Third Edition

For AQA Specification A

Understanding GCSE Geography

Ann Bowen

John Pallister

D0996103

Heinemann is an imprint of Pearson Education Limited, a
company incorporated in England and Wales, having its
registered office at Edinburgh Gate, Harlow, Essex, CM20 2JE.
Registered company number: 872828

Heinemann is a registered trademark of Pearson Education Limited

© Harcourt Education Limited, 2006

First published 2006

10 09 08
10 9 8 7 6 5 4 3

British Library Cataloguing in Publication Data is available from
the British Library on request.

ISBN: 978 0 435351 71 7

Edited by Jane Anson
Designed by hicksdesign
Typeset by HL Studios

Original illustrations © Harcourt Education Limited, 2006

Illustrated by HL Studios

Printed in China by South China Printing Co.

Cover photo: Corbis

Picture research by Elisabeth Lastchenko/Dan Sinclair/Zooid

Contents

Features of the book:	
Activities	Differentiated tasks catering for the needs of both foundation and higher tier students
Exam focus	Exam-style questions familiarizing students with exam vocabulary and giving further practice in exam technique
ⓘ	A panel giving extra information on a particular topic or theme
There are Case Studies in each chapter, both integrated within the text and as separate pages, providing further real examples of the topic.	

Acknowledgements

The authors and publisher would like to thank the following individuals and organizations for permission to reproduce copyright material.

Photographs

Albert McCabe/Hulton Archive/Getty Images p.137. Alexandra Carlile/Elvele Images/Alamy p.172. Alistair Berg /Alamy p.152. Andrew Bell/Alamy p.80. Andrew Woodley/Alamy p.5. APS UK p.79. Bill Wymar/Alamy p.144. Brian A. Vikander/Corbis UK Ltd p.223. Brian Shuel/Collections p.145. Cambridge Science Park p.179. Charlotte Thege/Das Fotoarchiv/Still Pictures pp.100, 111. Chris Howes/Wild Places Photography/Alamy p.25. Colin McPherson/Scottish Viewpoint p.147. Collections p.93. Corbis UK Ltd p.87. Danita Delimont/Alamy p.243. Dave G. Houser/Post-Houserstock/Corbis UK Ltd p.136. Dave G. Houser/Post-Houserstock/Corbis UK Ltd p.161. David Robinson/Snap2000 Images/Alamy p.56. Dean Conger/Corbis UK Ltd p.240. Dewitt Jones/Corbis UK Ltd p.7. Dick Makin/Alamy p.22. Dobson Agency/Rex Features p.74 Earth Satellite Corporation/Science Photo Library p.108. Geoscience Features Picture Library pp.30, 31, 65, 103. Janine Wiedel Photolibrary/Alamy p.146. Jeremy Hartley/Panos Pictures p.243. Jim Holmes/Panos Pictures p.175. John & Eliza Forder/Collections p.159. John D Beldom/Collections p.166. John Giles/PA/Empics p.67, 75. Jon Sparks/Alamy p.164. Jorgen Schytte/Still Pictures p.176 Joseph Sohm; Chromosohm Inc./Corbis UK Ltd p.181. Karen Kasmauski/Corbis UK Ltd p.127. Layne Kennedy/Corbis UK Ltd p.40. Les Stone/Sygma/Corbis UK Ltd p.50. Marion Kaplan/Alamy p.168. Mark Edwards/Still Pictures p.99, 176. Martin Bond/Still Pictures p.148. Matt Cardy/Getty Images p.68 Mike Kipling Photography/Alamy p.183. Neil Grant/Alamy p.170. Nigel Cattlin/Holt Studios International Ltd/Alamy p.161. Patrick Ward/Corbis UK Ltd p.218. Paul Freytag/Zefa/Corbis UK Ltd p.102. Rainer Kzonsek/Das Fotoarchiv./Still Pictures p.154. Rick Wilking/Reuters/Corbis UK Ltd p.98. Robert Harding World Imagery/Robert Harding Picture Library Ltd/Alamy p.26. Robert Holmes/Corbis UK Ltd p.51. Rodger Tamblyn/Alamy p.166. ROLAND SEITRE/Still Pictures p.100. Science Photo Library p.42. Shehzad Noorani/Still Pictures p.48. Simmons Aerofilms Ltd/Alamy p.135. Stan Gamester/Alamy p.145. Stan Gamester/Photofusion Picture Library p.145. The Lighthouse for Education p.74. Tony Wharton/Frank Lane Picture Agency p.11. ullstein/LS-PRESS/Still Pictures p.16. University of Dundee, www.sat.dundee.ac.uk pp.91, 93. Vincent Lowe/Leslie Garland Picture Library/Alamy p.34.

All other photographs are the copyright of John Pallister.

Maps, diagrams and extracts

Pages 31, 34, 44, 45, 57, 59, 82 Ordnance Survey. Page 134 *Daily Telegraph*. Pages 136, 137, 141 Ordnance Survey. Page 162 Reprinted from *A Course in World Geography: The British Isles* by J. Lowry, Hodder Arnold, 1984. Page 172 Reprinted by kind permission of the World Resources Institute. Page 176 Reprinted by kind permission of the *New Internationalist*. © *New Internationalist*. www.newint.org. Pages 187, 189 Ordnance Survey. Page 195 *Sunday Telegraph*. Page 197 *Daily Telegraph*. Page 204 *BP Statistical Review of World Energy*. Page 206 Reprinted by kind permission of WWF. Page 206 Reprinted by kind permission of the *New Internationalist*. © *New Internationalist*. www.newint.org. Page 211 *Financial Times* and British Wind Energy Association. Page 212 Guardian Newspapers Limited. Page 217 *Yorkshire Life*. Page 235 Reprinted by kind permission of the World Resources Institute. Page 237 Reprinted by kind permission of the World Resources Institute.

Every effort has been made to contact copyright holders of material reproduced in this book. Any omissions will be rectified in subsequent printings if notice is given to the publishers.

Websites

On pages where you are asked to go to Hotlinks to search for information or complete a task, please insert the code **1710P** at www.heinemann.co.uk/hotlinks.

Chapter 1
Tectonic activity

The great power of natural forces – the Soufrière Hills volcano on the island of Montserrat in the Caribbean erupted in 1997 causing massive damage to a large part of the island. Why do volcanoes occur in certain parts of the world?

Key Ideas

The Earth's crust is unstable and creates hazards:
- the tectonic plates may move together or move apart
- at plate margins earthquakes, volcanoes and fold mountains occur.

There are interactions between people and the environment and hazards created by tectonic activity:
- human activities in fold mountain areas are affected by the physical conditions
- areas of tectonic activity have advantages and disadvantages for settlement
- there are both primary and secondary effects for people from tectonic hazards
- tectonic activity has different effects in rural and urban areas and in rich and poor countries.

The Earth's tectonic plates

Figure 1 shows the structure of the Earth. At the centre there is the **core** surrounded by a large mass of molten rock called the **mantle**. At the surface there is a thin **crust** 'floating' on the mantle below. There are two main types of crust: oceanic crust which is denser and about 5 kilometres thick and continental crust which is lighter but about 30 kilometres thick.

The Earth's crust is not one continuous layer but is made up of seven large tectonic **plates** and many smaller ones. **Figure 2** shows the distribution of the main tectonic plates. The Earth's crust is unstable because the plates are moving in response to rising hot currents called **convection currents** within the mantle. The movement of the plates has greatest impact at the plate margins, where two tectonic plates meet. The centres of the plates, away from the margins, tend to be stable and away from major tectonic activity.

Plates may move apart, or closer together, or slide past each other. These movements lead to volcanoes, earthquakes and the formation of fold mountains.

Compressional plate margins

Plates that move together form a **compressional margin** or destructive plate boundary. **Figure 3** shows what happens when two plates are moving together. The Nazca plate is made of oceanic crust which is denser than the continental crust of the South American plate.

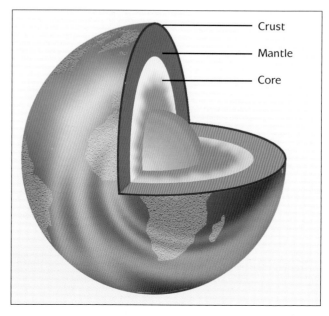

▲ **Figure 1** The structure of the Earth.

The Nazca plate is forced to sink below the South American plate. The oceanic crust sinks into the mantle where it melts in the **subduction zone**. Energy builds up in the subduction zone – at certain times this may

▼ **Figure 2** Earth's tectonic plates.

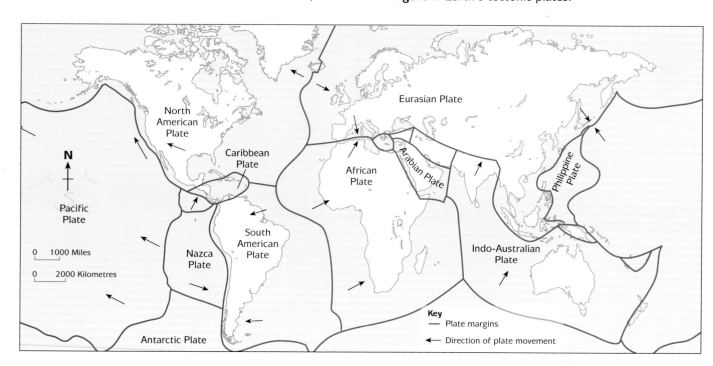

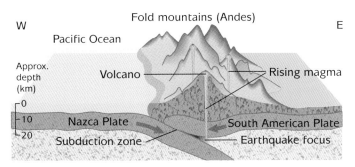

W　　　　　Fold mountains (Andes)　　　　　E

▲ **Figure 3** A compressional plate margin.

be released as an earthquake. The molten rock, called magma, may rise upwards, causing volcanic eruptions and leading to the creation of **composite volcanoes**. The lighter continental crust stays at the surface but sediment becomes crumpled into fold mountains. The Andes are the fold mountains that have formed along the west coast of South America.

Tensional plate margins

Some plates, like the North American and Eurasian plates, are moving in opposite directions, away from each other. This is called a **tensional margin** or constructive plate boundary. This type of movement mostly happens under the oceans. As the plates move apart the gap is filled by magma rising up from the mantle below. The rising magma creates **shield volcanoes** which, if they become high enough, form volcanic islands, such as Iceland. So much magma is poured out that ridges are built up from the sea bed, like the Mid-Atlantic ridge shown in **Figure 4**, upon which Iceland is located.

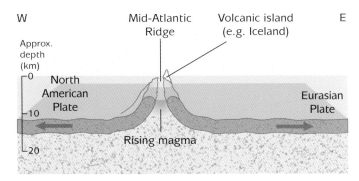

▲ **Figure 4** A tensional plate margin.

Passive plate margins

At the San Andreas fault in California, the North American plate and the Pacific plate are sliding past each other. They are moving in the same direction but the North American plate is moving slightly faster. Pressure builds up along the fault until one plate jerks past the other, causing an earthquake. The movement has also caused the land to become ridged and crumpled, as shown in **Figure 5**. This is called a **passive margin** or conservative plate boundary.

▲ **Figure 5** The San Andreas fault.

Activities

1 **(a)** Draw a labelled diagram to show the structure of the Earth.

   (b) Describe how the Earth is like an apple – mention the core, the 'fleshy' part and the skin.

2 On an outline world map mark and label the main tectonic plates and indicate the direction of movement of the plates.

3 Study **Figure 2**.

   (a) What does it show about the likelihood of tectonic activity in the UK?

   (b) Tectonic activity in Europe is concentrated in Iceland and southern Italy. State one tectonic similarity and one difference between Iceland and Italy.

4 Copy and complete a table like the one below:

Type of plate margin	Examples of plates	Features produced	Example country/area

5 Describe and explain the movements which take place at:

   (a) a compressional margin

   (b) a tensional margin.

   Draw diagrams to illustrate your answers.

Fold mountains

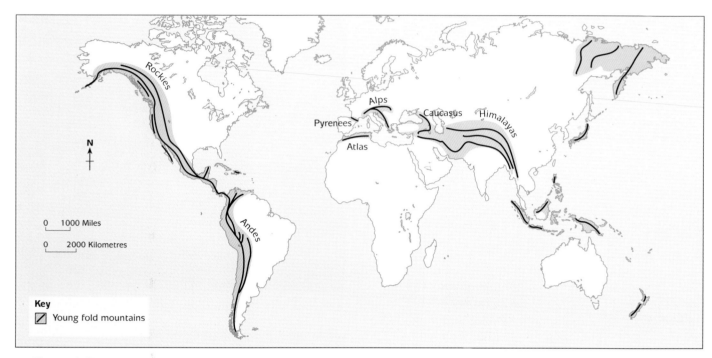

▲ **Figure 1** The world distribution of young fold mountains.

Young fold mountains are found in many parts of the world (**Figure 1**) and a glance back at **Figure 2** on page 6 shows that they form along the plate margins where great Earth movements have taken place.

Figure 2 shows the formation of fold mountains. There were long periods of quiet between Earth movements during which sedimentary rocks, thousands of metres thick, formed in huge depressions called **geosynclines**. Rivers carried sediments and deposited them into the depressions. Over millions of years the sediments were compressed into **sedimentary rocks** such as sandstone and limestone. These sedimentary rocks were then forced upwards into a series of folds by the movement of the tectonic plates. Sometimes the folds were simple upfolds (**anticlines**) and downfolds (**synclines**), as shown in **Figure 3**. In some places the folds were pushed over on one side, giving overfolds.

Fold mountains have been formed at times in the Earth's geological history called mountain-building periods. Recent mountain-building movements have created the Alps, the Himalayas, the Rockies and the Andes, some of which are still rising. For this reason many of these ranges are called young fold mountains.

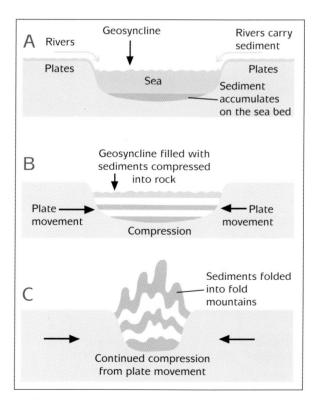

▲ **Figure 2** The formation of fold mountains.

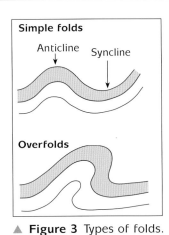

Figure 4 Evidence of ▶ folding can be seen where layers of rock are exposed.

▲ **Figure 3** Types of folds.

Opportunities and problems for human settlement

In general, high fold mountains are places with low densities of population, because of the limited opportunities for making a living there, and difficult communications.

Physical problems are immense and include the following:

- Relief – mainly high and steep. Rock outcrops are frequent and many mountain valleys are narrow and gorge-like. There is little flat land for farming and building settlements.

- Climate – with increasing height it becomes colder, windier and wetter, and more of the precipitation falls as snow. The growing season is short and it is often impossible to grow crops at high levels.

- Soils – mountain soils are typically stony, thin and infertile.

- Accessibility – roads and railways are expensive and difficult to build; travel on them is frequently disrupted by rock falls, avalanches and bad weather. High mountains in inland locations, such as the Himalayas, are the least accessible of all.

Valley floors are the most favourable locations for settlement within young fold mountain ranges. Physical problems tend to be less severe; soils are deeper, and more likely to be made of silt. Some of the highest population densities can be found in the mountains of Central America and in the Andes of Colombia, Ecuador and Peru (see **Figure 1** on page 114) where very fertile volcanic soils are present. Additionally some Andean countries have rich mineral resources including copper, tin, silver and gold. Most settlements at extreme heights in the Andes (above 4000 metres where the air is very thin) owe their existence entirely to mineral wealth; they lie well above the upper limit of cultivation.

◀ **Figure 5** Andean valley in Peru surrounded by mountain peaks more than 6000 metres above sea level.

Activities

1 Study **Figure 1**.

  (a) Describe the two main directions in which ranges of young fold mountains are aligned.

  (b) For one range of fold mountains, name the two tectonic plates responsible for its formation.

  (c) Explain why fold mountains are formed along compressional plate margins.

2 Draw a simple labelled sketch to show the evidence that the mountains in the photograph in **Figure 4** are fold mountains.

3 (a) Describe the changes in **Figure 5** from valley floor to mountain top in:

  (i) physical features

  (ii) human land uses.

  (b) Why are valleys important to people living in high mountainous areas?

Human activity in the Alps

The Alps are a range of young fold mountains formed 30–40 million years ago. They form the border between Italy and the neighbouring countries of France, Switzerland, Austria and Slovenia. The highest peak is Mont Blanc near the Franco-Italian border at 4810 metres, but there are many other peaks over 3000 metres. If you have flown over the Alps, you may have seen views similar to the one in **Figure 1**. From above, the Alps look like a wasteland of rock, ice and snow; only the valleys give a hint that human settlement may be possible.

Figure 1 Aerial view over the Alpine fold mountain range.

Human activities in the Alps vary according to height, which creates vertical zones of land use. **Figure 2** summarizes typical land uses in an Alpine valley. Many Alpine valleys are aligned west to east so that better opportunities for settlement, farming and other economic activities are created on sunny south-facing slopes. Land uses are higher and less varied on more shaded north-facing slopes. The main human activities in the Alps are connected with farming, forestry, tourism, hydro-electric power and industry.

Farming

Most farms are located on the sunnier and warmer south-facing slopes. The traditional farming is dairy farming using a system called transhumance, the seasonal movement of animals. In summer the cattle are taken up to the high Alp to graze, which allows hay and other fodder crops to be grown on the small fields on the flat land in the valley floor where summers are warmest. In winter the animals return to the farm on the valley floor, where they are kept in cattle sheds and stall-fed on silage and other fodder crops. Over the years there have been many changes to this system of farming.

- Cable cars and plastic pipes are now used to bring milk to the co-operative dairies down in the valley. In the past the farmers would have stayed with the cattle, and turned the milk into butter and cheese (which keep for longer) on the Alp.

- Farmers now use artificial feeds, so they, and their cattle, can stay at home all year.

- New roads, quad bikes and cable cars give easier access to the upland pastures.

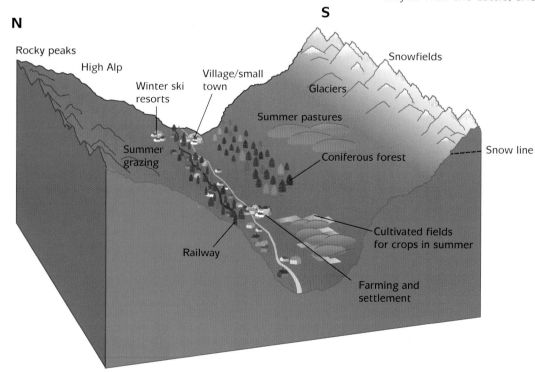

Figure 2 Human activities in an Alpine valley.

◄ **Figure 3** Alpine view – the Jungfrau region of the Lauterbrunnen valley.

There are good transport links from surrounding affluent European countries, where plenty of people can afford to take winter-sports holidays. Within the Alps is a dense communications network of railways and roads, as well as mountain railways, cable cars and ski-lifts.

The main worry is that Alpine winters are warming up and becoming less snowy than they used to be. More people are skiing on worn slopes, damaging the vegetation and the surface below and thereby increasing the number of bare surfaces and the risk of soil erosion on steep slopes.

Forestry

Coniferous trees cover many of the slopes, especially north-facing ones. Wood, as a plentiful local resource, has always been the main building material and winter fuel in Alpine lands. Sawmills are mainly located on the valley floors near to rivers; timber that cannot be used for construction is made into pulp and paper.

Tourism

The Alps have physical advantages for tourism all year round, attracting skiers, climbers and walkers as well as those who simply want to admire the spectacular scenery. These activities are a major industry.

A For winter tourism (examples of resorts are St Moritz, Chamonix):

- snow for skiing and other winter sports; in between the days with heavy snowfall are many sunny, crisp and clear days

- flatter land on the high-level benches for easy building of hotels, restaurants, ski-lifts and other facilities

- steep slopes above the resorts for ski runs amid great mountain views.

B For summer tourism (examples of resorts are Interlaken and Garda):

- large glacial lakes on valley floors

- beautiful mountain scenery with snow-capped peaks.

Hydro-electric power (HEP) and industry

The steep slopes, high precipitation and summer melting of the glaciers produce fast-flowing rivers that are ideal for generating HEP. The narrow valleys are easy to dam and there are lakes in which to store water. Some of the cheap HEP is used by industries which require a high input of electricity, such as sawmills, electrochemicals, fertilizer manufacture and aluminium smelting. Some of the electricity is also exported to other regions to supply towns and cities.

Activities

1 Study **Figure 3**.

   (a) Draw a labelled sketch of the photograph to show different land uses.

   (b) Explain why most human activities in the Alps are concentrated on valley floors.

2 **(a)** Describe the attractions of the area shown in **Figure 3** for tourists.

   (b) Give reasons why the Alps are one of Europe's major tourist regions.

Volcanoes

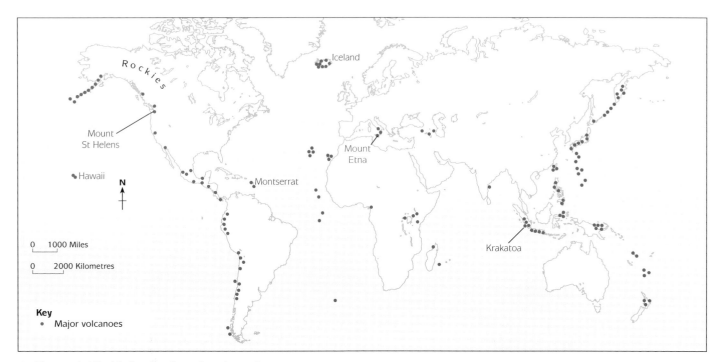

▲ **Figure 1** World distribution of active volcanoes.

A **volcano** is a cone-shaped mountain formed by surface eruptions of magma from inside the Earth. The magma that reaches the surface in an eruption is called **lava**, and is one of the many different products that can be thrown out, including ash, cinders, pumice, dust, gases and steam. The world distribution of active volcanoes (**Figure 1**) shows an almost perfect fit with the locations of the tectonic plate margins (see **Figure 2** on page 6).

How are volcanoes formed?

Volcanoes form where magma escapes through a **vent**, which is a fracture or crack in the Earth's crust. This

▶ **Figure 3** The Osorno volcano in Chile, an almost perfect cone shape.

happens most often at plate margins. Lava and other products are thrown out from the circular hole at the top called the **crater** (**Figure 2**). Each time an eruption takcs place, a new layer of lava is added to the surface of the volcano; since more accumulates closer to the crater during every eruption, a mountain that is cone-shaped is formed (**Figure 3**).

Different types of volcanoes

Volcanoes are divided into two main types, depending upon the material thrown out in an eruption and the form (height and shape) of the volcanic cone produced. These differences are shown in **Figure 4**. Basically the division is between volcanoes formed along tensional (constructive) plate margins and along compressional (destructive) margins, because of the different types

◀ **Figure 2** Looking into the crater of a dormant volcano.

	Tensional (constructive)	Compressional (destructive)
Plate margin	Movement of plate ← → Movement of plate Magma	Continental crust → ← Movement of plate Subduction zone — Oceanic plate Mantle
Formation	As the plates move apart, magma rises upwards from the mantle to fill the gap. This adds new rock to the spreading plates. Some of the magma may also be forced out to the surface through a vent. Some volcanoes grow high enough to form volcanic islands.	When the plates collide, the denser oceanic plate is pushed down into the mantle. Here the plate melts and is destroyed in the subduction zone. In the subduction zone the plate forms a pool of magma. The great heat and pressure may force the magma along a crack where it erupts at the surface to build up a volcano.
Form of volcano	**Shield volcano (basic lava)** Gentle slopes　Vent　Wide base Crater Magma chamber	**Composite cone volcano** Crater　Layers of ash and lava Subsidiary cone　Vent Magma chamber
Characteristics	• cone with wide base and gentle slopes • made of lava only • regular and frequent eruptions • lava pours out with little violence	• tall cone with narrow base and steep sides • made of alternate layers of lava and ash • irregular with long dormant periods • violent explosions possible
Examples	Hekla and Surtsey in Iceland Mauna Loa and Kilauea in Hawaii	Etna, Vesuvius and Stromboli in Italy Krakatoa in Indonesia

▲ **Figure 4** How different plate margins affect volcanoes.

of lava emitted. Along tensional margins the basic lava that has come from within the mantle has a low silica content: it pours out easily, is runny and flows long distances, building up shield volcanoes. However, along compressional margins the **acid lava** has a high silica content, which makes it more viscous so that it travels shorter distances before cooling; these are more explosive volcanoes. After an eruption the vent becomes blocked, which results in great pressure building up before the next eruption. During explosive eruptions lava is shattered into pieces so that bombs, ash and dust are showered over a wide area.

Exam focus

1 a Name the type of lava associated with **(i)** violent volcanic eruptions and **(ii)** gentle volcanic eruptions. (2 marks)

b State two reasons why some eruptions are more violent than others. (2 marks)

c What shows that the crater in **Figure 2** is a dormant volcano? (3 marks)

d (i) Describe the differences in size and shape between volcanoes such as Mauna Loa and Mount Etna (2 marks)

(ii) Is the Osorno volcano in **Figure 3** a shield or composite volcano? Explain your answer. (3 marks)

e With the aid of a labelled diagram, explain how a composite volcano such as Mount Etna is formed. (4 marks)

Effects of volcanoes on people

Natural hazards have both primary and secondary effects for people living in the surrounding area (see the Information Box). Volcanoes are no exception. Loss of life can be high if the eruption is large, explosive and happens with little warning; people may be hit or covered by falling debris, suffocated by poisonous gases or buried under mud flows after the heat of the eruption melts snow at the top of the volcano. Whilst the human and physical effects of most volcanoes are limited to the local area, a few massive eruptions have had global effects when volcanic dust has been sent high into the atmosphere, blocking out some of the sunlight and reducing world temperatures for a time.

No matter how gentle, however, a volcanic eruption is always destructive (**Figure 1**). Buildings in the path of the lava are destroyed and farmland is covered. It will be many years before the lava visible in **Figure 1** weathers into fertile soil that can be used again for farming. Homes, farms, animals and crops are lost; people lose their livelihoods and are forced to migrate, often with nothing. Roads are blocked and electricity and telegraph poles are brought down, further disrupting normal life and economic activities.

Loss of life is always more likely in volcanic eruptions along compressional plate margins. Living close to shield volcanoes is safer. Lives are rarely threatened when only lava is being erupted, even from a composite cone, because people have plenty of time to move out of the way of the lava flows. It may even be possible for people to reduce the damage caused, although the success of human efforts during the 2001 eruption of Mount Etna is open to question (**Figure 2**). Advance warning signs, such as small earthquake shocks and increased emissions of steam and gases, are often clearer for volcanoes than for earthquakes, which is why total loss of life is lower (see **Figure 2** on page 233).

Information

Primary and secondary effects of natural hazards

Primary effects – the immediate impact

- people injured and killed

- buildings, property and farmland destroyed

- communications and public services (transport, electricity, telephones etc.) disrupted.

Secondary effects – the medium- and long-term after-effects

- shortages of drinking water, food and shelter

- spread of disease from contaminated water

- economic problems from the cost of rebuilding and the loss of farmland, factories, tourism and other economic activities

- social problems from family losses and stress.

People against nature

When a volcano begins to erupt lava, ash and gas, the people living close by are forced to take note. In Catania, a city of 380 000 people located on the coast of Sicily, about 50km from the top of Mount Etna, there were two effects of the volcano: a fine ash settled on everything, and the explosions and flows of glowing lava provided spectacular evening entertainment which made the city a magnet for visitors.

The threat was much greater for the 6300 people who lived in the town of Nicolosi, higher up on the side of the volcano, only 20km from the start of the lava flows. One lava flow was heading straight for the town. The Italian government declared a state of emergency and provided US$ 7 million of help. Thirty bulldozers worked night and day building walls of earth on the higher slopes above Nicolosi to try to divert the lava flow away from the town. Two aeroplanes and a helicopter also dropped water to cool the lava and decrease its speed of flow, although they couldn't stop it destroying the ski-lifts.

Meanwhile the people of Nicolosi prayed. In the end their prayers seem to have been answered, because the lava flow stopped 4km from the town. A new crack opened up on the side of the volcano, which took some of the lava away from the flow that was moving towards Nicolosi. The volcanic activity decreased, the lava flow became wider and the lava itself became more dense. Each of these three things helped to reduce the speed of flow of the lava. 'We have a love affair with this volcano,' said the mayor of Nicolosi. 'Even in the past when eruptions have ruined some property, we just start again.'

◀ **Figure 1** Results of a lava flow from the 2001 eruption of Mount Etna, which covered vineyards and orchards.

Figure 2 ▶ Newspaper report from July 2001 about the eruption of Mount Etna.

Volcanoes: hazard or blessing?

Why do people choose to live near active volcanoes when an eruption could happen at any time? The answer is that living in volcanic areas has its attractions. After the lava weathers, volcanic soils are some of the world's most fertile soils. Often there are sharp contrasts in quality compared with the soils in surrounding mountainous areas. Soils from the many eruptions of Vesuvius on the Plain of Campania are the best in southern Italy (**Figure 3**). Volcanic areas offer a variety of attractions for tourists, including bathing in the hot springs and mud pools, watching geysers perform and volcano walking up to and around the crater (**Figure 4**). The supplies of hot water have economic uses, either as domestic heating or for generating electricity (geothermal power) as in Iceland. In addition, there are valuable minerals, notably sulphur, borax and pumice.

▼ ▲ **Figure 4** Vulcano, a volcano on one of the Lipari Islands off the north coast of Sicily. **A:** Monitoring equipment in the crater. The yellow deposits are sulphur. **B:** Italian tourists in a hot mud pool at the foot of the volcano.

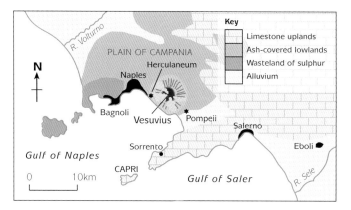

▲ **Figure 3** Vesuvius and the Plain of Campania.

Campania is a region in southern Italy. It includes the major city of Naples and the famous volcano, Mount Vesuvius (1198 metres).

Vesuvius is a composite cone volcano which is dormant at present but has been very destructive in the past. Its most notable eruption was in AD 79 when the towns of Pompeii and Herculaneum were destroyed. Thousands of people were killed by the poisonous gases and the area was buried under metres of ash. So why do so many people choose to live close by?

Excavations at Pompeii and Herculaneum, trips to the crater on Vesuvius and visits to hot springs have brought a thriving tourist industry employing many local people. The fine ash is very fertile. Wheat, maize, peaches, almonds, vines and especially tomatoes are all grown intensively. Yields are five times higher than the national average. The fertility of the plain contrasts with the dry, arid, thin soils on the limestone uplands. West of Naples is the Phlegraean Fields, a wasteland with hot springs, geysers and sulphur domes, which is useless for farming, although the sulphur has industrial uses.

Activities

1 Draw two spider diagrams to show the advantages and disadvantages of volcanoes.

2 **(a)** From **Figures 1** and **2**, describe the damage caused by the eruption of Mount Etna in 2001.

   (b) Describe how local people responded to try to reduce the effects of this eruption.

   (c) Which was more important for limiting the effects of this eruption – nature or the actions of people? State and explain your view.

3 **(a)** Explain the dangers for people living near Mount Vesuvius, a dormant composite volcano.

   (b) Why, despite the dangers, is the area around Vesuvius one of the most densely populated parts of Italy?

A volcanic eruption – the Soufrière Hills volcano on the island of Montserrat in the Caribbean

◀ **Figure 1** The location of Montserrat.

▶ **Figure 2** Plymouth, a ghost town.

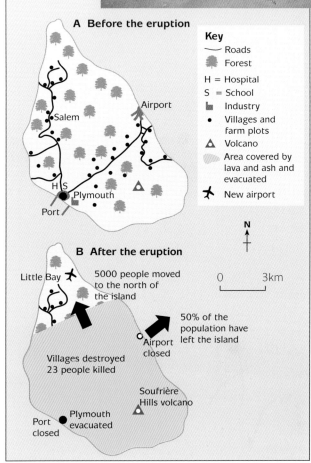

Figure 3 The impact of the eruptions on Montserrat.

Montserrat (**Figure 1**) is a small island in the Caribbean which is still a British colony. The island is mountainous and wooded, earning it the name 'Emerald Island of the Caribbean'. It has been popular with many wealthy British people including Paul McCartney and Sting – exclusive villas and hotels line the coast. However, many of the residents of Montserrat are quite poor, living in small villages and practising subsistence farming. Before the eruption the population was 12 000, 50 per cent of whom lived in the capital city, Plymouth, in the south of the island.

In July 1995 the Soufrière Hills volcano erupted for the first time in 350 years. One month later 50 per cent of the population were evacuated to the north of the island away from the danger zone. In April 1996, as the eruptions continued, Plymouth became a ghost town (**Figure 2**) as more and more people were evacuated. The eruptions became more explosive and the lava and ash caused great damage to the island. In June 1997 another eruption destroyed villages in the centre of the island, killing 23 people. Of the island's 40 square miles only 15 square miles in the north of the island were considered safe. Over 5000 people left the island, most to settle on nearby islands such as Antigua or to move to the UK. Study **Figure 3**, which shows the impact of the eruptions.

Those people who have stayed on the island are suffering very harsh conditions. The south of the island was the most developed, with the main towns, communications and services. In the north there were few roads and settlements. Many of the evacuees are forced to live in makeshift shelters with inadequate sanitation; there are few schools and no proper hospital, and living conditions are very poor. The country's tourist industry has stopped

with the closure of the airport, and other industries are suffering with the restricted port activities. The processing of imported rice and the assembly of electronics products have both declined. The country now relies upon aid from London.

After the eruptions which forced the evacuation of most of the island, the people called for the British government to pay compensation and to rebuild the island. Aid totalling £41 million has been offered to redevelop the north of the island and £10.5 million to relocate refugees. In 1997, £2400 was offered to each adult over eighteen wanting to leave the island. There was rioting on the island when the local people felt the British government was not offering enough help to the people. The Montserratians were demanding £20000 per person. Can the rebuilding of the island be justified for a population of only about 4000 people? Also, scientists cannot predict if and when the volcano may erupt again. Money may be invested in rebuilding only to be wiped out by another eruption which destroys the whole island.

Has the volcano gone quiet?

Although the last major eruption was in July 2003, scientists at the Observatory are unable to give any guarantees about the volcano. In early 2006 lava was still being emitted and there were signs of increased activity. Check the Observatory website (see Hotlinks, page iv) for up-to-date information. Instead of waiting for the volcano to stop erupting, government policy is to concentrate on rebuilding in the 'safe' north of the island. The government is pushing ahead with plans to replace Plymouth as the capital with a new town at Little Bay in the north west of the island. In 2005 the new airport opened, with regular links to the international airport in Antigua. Despite this vital improvement in communications, attracting tourists back to the island and restarting profitable economic activity is proving to be an uphill struggle. It must be remembered that few people lived in the north before the 1995 eruption, because natural resources for the two main economic activities, agriculture and tourism, were inferior to those in the south.

▲ **Figure 4** Northern limit of the exclusion zone which covers about two-thirds of the island.

▲ **Figure 5** Housing hurriedly built on unused land in the north after 1997 by families forced to leave Plymouth.

Activities

1 Put together information for the case study of a volcano – Montserrat.

(a) Draw a sketch map to show the location of Montserrat.

(b) Explain with the aid of diagrams why the volcanic eruption occurred on Montserrat. (**Figures 2** and **3** on pages 6–7 and **1** and **4** on pages 12–13 will help.)

(c) Describe **(i)** the primary effects and **(ii)** the secondary effects of the eruption.

(d) Describe the responses:

(i) short term (until 1997)

(ii) medium term and

(iii) longer term (since 2003).

(e) Many of the remaining islanders are asking the government questions like 'When is Plymouth going to be opened up?' and 'When can we go back to live in the south where we used to live?' The government is not keen to respond. Explain these different points of view about what should happen next.

The earthquake hazard

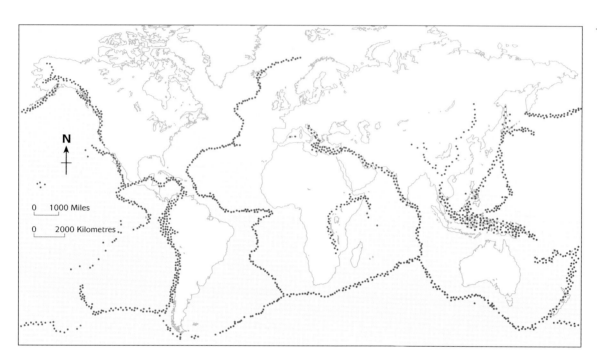

◀ **Figure 1**
Distribution of
earthquakes.

Earthquakes are vibrations in the Earth's crust which shake the ground surface. They are sudden, and, because they happen without warning, often lethal. Up to 2 million people have died as a result of earthquakes since 1900. Earthquakes are common events; there are around 50 000 detectable earthquakes each year, although most are too small for people to notice. They occur in well-defined zones (**Figure 1**). How good is the fit with the tectonic plate margins shown in **Figure 2** on page 6?

Between 15 and 20 earthquakes a year are magnitude 7 or more on the **Richter scale**, which are strong enough to have devastating effects on life and property. The magnitude of an earthquake is measured by an instrument called a **seismograph** and given a value between 1 and 10. The Richter scale is logarithmic: an earthquake measured at 7 is 10 times stronger than one measured at 6 and 100 times stronger than one measured at 5.

Why do earthquakes happen?
Over 90 per cent of earthquakes occur where plates are colliding at compressional plate margins. Great stresses build up in the subduction zone as one plate is forced down below the other. Energy builds up, and is released in an earthquake. The point at which the earthquake happens below the ground surface is called the **focus** (see **Figure 3** on page 7).

The **epicentre** is the point on the surface directly above the focus, where the greatest force of the earthquake is felt. Earthquakes also occur along passive plate margins like the San Andreas fault (see **Figure 5** on page 7).

The effects of earthquakes
Primary effects are immediate damage caused by the earthquake, such as collapsing buildings, roads and bridges. People are killed by being trapped in their homes, places of work and cars. The severity of the primary effects is determined by a mixture of physical and human factors (**Figure 2**). The chance element is time of day – were there fewer people close to the epicentre when the earthquake struck than at other times?

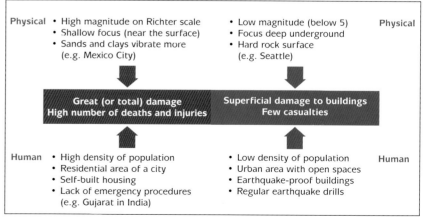

Physical	• High magnitude on Richter scale • Shallow focus (near the surface) • Sands and clays vibrate more (e.g. Mexico City)	• Low magnitude (below 5) • Focus deep underground • Hard rock surface (e.g. Seattle)	Physical
	Great (or total) damage **High number of deaths and injuries**	**Superficial damage to buildings** **Few casualties**	
Human	• High density of population • Residential area of a city • Self-built housing • Lack of emergency procedures (e.g. Gujarat in India)	• Low density of population • Urban area with open spaces • Earthquake-proof buildings • Regular earthquake drills	Human

▲ **Figure 2** Factors controlling the effects of an earthquake.

Secondary effects are the after-effects, such as fires, tsunamis, landslides and disease.

- Fires are caused by earthquakes fracturing gas pipes and bringing down electricity wires. Fires spread quickly in areas of poor-quality housing.

- Tsunamis are giant sea waves caused by an earthquake on the sea floor. The 8.9 magnitude earthquake off the coast of Sumatra in Indonesia on Boxing Day 2004 was the fifth-strongest earthquake ever recorded; it produced a wave of awesome size and speed which devastated many densely populated coastal communities in Indonesia, Thailand, India and Sri Lanka (see the Information Box).

ⓘ Information

Asian Tsunami, December 2004

Speed	500km/h
Highest recorded wave	34m
Distance travelled from epicentre	4000km
Number of people dead or missing	220000+
Number of people displaced	about 2 million
Houses destroyed	over 500000

- Landslides are most likely on steep slopes and in areas of weak rocks such as sands and clays.

- Diseases such as typhoid and cholera spread easily when burst pipes lead to shortages of fresh water and to contamination from sewage.

The impact of earthquakes is often much more severe in LEDCs, where earthquake-resistant buildings are often considered too expensive to build. Even where building regulations exist, they are frequently ignored, because builders want to make more money and people can only afford cheap houses. People and authorities are also less well prepared.

The responses to earthquakes

Irrespective of whether people are rich or poor, immediate emergency aid (pages 242–3) is desperately needed everywhere after a strong earthquake strikes. Specialist rescue and medical teams can be expected to be on the scene within hours in MEDCs, thanks to advance preparations and modern communications. The poorer the country, the greater its reliance upon short-term aid from overseas.

In the medium term, the need is for a quick return to normal life (or as near normal as possible) by repairing and replacing what has been lost and restarting economic activity. The focus needs to be switched from disaster aid to development aid.

However, without any method of earthquake prediction, the only effective long-term response for people living in known earthquake zones is to be prepared. Tall buildings are more likely to remain standing if built using the methods shown in **Figure 3**.

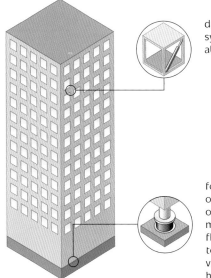

damping and bracing systems to help absorb shocks

foundation piles made out of alternative layers of steel and rubber to make the skyscraper flexible in an earthquake – to make them stiff vertically, but flexible horizontally

▲ **Figure 3** How to make a skyscraper resist earthquake shocks.

Activities

1 **(a)** Describe the global pattern of earthquakes shown in **Figure 1**.

  (b) Explain why many strong earthquakes occur along compressional plate margins.

  (c) Why are tectonic stresses lower along tensional plate margins (page 7)?

2 **(a)** What is the difference between the primary and the secondary effects of an earthquake?

  (b) Explain why the primary effects of earthquakes are usually more severe in urban than in rural areas.

  (c) (i) Name one earthquake which had more serious secondary than primary effects.

   (ii) Give reasons why the secondary effects were so great.

3 Two earthquakes of similar magnitude (Richter scale 8.3–8.4) were:

  1964 Anchorage, Alaska   Estimated deaths 130
  1985 Mexico City   Estimated deaths 7000

  Name and explain two factors that could have accounted for the different number of deaths.

Case Study – Indian earthquake 2001

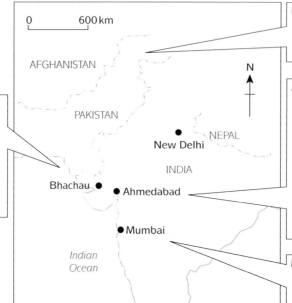

▶ **Figure 1** Effects of the Gujarat earthquake in 2001.

0   600 km

AFGHANISTAN

PAKISTAN

N

New Delhi   NEPAL

INDIA

Bhachau ● ● Ahmedabad

● Mumbai

Indian Ocean

Pakistan
• Mild rocking of the ground in the south led to 15 deaths being reported

Bhachau
• Population 25 000
• Rescuers could only find 10 000 people alive
• All buildings were destroyed along with 90% of the streets
• One survivor said 'Within half a minute thousands were killed, covered by massive piles of rubble'

Ahmedabad (the capital of Gujarat)
• Population 5 million after a decade of rapid growth
• Main tremors lasted only 45 seconds, but at least 40 high-rise buildings swayed so much that they collapsed
• Much of the rest of the city was still intact because the earthquake was selectively destructive, with apartment blocks less than ten years old the worst affected
• 1000 or more were killed

Mumbai (Bombay)
• Population 16 million
• High-rise buildings swayed during the mild tremors
• People rushed out on the streets but there were no reported deaths

The state of Gujarat in northern India is an active earthquake zone. In January 2001, a devastating earthquake occurred here, measuring 7.9 on the Richter scale. Tectonic stresses across northern India result from the pressure of the Indo-Australian plate pushing northwards against the Eurasian plate. The epicentre was close to Bhachau (**Figure 1**). As much as 48 hours elapsed before rescue efforts began; there were many complaints about the authorities' lack of preparedness for a natural disaster of this kind. In some places, rescue workers had no better equipment than shovels. The earthquake left at least 20 000 dead, 160 000 injured and 600 000 homeless. Direct economic losses were estimated at US$ 2–4 billion.

Earthquake rocks Seattle

An earthquake measuring 6.8 on the Richter scale rocked the Pacific Coast American city of Seattle last night, sending workers onto the streets in panic. Office workers used fire escapes to evacuate skyscrapers in downtown Seattle. The hotel where Microsoft founder Bill Gates was addressing a conference was evacuated. In the city, one woman died of a heart attack and more than 250 were injured. Extensive damage to property was caused in the centre, which could cost billions of dollars to repair. It was the biggest earthquake to hit the area in 52 years.

Experts said that one of the reasons for the lack of deaths and serious injuries was that the earthquake's focus was 50km underground, in solid rock. Also a lot of money had been spent in the last 50 years to make sure that buildings could withstand major shocks. One of the city's best-known landmarks, the Space Needle, was built to withstand a 9.1 magnitude earthquake. It shuddered and rolled violently during the tremor, but remained undamaged.

▲ **Figure 2** Another earthquake in 2001 (for comparison).

Activities

1 Make notes for a 'Case study of an earthquake – India 2001'. Suggested headings: Earthquake information (such as strength, location of epicentre and areas affected); Cause of earthquake; Effects of earthquake; Variations in effects with distance from the epicentre.

2 State two reasons why loss of life and injuries were lower in the Seattle earthquake than in the Indian quake.

3 View of one scientist: 'Earthquakes are inevitable, but death from earthquakes is not'.

  (a) Describe more fully this scientist's view about earthquakes and their effects.

  (b) Do the two examples of earthquakes on this page support this view? Explain your answer.

Chapter 2
Rocks and landscape

Limestone pavement near Malham in the Yorkshire Dales National Park. This landscape feature forms only in areas of Carboniferous limestone rocks.

Key Ideas

The Earth's crust is composed of different rock types and is modified by weathering:

- the three rock groups are igneous, sedimentary and metamorphic
- weathering processes that modify rocks include physical processes (e.g. frost shattering) and chemical processes (e.g. limestone solution)
- rock types such as granite, Carboniferous limestone, chalk and clay form distinctive landscapes.

The interaction between people and landscapes of different rock types leads to:

- different land uses according to rock type
- activities such as quarrying because rocks have economic uses
- local issues associated with quarrying and its impacts.

Types of rocks and their distribution within the UK

Types of rocks

Although there are many different types of rock on the Earth's surface, there are only three groups of rocks. Rocks are either igneous, sedimentary or metamorphic.

Igneous rocks are 'formed by fire'; they begin as magma in the interior of the Earth. Some are formed by lava cooling on the Earth's surface after being thrown out by a volcanic eruption. For example, the basic lava that flows from constructive margins and forms shield volcanoes cools to form *basalt* rock. This has been eroded into hexagonal blocks at the Giant's Causeway in Northern Ireland (**Figure 1**). Others are formed by magma cooling underground after having been intruded into other rocks without reaching the surface. *Granite* is an example of this type of igneous rock. It is often intruded during the building of fold mountains along destructive plate boundaries. Granite outcrops on the surface after erosion of the rocks above it over millions of years. Today it is exposed in many places in Scotland (**Figure 4**) and forms most of the moorlands of Devon and Cornwall as well the dramatic cliffs at Land's End (**Figure 2**).

Sediments are small particles of rock transported by water, ice and wind. Most eventually reach the sea bed where over the years successive layers of sediments accumulate. The weight of materials above compresses the sediments below into **sedimentary rocks**. These rocks are laid down in layers, or beds, with lines of weakness, or bedding planes, between layers (**Figure 3**). When sand is compressed, *sandstone* rock is formed. *Clay* forms from the accumulation and compression of deposits of mud. *Limestone* and *chalk* consist of calcium carbonate which comes from the remains of plants and animals. For example, the shells of sea creatures are made of calcium carbonate; when these animals die, masses of shells accumulate on the sea floor, building up layers of limestone rock. A lot of limestone was formed during the Carboniferous period (280–345 million years ago) because at that time much of Britain was a warm shallow sea, rich in plant and animal life.

Metamorphic rocks are those which have been changed in shape or form. They begin as either igneous or sedimentary rocks but are later altered by heat or pressure. This happens, for example, along destructive plate boundaries and fault lines. Heat and pressure change limestone into *marble* and clay into *slate*. Both marble and slate are harder forms of the original rocks, and have greater economic value. Marble is widely used in building and for floors in Mediterranean countries such as Italy. Slate splits easily into sheets and, until recent times, was the main roofing material used in the UK.

◀ **Figure 1** The Giant's Causeway is made of basalt.

▼ **Figure 2** Land's End is made of granite. Notice the many vertical joints.

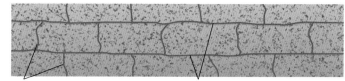

joints – vertical weaknesses within the layers of rock   *bedding planes* – horizontal weaknesses between the layers

fault – earth movements have broken up the beds of rock

▲ **Figure 3** Rock weaknesses. These are important because they are the first points to be attacked by processes of weathering and erosion.

Distribution of rock types within the UK

The distribution of rocks reflects the geological history of the UK. It is customary to divide the country into two parts using a line running from the mouth of the River Tees to the mouth of the River Exe, separating Highland from Lowland Britain (**Figure 4**).

The geology of Highland Britain to the north and west of the Tees–Exe line is dominated by old and hard rocks. The majority are igneous and metamorphic rocks which have resisted erosion and therefore form the upland and mountainous parts of the country.

In Lowland Britain, to the south and east of the Tees–Exe line, the geology is dominated by younger sedimentary rocks. There is much low-lying and flat land, such as in the clay vales. Chalk, however, is more resistant to erosion than many of the other sedimentary rocks which surround it. This is why chalk ridges and scarps, such as the North and South Downs, appear to be high and steep, but they form lower, gentler and more rounded landscapes than the rocks in the uplands of Highland Britain.

Activities

1 Use **Figure 4** and a map of the British Isles.

   (a) On an outline map of the British Isles, shade and name two areas covered by each of the following rock types:

   • granite

   • Carboniferous limestone

   • chalk

   • clay.

   (b) Describe the differences between Highland and Lowland Britain using these headings:

   (i) Rocks

   (ii) Relief (height and shape of the land).

2 Draw a larger version of the table below and fill it in using information from pages 22–23.

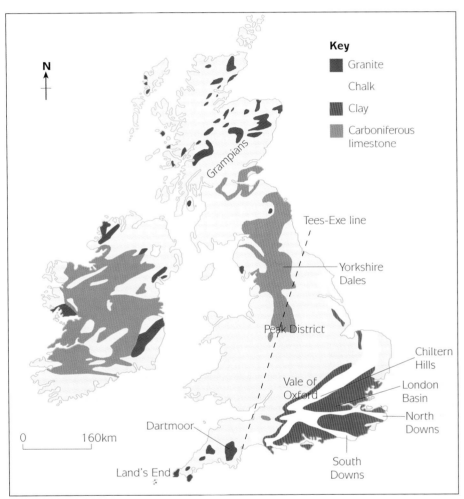

▲ **Figure 4** The distribution of some of the rocks found in the British Isles.

Heading	Igneous	Sedimentary	Metamorphic
Brief definition			
Where they were formed			
How they were formed			
Rock types found in the UK			
Main areas in the UK where rock outcrops occur			

Weathering

Weathering is the breakdown of rock at or near the surface, for which the weather, such as changes in temperature, is mainly responsible. The rocks are broken down *in situ*, which means that no movement is involved (unlike erosion, which is caused by the movement of water, ice and wind). There are two main types of weathering: **physical weathering** and **chemical weathering**.

Physical weathering

This leads to the break-up of rock without any change in the minerals that form the rock.

In cold climates the most widespread type is frost shattering or **freeze–thaw** (**Figure 1**). The volume of water expands when it freezes, so each time water freezes and expands within a crack or joint in the rock, more pressure is put on the surrounding rock and the crack widens. The more often the temperature fluctuates above and below freezing point during the year, the more effective the frost shattering is at breaking off pieces of rock. The sharp-edged (or angular) pieces of rock that are broken off form **scree**, which can be seen below rock outcrops in all upland areas (**Figure 2**). Some of the largest scree slopes in the UK are on the side of Wastwater in the Lake District (see **Figure 5** on page 59).

Chemical weathering

This happens when the minerals of which the rock is composed are changed, leading to the disintegration of the rock. Granite (pages 26–27) is one type of rock that is vulnerable to chemical weathering. Feldspar, one of the minerals that make up granite, is converted into clay minerals such as kaolin (china clay).

The distinctive landforms both above and below the ground in areas of Carboniferous limestone (pages 28–29) owe their origins to **limestone solution**. This type of chemical weathering is also called carbonation, because the dissolving of the limestone changes calcium carbonate into calcium bicarbonate.

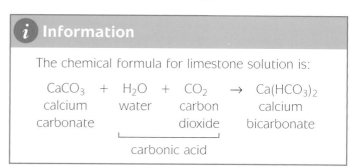

> ### 𝒊 Information
>
> The chemical formula for limestone solution is:
>
> $$CaCO_3 + H_2O + CO_2 \rightarrow Ca(HCO_3)_2$$
>
> calcium carbonate   water   carbon dioxide   calcium bicarbonate
>
> carbonic acid

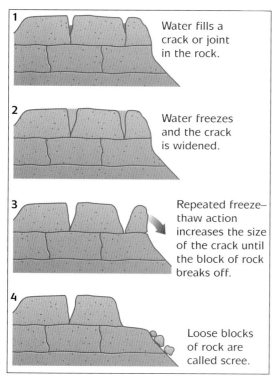

1   Water fills a crack or joint in the rock.

2   Water freezes and the crack is widened.

3   Repeated freeze–thaw action increases the size of the crack until the block of rock breaks off.

4   Loose blocks of rock are called scree.

▲ **Figure 1** How frost shattering/freeze–thaw weathering operates.

▶ **Figure 2** Scree slopes in the South Orkney islands (near Antarctica).

Limestone is little affected by pure water, but rain water is slightly acidic and contains some carbon dioxide. Rain water and carbon dioxide in the atmosphere combine to form carbonic acid, in which calcium carbonate (of which limestone is made) slowly dissolves. Limestone is changed into calcium bicarbonate, which is removed. The limestone is very vulnerable to attack from chemical weathering because of its many lines of weakness, both horizontal (bedding planes) and vertical (joints). The joints visible between the surface blocks of the limestone pavement shown on page 21 were widened by limestone solution to form **grykes** (page 28).

Surface streams disappear underground down joints widened by limestone solution. Underground streams follow the lines of joints and bedding planes. Fresh supplies of acidic water continue the work of solution until a labyrinth of caves is dissolved out of the limestone. Loosened blocks of rock fall from the roofs which have been weakened by solution, turning caves into **caverns**. There is a slow seepage of water charged with lime into the roofs and walls of caves. Lime (calcium carbonate) is deposited when water evaporates or loses its carbon dioxide. The chemical formula shown in the Information Box is reversed. The lime builds up to form **stalactites**, **stalagmites** and **pillars** within caves and caverns (**Figure 3**).

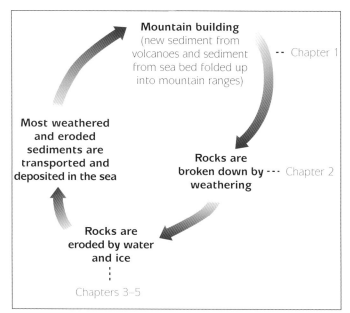

▲ **Figure 4** The Earth's sediment cycle.

Why are weathering processes important?

Weathering loosens and breaks off pieces of rock. It is the first stage in wearing away the rocks of the Earth's surface after mountains have been formed. It speeds up rates of erosion because the loose pieces of rock can be picked up and carried away by rivers, waves and glaciers, which use them as tools for wearing away rock surfaces over which they pass. This is part of the sediment cycle (**Figure 4**).

◀ **Figure 3** A different world awaits exploration underground in areas of Carboniferous limestone.

Activities

1 **(a)** Define 'weathering'.

   (b) Explain the difference between physical and chemical weathering.

2 **(a)** Draw a labelled sketch of **Figure 2**.

   (b) Explain how it shows the results of weathering.

3 Study **Figure 3**.

   (a) Describe the underground features shown.

   (b) Explain how they are formed by chemical changes in Carboniferous limestone rock.

Granite

All the granite rocks in the UK are found to the north and west of the Tees–Exe line (see **Figure 4** on page 23).

Landscape features

In south-west England granite gives relatively flat-topped moorland plateaus with frequent rock outcrops, which from time to time form rock blocks called **tors** (**Figure 1**). Tors are some 5–10 metres high and are surrounded by weathered materials of all sizes from boulders to sand. On the higher parts of the moorlands there are many areas of standing surface water forming marshes and bogs. The many surface streams have cut deeply into the upland block of Dartmoor to form deep and steep V-shaped valleys, especially where rivers such as the Dart go over the edge of the plateau. Dartmoor has a radial pattern of drainage.

▼ **Figure 1** Bowerman's Nose, a tor on Dartmoor.

Dramatic coastal scenery occurs where granite and Atlantic breakers meet, as at Land's End (see **Figure 2** on page 22). In Scotland the granite peaks in the Grampians and on Goat Fell in Arran are rocky and frost-shattered, although where the land is relatively flat, such as on Rannoch Moor, extensive bogs occur.

Granite is a hard rock, resistant to erosion, which is why it forms areas of high relief inland and cliffs along the coast. It is an impermeable rock, which explains why there is so much surface water. Another reason for the presence of so many bogs is the high precipitation in western upland areas.

Formation of tors

The rock which forms tors is that which remains after the surrounding rocks have been weathered and carried away. Where tors occur, the joints in the granite are wider apart

than in the rock around them. Freeze–thaw weathering (see **Figure 1** on page 24) can operate more effectively and blocks of rock break off more quickly where the joints are close together, because there are more cracks in the rock for the water to fill (**Figure 2**). Each time the water freezes and expands within a joint, more pressure is put on the surrounding rock and the crack widens. Where there are fewer joints, it takes longer for the blocks of rock to be broken off and the blocks are left upstanding as tors.

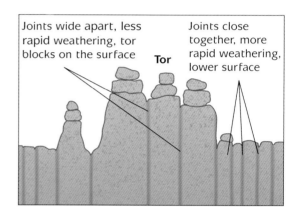

▼ **Figure 2** Effects of joints upon tor formation.

Land use and economic uses

On the higher areas, bog, marsh and moorland produce some of the least useful land in the UK. In some places there may be opportunities for water storage. At lower levels there may still be nothing better than poor grazing land suitable only for sheep and cattle (and, on Dartmoor, also for ponies). Soils are acidic and infertile; it is only around the edges of the uplands that the pastures improve sufficiently to allow grazing by dairy cattle.

Granite is a fine building stone. Aberdeen is known as 'the granite city' since so much use was made by builders of locally available supplies of stone. It is also often used for headstones in graveyards.

Granite rock is susceptible to attack by chemical weathering and in some places it has decomposed. This has resulted in the feldspar in the granite being converted into clay minerals, such as china clay (kaolin). China clay is best known as the raw material for the pottery and porcelain industries, and much is sent to the Potteries region around Stoke-on-Trent. It is also used in the manufacture of paper and is an ingredient in paint, toothpaste, skin creams and many other products.

▼ **Figure 3** Map of Dartmoor.

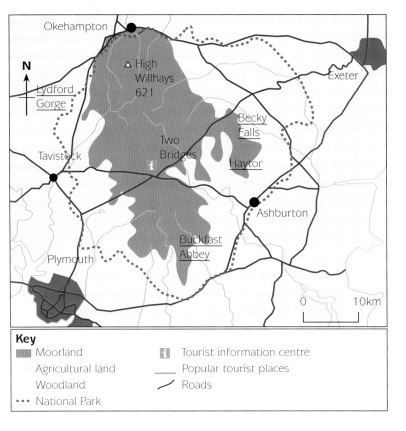

Key

▨ Moorland
☐ Agricultural land
☐ Woodland
••• National Park

ℹ Tourist information centre
⎯ Popular tourist places
╱ Roads

ℹ Information

Dartmoor

- Dartmoor has a high rainfall and is known for its mists and fogs.

- Much of the land is covered by heather.

- The many boggy areas contain a rich variety of plant life.

- The central upland block was enclosed within a National Park in 1951.

- The Park covers almost 100 000 hectares and over 30 000 people live inside it.

- Up to eight million people visit or pass through the Park each year.

- Most of the towns, such as Tavistock, Okehampton and Ashburton, are located around the edges of the central block.

- Places popular with visitors include Buckfast Abbey, Haytor, Becky Falls and Lydford Gorge.

- Some of the remains of old woodlands have been preserved as nature reserves.

Activities

1 Draw a frame the same size as **Figure 1**. Draw a sketch from the photograph and label the landscape features and land uses shown.

2 The tor is the most distinctive granite landform.

   (a) Name and locate an example of a tor.

   (b) With the aid of a diagram, describe the features of a tor.

   (c) Explain how a tor is formed.

3 Answer the questions below using information from **Figure 3**.

   (a) *Relief and drainage*

   (i) Name and give the height of the highest point on Dartmoor.

   (ii) Describe where the moorland is located.

   (iii) In which directions do most of the rivers flow?

   (b) *Land uses*

   (i) Describe and explain the pattern of land use shown in **Figure 3**.

   (ii) Suggest as many reasons as you can why the tourist information centre has been located in Two Bridges.

4 Use the information provided and do some research of your own, including using websites (see Hotlinks, page iv). Produce a short case study on one side of A4 paper with the title 'Tourism on Dartmoor'. Make it interesting and informative; try to use a mixture of maps, sketches, diagrams and written information.

Carboniferous limestone

Although outcrops of Carboniferous limestone occur widely in the uplands of England and Wales, the greatest number and variety of distinctive landforms can be seen in two of the English National Parks – the Yorkshire Dales around Malham and Ingleton, and the Peak District near Castleton. Outcrops of Carboniferous limestone cover much more extensive areas around the Mediterranean – you may have visited caves and caverns while on holiday in places such as Majorca.

Landscape features

Carboniferous limestone weathers by the solution process described on page 24 to produce distinctive landforms both above and below ground. **Limestone pavements** (page 21) are flat surfaces of bare rock broken up into separate blocks. The flat surfaces of the blocks are **clints** and the gaps are **grykes**. Rivers disappear underground either through small holes in the rock, called **sink holes**, or down larger holes with a funnel shape above, called **swallow holes** (**Figure 1**). Underground the limestone is full of holes: small passageways, or cave systems, which from time to time open out into large chambers, or **caverns**. **Stalactites** made of lime hang down from the roofs like long icicles, whilst **stalagmites** are the thicker columns built up from the floor. In places the two meet to form a **pillar** of limestone. Rivers reappear on the surface once they have passed through the limestone outcrops and reach impermeable rocks. In a few places there are surface rivers across the limestone flowing at the bottom of a **gorge**. When many limestone landforms

▲ **Figure 1** The swallow hole at Gaping Gill. Fell Beck is the surface stream seen disappearing underground. It drops 110m as a waterfall into a giant chamber more than 150m long and 30m high (i.e. large enough to fit a cathedral into). The stream flows several kilometres underground through a complex system of caves before it reappears on the surface through Ingleton Cave and forms Clapham Beck.

like those described above occur together in an area, they form **karst scenery** (**Figure 2**).

Occasionally the holes that are formed by solution become so large that the roof collapses. When the roof of a long underground passageway falls in, a deep steep-sided valley, or limestone gorge, forms with the river flowing at the bottom of it. A possible example is Gordale where the blocks which may have formed the cavern roof can be seen as debris on the floor (**Figure 3**).

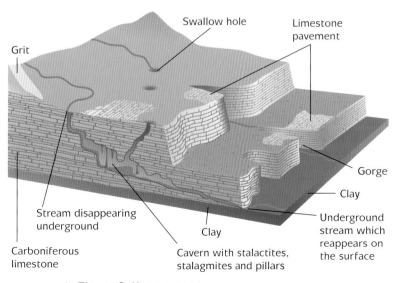

▲ **Figure 2** Karst scenery.

Labels: Grit; Swallow hole; Limestone pavement; Stream disappearing underground; Carboniferous limestone; Clay; Cavern with stalactites, stalagmites and pillars; Gorge; Clay; Underground stream which reappears on the surface

▲ **Figure 3** The gorge at Gordale, a possible collapsed cavern.

Carboniferous limestone scenery near Malham

The stream which disappears underground at A¹ in **Figure 4** reappears at A² at the bottom of Malham Cove. The stream which disappears underground at B¹ in **Figure 4** is shown in **Figure 5**. It reappears on the surface as a spring at B² in **Figure 4**.

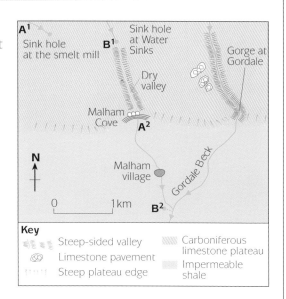

▶ **Figure 4** Location of Carboniferous limestone features near Malham.

Key
- Steep-sided valley
- Limestone pavement
- Steep plateau edge
- Carboniferous limestone plateau
- Impermeable shale

▶ **Figure 5** Sink hole at Water Sinks.

◀ **Figure 6** Dry-stone walls, made of limestone, on a farm in the Yorkshire Dales.

Land use and economic uses

Carboniferous limestone lies on or close to the ground surface, so the soil is too thin to be used for cultivation, and also dry. However, a turf-like grass covers the surface. This is good for sheep farming because they graze short grass. Population density in limestone areas is low, but the limestone landforms are attractive to visitors. Service sector employment has been boosted in the villages and small towns, while some farmers earn a supplementary income from camping and caravan sites and bed and breakfast.

Limestone is of great economic importance. It is widely used as building stone. It is more easily worked than a hard rock such as granite. Limestone has been used in well-known buildings such as St Paul's Cathedral, the Houses of Parliament and the front of Buckingham Palace. When crushed up it is used for chippings for drives or for making concrete and cement. Farmers spread lime on their fields as fertilizer. Limestone is also used as a cleanser in many industries such as smelting steel, to absorb harmful sulphur dioxide from coal-fired power stations, and to purify water. With so many and varied uses there are many quarries in limestone areas, and some visitors feel that these ruin their scenic beauty (pages 32–34).

Activities

1 Describe the physical features of each of the following on a labelled sketch. Explain how it was formed.

  (a) The sink hole at Water Sinks.

  (b) The limestone pavement above Malham Cove.

  (c) The gorge at Gordale.

2 What underground features are likely to be present between Water Sinks and the bottom of Malham Cove? Explain your answer with the help of diagrams.

3 **(a)** Draw a spider diagram to show the economic uses of limestone.

  (b) Why does stonework on buildings made of limestone 'rot away' over time?

Chalk and clay

◀ **Figure 1** Chalk and clay meet at the foot of the South Downs near Fulking. In many places in England chalk (on the right) and clay (on the left) outcrop next to one another; together they form a distinctive but contrasting landscape of chalk escarpment and clay vale. Chalk and clay are both sedimentary rocks and only outcrop in Lowland Britain; they have little else in common.

▲ **Figure 2** Devil's Dyke dry valley east of Fulking.

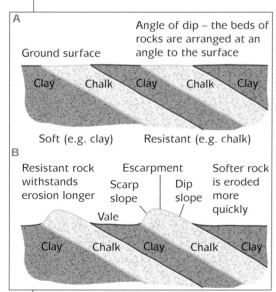

▲ **Figure 3 A:** Rock arrangement needed for the formation of an escarpment. **B:** Formation of an escarpment.

Landscape features

The **chalk escarpment** (also known as a cuesta) is the most distinctive feature of chalk scenery in England. It consists of two parts – the **scarp slope**, which is steep, and the **dip slope**, on which the land falls away more gently. The top of the escarpment has gently rolling hills with rounded summits. There is little surface drainage and rivers are few and far between; however, in places the dip slope has been cut by deep, steep-sided, V-shaped **dry valleys**, which are marked landscape features (**Figure 2**). After spells of wet weather temporary streams may flow in the valleys; these are known as **bournes**, and this term is used in place names such as Bournemouth and Eastbourne.

Chalk outcrops along the coast often lead to high cliffs such as the famous 'white cliffs of Dover', and to prominent headlands, such as Beachy Head in Sussex and Flamborough Head in Yorkshire. Erosion around headlands can lead to the formation of caves, arches and stacks. The Needles off the north-west corner of the Isle of Wight are examples of stacks (Chapter 5).

In contrast, the **clay vale** is a wide and often almost flat area of land. Surface drainage is abundant and the vale is crossed by meandering streams. At the coast, clay forms weak cliffs which slide and collapse.

Formation of the chalk escarpment

There are two requirements before an escarpment can be formed (Figure 3A):

1 Alternate outcrops of different types of rocks. One rock needs to be soft and the other needs to be more resistant to erosion.

2 Beds of rock dip at an angle to the ground surface. Instead of being horizontal, the beds were tilted by earth movements so that they lie at an angle to the surface.

These two needs are commonly met in eastern and southern England. The clay is eroded more quickly than the chalk. As the clay is eroded down into a vale, the chalk is left standing up because of its greater resistance. The scarp slope forms a prominent feature where the layer of chalk reaches the surface. The dip slope is more gentle following the tilt of the beds of rock.

Land use and economic uses

There are great differences in settlement and other land uses between areas of chalk and clay.

Settlement

Some of the earliest human arrivals to the British Isles settled on chalk escarpments. Above the village of Fulking in **Figure 4** there are signs of burial mounds (tumulus) and old defences (fort and motte and bailey). The chalk escarpments were drier than the wet clay vales and contained flint which could be used for tools and weapons by the early settlers. The main problem for settlement in areas of chalk was shortage of water. **Springs** form at the junction of the chalk and clay; water seeping down through the spaces in the porous chalk meets the impermeable clay and reappears on the surface as a flow of water. Settlements grew along the spring line and Fulking is a classic example.

▲ **Figure 4** OS map of the area around Fulking at a scale of 1:50000 (2cm = 1km).

Land uses

The main land use on chalk is pasture. The short but rich turf is good for grazing sheep and training racehorses. Famous racecourses such as Epsom are situated on the Downs. High cereal prices and intensification of farming have led some farmers to plough up gentler and lower slopes for wheat and barley, despite the dry and stony nature of the chalk soils. The main land use on clay is also pasture. In general the soils are too wet and heavy to plough. The grass grows longer than that on the chalk and is more likely to be grazed by dairy cattle.

Economic uses

Chalk with flint provides a strong and attractive building material. Chalk has many of the same economic uses as limestone, such as in the manufacture of cement. The underground stores of water, known as **aquifers**, are widely used for water supply in south-east England. Clay, taken from pits, is a raw material for making bricks.

Activities

1 Describe the differences between the areas north and south of Fulking using these headings.

   (a) Relief (height and shape of the land)

   (b) Drainage (number and density of surface streams)

   (c) Land uses and economic activities (woodlands, farms, settlement, quarrying, etc.).

2 Using **Figure 4**, draw a sketch cross-section from grid reference 226140 in the north (height 10m) to 226080 in the south (height 100m). Add labels for the clay vale, scarp slope and dip slope.

3 **(a)** In which direction was the camera pointing when the photograph for **Figure 1** was taken?

   (b) Describe the additional information about land uses given on the photograph compared with the OS map.

Quarrying

Quarrying is almost as old as settlement in the British Isles. Early settlers used stone for building shelters, defensive works and burial mounds. Demand for stone increased over the centuries as populations grew and economic development occurred. Quarrying limestone, for example, is big business in the Yorkshire Dales today, because limestone has so many uses (page 29). About 5 million tonnes of the rock are extracted each year from quarries located within the Yorkshire Dales National Park (**Figure 1**).

▲ **Figure 1** Horton-in-Ribblesdale quarry.

Quarrying has two major advantages:

• it provides the raw materials for the huge (and increasing) demand for building, road construction, cement manufacture and sea defences

• it is a major source of employment for people in upland areas where few other employment opportunities exist.

Unfortunately, quarrying is an example of a **non-sustainable** activity. It involves removing natural resources from the ground to be used up and not replaced; they will not be available for use by future generations. The disadvantages of quarrying are a mixture of environmental, economic and social. They are illustrated in **Figure 2**. Can you distinguish the environmental disadvantages from the social and economic? Some negative impacts continue well after quarrying has ceased.

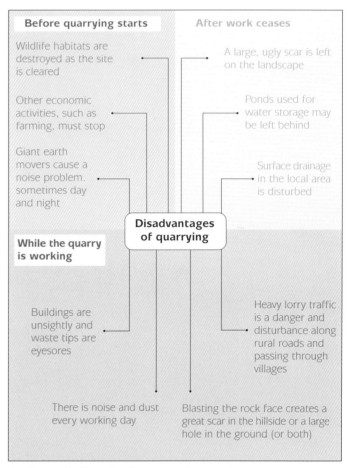

Before quarrying starts

Wildlife habitats are destroyed as the site is cleared

Other economic activities, such as farming, must stop

Giant earth movers cause a noise problem, sometimes day and night

While the quarry is working

Buildings are unsightly and waste tips are eyesores

There is noise and dust every working day

Disadvantages of quarrying

After work ceases

A large, ugly scar is left on the landscape

Ponds used for water storage may be left behind

Surface drainage in the local area is disturbed

Heavy lorry traffic is a danger and disturbance along rural roads and passing through villages

Blasting the rock face creates a great scar in the hillside or a large hole in the ground (or both)

▲ **Figure 2** Negative impacts of quarrying.

Reducing the negative effects of quarrying

Planners and local authorities have an important role to play in placing controls upon the commercial quarry companies, who are really only interested in making the largest profits. A full consideration of likely environmental effects is needed before planning permission is given and work begins. Planning authorities can restrict the size of the quarry, insist that buildings and waste tips are screened by trees in areas of great scenic beauty, limit noisy operations like blasting to certain times and impose binding commitments for cleaning up the site after work finishes. After several years of working, it is common for extraction companies to seek to extend the quarry, driven by strong commercial interests, as well as economic need due to continuing demand for the natural resource. The authorities need to be vigilant and check regularly that the planning restrictions are not being 'overlooked' by these companies.

▲ **Figure 3** Quarry lorry in the centre of Horton-in-Ribblesdale, just below the train station. This was one of nine counted in a 20-minute period on a weekday. How could this negative impact be reduced?

Alternatives after quarrying stops

The most environmentally friendly action is for the quarrying company to fill the hole and replace the topsoil – in order to leave the land looking similar to the way it looked before the work began. Trees, grass and shelter belts can be planted to landscape the land. This is an example of **reclamation**: the land has been reclaimed from quarrying to be used again for farming and other rural land uses. Usually this is only possible where the quarry is small and the company is forced to abide by strict planning rules. It is an expensive option.

Large holes and old quarries are convenient places for the disposal of waste. This is **landfill**, a cheap and easy way to dispose of waste. From time to time large machines are used to level it off and compact the waste that has been tipped into the hole. When full, the land can be reclaimed for other uses such as forestry, farming or

recreation, although it is rarely very productive. The land never seems to look natural. Disposal of waste in landfill sites needs to be managed carefully, otherwise it may contaminate the land and water courses, and become a hazard to the health of people living in the area.

Two large quarries in the UK have attracted special uses. The glass domes of the Eden Centre, which house plants from different world environments, are in a china-clay quarry in Cornwall (**Figure 4**). The large Bluewater out-of-town shopping centre, east of London, is built in an old quarry, in a part of the UK where new building land is in short supply.

Activities

1 (a) Why is quarrying a much-needed and widespread economic activity in the British Isles?

   (b) Draw a labelled sketch based on **Figure 1** to highlight differences in land uses between the quarry and the rural area around it.

2 (a) Rearrange the disadvantages (negative impacts) of quarrying from **Figure 2** in a table, using the headings Environmental, Social and Economic.

   (b) Choose one disadvantage from each heading. For each one:

      (i) explain why it is a disadvantage of quarrying

      (ii) describe ways to reduce its negative effects.

3 It is usual to call a local meeting when an extraction company is seeking permission to increase the size of a quarry. Among the people likely to attend will be:

- retired residents
- young married couples
- the manager of the construction company
- activists from an environmental group
- chairman of the parish council
- an officer from the local planning authority
- the secretary of the local tourist board.

   (a) Write down a list of points that one of these people might use if given the opportunity to speak at the meeting.

   (b) In a debate, how would you expect some of the others to reply to your points? Make a brief list of their likely counter-arguments.

▲ **Figure 4** The Eden Centre near St Austell in Cornwall.

Case Study – Peak District National Park near Castleton

Limestone quarrying and Hope cement works in the Peak District

The village of Castleton (**Figure 1**) grew due to its closeness to lead mines. Some of the limestone caves and caverns in the region that now attract thousands of visitors each year, such as Blue John and Speedwell Caverns, had previously been used by miners. The railway line, which gives access with a station in Hope, was also part of Britain's mining history because it was built to transport salt from Cheshire to Sheffield just before the end of the 19th century. Tourists first arrived in the area by train, but road access is now much more important. Although there are still several farms in the area, many people make at least a part-time living from tourism, in hotels, bed and breakfasts, camp sites, cafés, pubs and shops. The Hope quarry and cement works, however, is the largest single employer in this area (**Figure 2**). Some 300 people are employed, nearly all of whom live locally. Without the works, many more people would be forced to commute by road to towns and cities outside the Park, particularly Manchester and Sheffield.

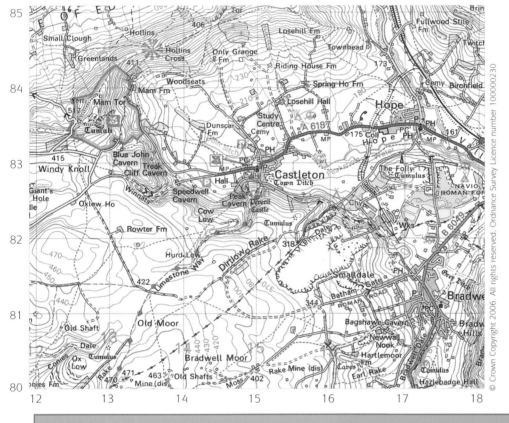

▲ **Figure 2** Hope quarry and cement works.

◄ **Figure 1** OS map of Castleton at a scale of 1:50000 (2cm = 1km).

Activities

1 From the OS map and other information, state the evidence for:

  (a) settlement for many centuries in the area

  (b) the long history of mining in the region.

2 Hope cement works is located in square 1682, north-east of the limestone quarry.

  (a) From the OS map measure the length and width of the quarry in metres and state the approximate area covered by it.

  (b) From **Figures 1** and **2**, describe the other ways in which the presence of the cement works is affecting the landscape.

  (c) Why might transport of the limestone and cement to market be less of a problem here?

3 Castleton attracts many tourist visitors. From the OS map:

  (a) describe the attractions for visitors in and around Castleton

  (b) state the facilities provided for them.

4 Local people in Castleton have differing views about both the quarry and tourist visitors.

  (a) Suggest some of the different views that may exist among the villagers.

  (b) Explain why they have arisen.

Chapter 3

River landscapes and processes

The Iguaçu Falls on the border between Brazil and Argentina are a stunning example of the erosive power of rivers.

Key Ideas

The Earth's crust is modified by river processes that result in distinctive landforms:

- the processes of erosion help to create waterfalls and gorges
- rivers transport and deposit material, helping to form flood plains, deltas and levées
- meanders and ox-bow lakes are formed by both erosion and deposition.

The interaction between people and fluvial environments:

- river flooding has different causes; most are natural, but some are human
- the effects of flooding in both LEDCs and MEDCs can be devastating
- river basin management can reduce the effects, although it raises other issues.

River basins

Rivers begin in upland areas and flow downhill, becoming wider and deeper, until they enter the sea. Where a river begins is called the **source** and where it ends is the **mouth**. Along a river's journey to the sea other smaller rivers called **tributaries** may join the main river at a **confluence**. A river and its tributaries obtain their water from the surrounding land. The area drained by a river and its tributaries is called the **drainage basin** (**Figure 1**). The boundary of the drainage basin is called the **watershed** and it is usually a ridge of high land.

Some of the river's energy is used in transporting loose material downstream. The material is transported in one of four ways (**Figure 2**). The amount of load being carried depends on:

- the volume of water – the greater the volume, the more load it can carry

- the velocity – a fast-flowing river has more energy to transport and can move larger particles

- the local rock types – some rocks, e.g. shales, are more easily eroded than others, e.g. granite.

The water in a river flows within a **channel** unless the river floods and spills out onto the surrounding land. The size and shape of the channel changes as the river flows downstream, becoming wider and deeper. A river also flows within a valley; the size and shape of the valley changes downstream from a steep V-shaped valley to a broad, almost flat V-shaped valley. **Figure 3** shows some of the changes which take place downstream in a river valley.

Many of these changes are caused by changes in the river's energy. In the uplands, close to the source, the river is high above its base level (usually sea level). This gives the river a lot of potential energy. The river is also trying to reach its base level, so the main processes at work are erosional. The river mainly erodes in a downwards direction (vertical erosion) to try to reach its base level. This helps to create V-shaped river valleys in upland areas. As the river moves downstream it uses a lot of energy to transport the material or load it has eroded. Surplus energy is now used to erode sideways (lateral erosion) because the river is much closer to its base level, and so the river valley becomes wider and flatter.

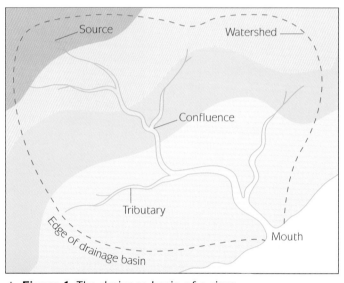

▲ **Figure 1** The drainage basin of a river.

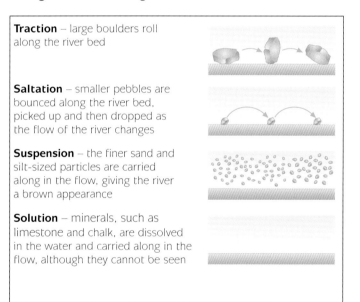

Traction – large boulders roll along the river bed

Saltation – smaller pebbles are bounced along the river bed, picked up and then dropped as the flow of the river changes

Suspension – the finer sand and silt-sized particles are carried along in the flow, giving the river a brown appearance

Solution – minerals, such as limestone and chalk, are dissolved in the water and carried along in the flow, although they cannot be seen

▲ **Figure 2** Transporting the river's load.

The long profile of a river shows an irregular steep gradient at the source, gradually becoming lower and less steep until the gradient is almost nil near to sea level. The normal profile is smooth and concave. The changes in the river valley and features along the profile allow the river to be subdivided into three sections – the upper, middle and lower courses. Study **Figure 3** to find out more about how a river and its valley change downstream.

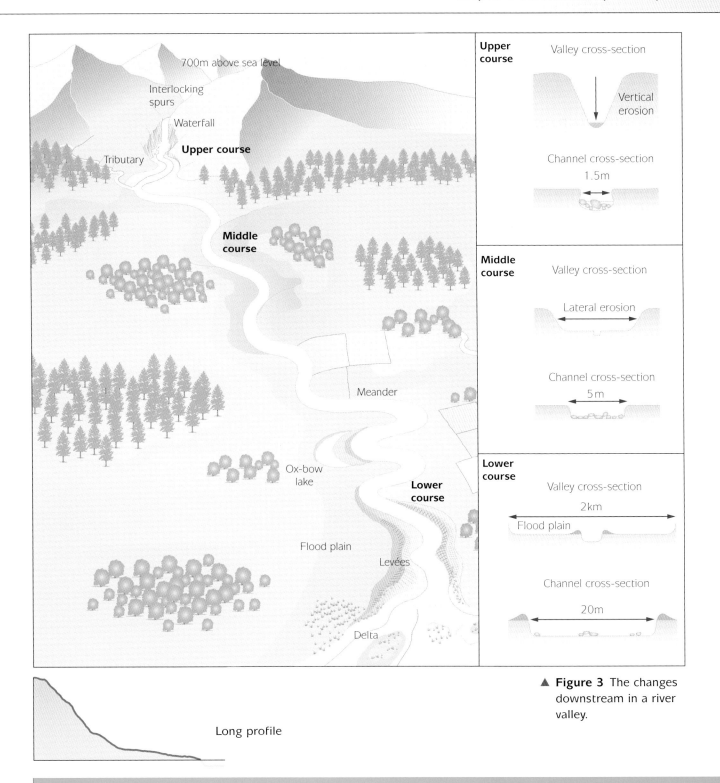

Long profile

▲ **Figure 3** The changes downstream in a river valley.

Activities

1 Write a definition for each of the terms used in **Figure 1**.

2 Draw a side view of a flowing river. Show the different ways in which boulders, pebbles, sand and silt are transported.

3 Fill in a copy of the table on the right using sketches and information from these two pages.

	Upper course	Middle course	Lower course
Long profile			
Valley cross-section			
Channel cross-section			
Examples of landforms			

The upper course of a river

In the upper course of a river, erosion is the dominant process. A river may erode by one of the four processes shown in **Figure 1**.

Hydraulic power

This is the force of the water on the bed and banks of the river. It is particularly powerful when the river is in flood. The force of the water removes material from the bed and banks of the river.

Corrasion

The river carries with it particles of sand and silt and moves pebbles and boulders at times of high flow. This material rubs against the bed and banks of the river and wears them away. This process is also called **abrasion**.

Corrosion

Some rock minerals, such as calcium carbonate in limestone and chalk, slowly dissolve in river water, which is sometimes slightly acid.

Attrition

The load being carried by the river collides and rubs against itself, breaking up into smaller and smaller pieces. The rough edges also become smooth, forming smaller, rounded material. Eventually the particles are reduced to sand and silt-sized particles.

▲ **Figure 1** Processes of erosion.

Landforms in the upper course
V-shaped valleys and interlocking spurs
The vertical erosion in the upper course creates the **V-shaped valley** (**Figure 2**) which is steep-sided and narrow. As the river erodes downwards, soil and loose rock on the valley sides are moved downhill by slopewash or soil creep. The river also winds its way around **interlocking spurs** of hard rock (**Figure 2**), which should not be confused with meanders! There is no flat valley floor and the valley gradient is steep.

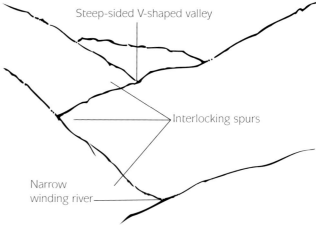

Steep-sided V-shaped valley

Interlocking spurs

Narrow winding river

Sketch from the photograph

▲ **Figure 2** A V-shaped valley with interlocking spurs.

The river channel
The river channel is narrow and shallow; it is often lined with large angular boulders. The gradient of the river may be quite steep and waterfalls and rapids may be found along the river. The velocity of the river is high at waterfalls and rapids but may be quite low in other stretches because so much energy is used in overcoming friction with the rocky bed and banks of the river. The water is often quite clear because the river is not carrying much load in suspension. The river has not had time to grind down the boulders into fine sand and silt-sized particles by abrasion and attrition.

Waterfalls and gorges

A waterfall is a steep drop in the course of a river. It has a high head of water and a characteristic plunge pool at the base. The rocks at the top of the waterfall are often hard and resistant, forming a cap rock, and softer rocks below are undercut (**Figure 3**). The waterfall may lie within a gorge.

Waterfalls often form when a band of resistant rock lies over softer, less resistant rocks. The softer rock is eroded more quickly, causing undercutting of the hard rock. The hard rock overhangs until it can no longer support its weight. The overhang then collapses, adding large blocks of rock to the base of the waterfall. The great power of the water falling to the base moves the material around, eroding the base into a deep plunge pool. The bed of the river below a waterfall contains boulders eroded by splashback from behind the waterfall, and some blocks of rock from the collapse of the hard cap rock.

Over a very long time the process of undercutting and collapse is repeated many times, causing the waterfall to retreat upstream. The retreat creates a steep-sided **gorge**. At the same time chips of the hard cap rock are eroded away, which reduces the height of the waterfall.

▲ **Figure 4** The Iguaçu Falls.

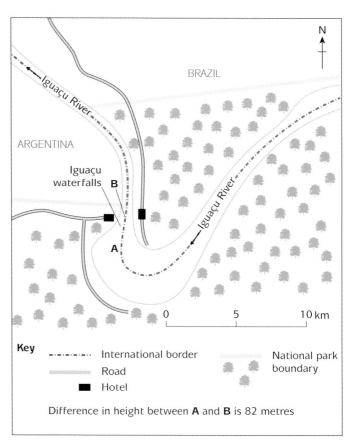
▲ **Figure 5** Map of Iguaçu Falls.

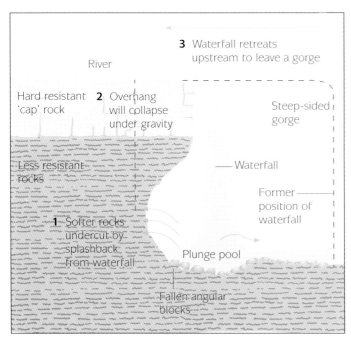
▲ **Figure 3** The formation of a waterfall.

Exam focus

1 a Make a frame and draw a sketch of **Figure 4**. Label features of the river, channel and valley. (5 marks)

b Explain the types of rocks needed for the formation of waterfalls. (3 marks)

c (i) What is the main land use shown in **Figure 4** in the rest of this area? (1 mark)

(ii) State the evidence from **Figures 4** and **5** for the importance of tourism in this area. (2 marks)

d (i) Why would **B** make a good site for a hydro-electric (HEP) power station? (2 marks)

(ii) Suggest reasons why no power station has been built here. (2 marks)

The middle and lower courses of a river

The middle course

As the river flows downstream, the gradient over which it flows becomes less steep and the river is not as high above its base level. The river continues to erode vertically, but lateral or sideways erosion becomes more important. When the river emerges from its upland area it begins to **meander** in order to use up surplus energy. The erosion on the outside of meanders removes the ends of the interlocking spurs and the valley becomes wider and has a more recognizable valley floor.

Meanders are bends in the river's course (**Figure 1**). On the outside of a meander the water is deeper and the current flows faster. The force of the water erodes and undercuts the outside bend by corrasion, forming a steep bank called a **river cliff**. On the inside bend there is slack water and the current is less strong, which encourages deposition. Sand and small pebbles are deposited creating a gentle **slip-off slope**. An underwater current spirals down the river, carrying the eroded material from the river cliff to the slip-off slope.

The lower course

In the lower course the river channel becomes wider and deeper. The velocity is often greater than in the upper course because the channel is more efficient with less friction. The channel is almost semi-circular and much smoother because of deposits of sand and mud. In the lower course a river flows through a wide, flat valley called the flood plain (**Figure 2**).

The meander bends become even larger in the lower course as the river meanders more vigorously (**Figure 3**). Continued erosion on the outer bends and deposition

▲ **Figure 1** Meander formation. **A:** Plan. **B:** Cross-section.

on the inside of the bends may eventually lead to the formation of an **ox-bow lake** (**Figure 4**). The neck of the meander narrows as erosion continues on the outside bends. Eventually the neck is broken through, creating a straight channel. This often happens during a flood when the river is particularly powerful. As the flood waters fall, and at times of low flow, alluvium is deposited which seals off the old meander and forms an ox-bow lake. Gradually the ox-bow lake dries up, forming a **meander scar**.

The river carries a large load of suspended material. Deposition of the sand and silt, called **alluvium**, becomes the most important process. Alluvium is found in great thicknesses on the flood plain, especially where

▶ **Figure 2** Landforms in the lower course.

◀ **Figure 3** The features of a meander.

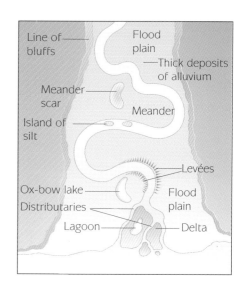

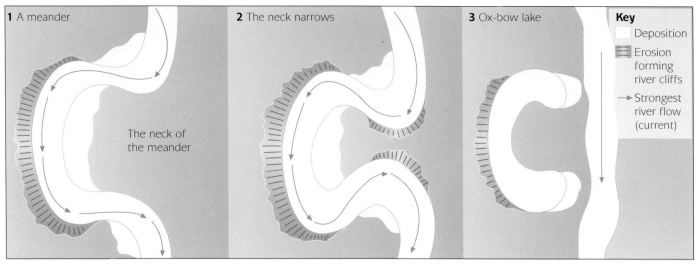

▲ **Figure 4** Formation of an ox-bow lake.

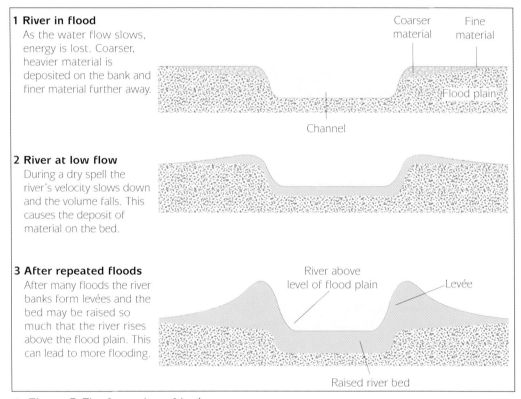

1 River in flood
As the water flow slows, energy is lost. Coarser, heavier material is deposited on the bank and finer material further away.

Coarser material   Fine material

Flood plain

Channel

2 River at low flow
During a dry spell the river's velocity slows down and the volume falls. This causes the deposit of material on the bed.

3 After repeated floods
After many floods the river banks form levées and the bed may be raised so much that the river rises above the flood plain. This can lead to more flooding.

River above level of flood plain   Levée

Raised river bed

▲ **Figure 5** The formation of levées.

Levées

Levées are natural embankments of silt along the banks of a river, often several metres higher than the flood plain. Levées are formed along rivers that flow slowly, carry a large load and periodically flood (**Figure 5**). Every time the river leaves its channel, the greatest amount of sediment is deposited on the edge of the channel where the loss of flow due to increased friction is most pronounced. Big rivers like the Mississippi have built up huge natural levées of 10 metres and higher in places.

levées are formed. Deposition is encouraged by several factors:

- a river carrying a large load, providing a great deal of material for deposition
- a reduction in velocity such as at the inside bend in a meander
- an obstruction, e.g. a river enters a lake and velocity falls, or it meets waves and currents, or bridge supports interrupt flow
- a fall in the volume of river water, e.g. at times of low flow during a period of drought.

Activities

1 (a) What is meant by lateral erosion?

(b) Where and when does it occur?

(c) Which landforms are formed by it?

2 (a) Draw a labelled diagram to show the differences between a river cliff and a slip-off slope.

(b) Explain why they are found on different sides of the channel.

3 Explain how river flooding leads to the formation of **(a)** levées and **(b)** ox-bow lakes.

Flood plains and deltas

The flood plain

The flood plain is the wide, flat area of land either side of the river in its lower course. The flood plain is formed by both erosion and deposition. Lateral erosion is caused by meanders and the slow migration downstream to widen the flood plain (**Figure 1**). The deposition on the slip-off slopes provides sediment to build up the valley floor. This is added to during a flood when the river spills over its banks onto the surrounding land. The river carries with it large quantities of suspended load. As the water floods onto the flood plain there is greater friction, the water is shallow and the river's velocity falls so its load is deposited onto the flood plain as alluvium. Over many thousands of years these deposits build up into great thicknesses of alluvium.

Deltas

A delta is the landform that is created when a river splits up and flows into the sea through more than one channel. The many separate channels are known as distributaries; the land between them is flat and made of river silt. A delta is new land built into the sea and is usually triangular in shape. The shape shows up well on the satellite image of the Nile Delta (**Figure 2**). The black areas near the sea are swampy lagoons of trapped sea water, which are also common delta features.

Key
| ■ | Clear water | ■ | Settlement |
| ■ | Vegetation | ■ | Desert/bare land surface. |

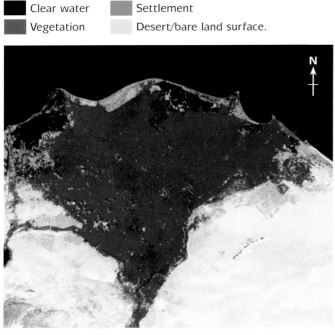

▲ **Figure 2** Satellite image of the Nile Delta in Egypt.

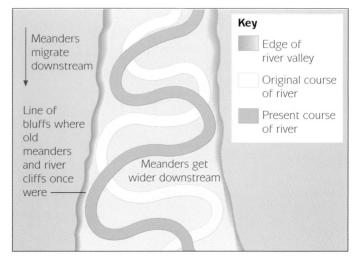

Meanders migrate downstream

Line of bluffs where old meanders and river cliffs once were

Meanders get wider downstream

Key
■ Edge of river valley
□ Original course of river
■ Present course of river

▲ **Figure 1** Meander migration.

For a large delta to form, the following conditions are necessary:

- the river carries a large load of sediment – from active erosion upstream and from having a large drainage basin

- the river is slow-flowing over a large flood plain – and it is slowed down more when it reaches the sea because sea water containing salt is denser than fresh water and holds back its flow

- the sea lacks strong tides or currents that would wash away the sediments faster than they could accumulate to build up the delta.

Where favourable conditions for delta formation exist, the river's flow is blocked by deposition, which forces the river to split up into distributaries. Silt is deposited by each of the distributaries and they spread it out over a wide area. More and more silt is deposited into the sea until a sufficient thickness accumulates to form new land.

There are no large deltas around the coast of the British Isles – the rivers are not big enough and many coastal stretches are noted for strong tides and currents. However, there are many in the enclosed Mediterranean Sea, such as the Nile and Rhone, where the tidal range is low and currents are weak. The Mississippi, with its huge drainage basin draining almost half of the USA (see **Figure 2** on page 50), has a huge delta in the Gulf of Mexico.

Flooding

Flooding is a normal occurrence in the lower course of a river, and is why a flood plain is created by a river. A flood occurs when the water in a river overtops its banks and leaves the channel. Most floods occur because of the weather, for example:

- long, continuous periods of rainfall, as happened in many parts of the UK in 2000 (page 52)

- a cloudburst in a thunderstorm which causes large amounts of run-off

- a sudden increase in temperature that rapidly melts snow and ice. In winter the water often cannot seep into the ground because it is still frozen.

Sometimes human activities can also make the flooding worse:

- building new towns or increasing an urban area makes surfaces impermeable so that more water runs off the surface

- deforestation reduces interception and increases run-off

- occasional disasters occur, such as a burst dam.

After a rain storm, the discharge of a river usually increases. However, two drainage basins may react very differently from a rain storm producing identical amounts of rainfall. The River Tees, for example, is a 'flashy' river. This means that it can rise very quickly after a rainfall event and also fall equally fast following a dry spell. The River Tees and other similar rivers are therefore more likely to flood. The discharge in other rivers may be much more even and the flood threat is then much less.

▲ **Figure 3** River in flood.

Activities

1 **(a)** Describe the main physical features of flood plains.

   (b) Why are they called flood plains?

2 Study **Figure 2**.

   (a) Draw a sketch to show the shape and land uses in the Nile Delta.

   (b) How and why is the delta different from the surrounding land to the south and west?

3 Study **Figure 4**.

▲ **Figure 4** River Cuckmere in Sussex near to the sea.

   (a) Make a frame and draw a labelled sketch to show the features of the river and its valley.

   (b) Explain what could happen in the future at the point marked **X**.

   (c) In times of flood, will all the land on the photograph be under water? Explain your answer.

4 'The larger the river, the more likely it is to form a delta.' Explain this statement.

Changes downstream in the Tees Valley

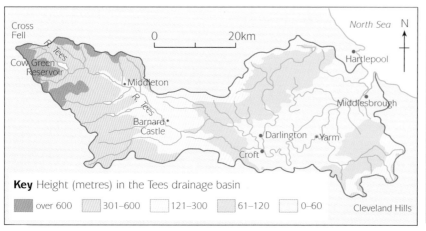

▲ **Figure 1** The drainage basin of the river Tees.

Key Height (metres) in the Tees drainage basin

over 600   301–600   121–300   61–120   0–60

▲ **Figure 3** High Force.

The River Tees (**Figure 1**) is located in north-east England. Its source area is high in the Pennines in the west and the river flows eastwards into the North Sea.

In the uplands

The source of the River Tees lies on Cross Fell in the Pennines, 893 metres above sea level, where rainfall is over 2000 milimetres a year. Run-off is high because of the impermeable rocks and the steep slopes. **Figure 2** shows part of the upper course of the River Tees. The valley cross-section is steep-sided and V-shaped and the long profile has a steep gradient. The river occupies the whole of the valley

floor. The river is turbulent and clear, although often stained brown by the peat which covers much of the moorlands. The river bed is rocky and there are many rapids and a waterfall at High Force (**Figure 3**).

High Force in grid square 8828 is one of the largest waterfalls in England, with a very deep plunge pool at its base. The cap rock is made of a very resistant igneous rock called **whinstone**. Below the whinstone there are bands of sandstone and shales as well as some very thin coal seams. These rocks are less resistant and erode more easily, creating an overhang in the cap rock. Over many thousands of years the waterfall at High Force has retreated upstream, creating an impressive gorge below it.

▼ **Figure 2** OS map of part of the upper course of the River Tees at a scale of 1:50 000 (2cm = 1km).

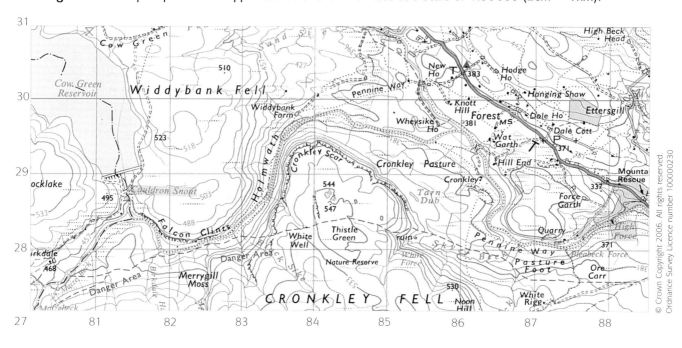

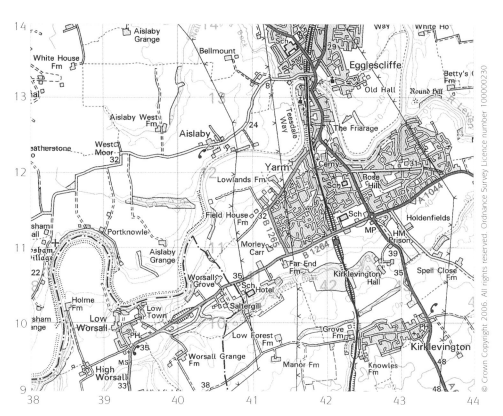

▲ **Figure 4** The River Tees meandering through Yarm. Scale 1:50 000 (2cm = 1km).

In the lowlands

Moving downstream the valley begins to widen and the river starts to meander (**Figure 4**). There are more bridging points and larger villages and towns, such as Yarm. Nearing the river mouth the river meanders in large loops across its flat flood plain. It is 30 kilometres as the crow flies from Darlington to Teesmouth but the river travels 75 kilometres. It used to be longer but several of the meanders were cut off in the nineteenth century to shorten the journey for boats navigating the river up to Stockton and Yarm.

The mouth of the River Tees is shown on the OS map in **Figure 3** on page 187. The river channel is at its widest here. At Teesmouth the river flows between huge areas of mud flats and sands such as Seal Sands, which are important wildlife areas for migratory birds and seals. There is plentiful map evidence of the importance of the estuary for shipping and large expanses of flood plain are covered by heavy industries.

Activities

1 From **Figure 1**, describe how the following features change between upland and lowland areas in the Tees drainage basin:

  (a) the height of the land

  (b) the channel shape of the River Tees

  (c) the number and appearance of the tributaries.

2 In the upper course of the River Tees:

  (a) describe the shape of the Tees Valley at Holmwath named in squares 8328 and 8329

  (b) explain the irregular long profile at High Force

  (c) state the land uses and explain the low density of population.

3 In the lower course of the River Tees (see **Figure 3** on page 187 and **Figure 4** on this page):

  (a) describe the channel and valley features of the River Tecs

  (b) state the land uses and explain the high density of population.

River-basin management

▼ **Figure 1** Dealing with floods: what are the options?

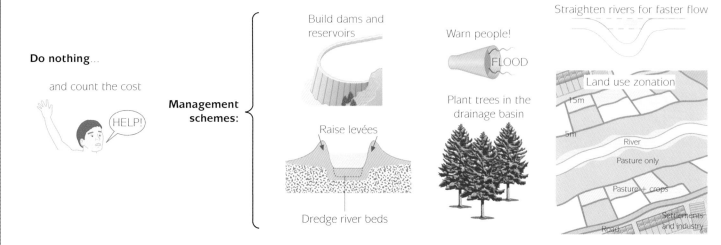

Do nothing…

and count the cost

HELP!

Management schemes:

Build dams and reservoirs

Raise levées

Dredge river beds

Warn people!

FLOOD

Plant trees in the drainage basin

Straighten rivers for faster flow

Land use zonation

15m

5m

River

Pasture only

Pasture + crops

Settlements and industry

Road

Management means planning ahead and controlling development and change. Rivers have many uses for people – how many different ones can you name? Therefore people attempt to manage rivers for their own purposes. In addition, a lot of people throughout the world live close to rivers and need to use management methods to try to reduce the flooding that occurs naturally on the sides of all rivers, especially in their lower courses.

Methods of river management

Some of the options for dealing with floods are shown in **Figure 1**. Building retaining dams, increasing the height of levées with concrete embankments and building walls to confine the channel course through urban areas are examples of hard engineering. In the short and medium term these usually afford the greatest protection, but they are expensive to build and require costly maintenance. In the long term, they offer less security than more sustainable methods such as zoning land uses so that no permanent structures are built on the flood plain.

The River Tees – a case study of river management

▲ **Figure 2** High walls and metal gates that can be closed in times of flood protect expensive housing on the banks of the River Tees in Yarm.

The River Tees has a long history of flooding. The first documented flood was at Croft (see **Figure 1** on page 44) on the lower Tees in 1356. The Tees valley is also home to a large population and many industries, all requiring a reliable water supply. The river is managed to provide a water supply and to control flooding. In recent years there have also been developments to increase its potential for recreation and tourism.

Cow Green reservoir was built in 1970 to provide water for the growing industries on Teesside. It is a regulating reservoir, storing water in times of plenty and releasing enough for the needs of industry in times of low flow. In times of severe summer drought, water can be added to the River Tees via a tunnel which connects it to the River Tyne and Kielder reservoir.

Yarm's flood defence scheme

Yarm, a historic market town and once an inland port, is located on the inside bend of a large meander (see **Figure 4** on page 45). Yarm is particularly prone to flooding. The most recent serious flood was in January 1995. Since then a new flood defence scheme costing £2.1 million has been built with:

- reinforced concrete walls with flood gates for access by people and vehicles

- earth embankments

- gabions (baskets filled with stones) to protect the walls and embankments from erosion

- fishing platforms, street lighting and replanting to improve the environment

- building materials approved by English Heritage to be in keeping with existing architecture.

Cutting of meanders

The tidal mouth section is the most heavily managed part of the Tees Valley. As early as 1810, the Tees Navigation Company cut across the neck of the Mandale Loop, a large meander near Stockton. The new route shortened the river by 4 kilometres. Other stretches of the river have been artificially straightened. This allows the water to move faster along the channel, reducing the flood risk.

Dredging

Ever since, the lower stretches of the Tees estuary have been dredged periodically to improve navigation by maintaining a deep-water channel.

The Tees Barrage

In 1995 the Tees Barrage was completed at a cost of £54 million. The 22 kilometre stretch of river between Yarm and Stockton is now kept permanently at high tide. The water is fresher and cleaner as it does not mix with the tidal, salt water in the lower estuary. The barrage also reduces the risk of flooding at very high tides or during a storm surge. The barrage has acted as a catalyst for £500 million of investment in offices, housing, educational, leisure and shopping facilities.

Activities

1 Make a copy of the table below and complete it for the management schemes shown in **Figure 1**. The first one has been done for you.

Option	Impact	Advantages	Disadvantages
Do nothing	None, may discourage people from settling	Cheap River naturally floods Fertile silt/water supply for farming	Floods homes, fields, roads, services, etc. Costly to clean up and repair damage

2 (a) Arrange the different methods used to manage the River Tees under the headings hard and soft strategies for river management.

(b) Choose one hard and one soft strategy used to manage the Tees. For each one, state where and why it was used.

(c) Explain why the attitudes of the following groups of people to methods of river management are often different:

(i) residents

(ii) Local Authorities and agencies paying for the works

(iii) environmental groups based in the large cities (e.g. London).

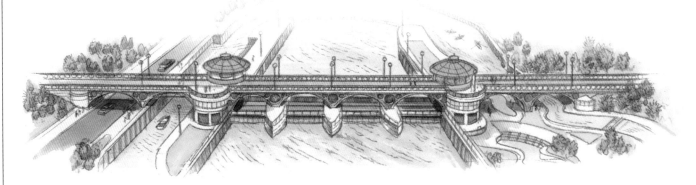

▲ **Figure 3** The Tees Barrage.

The 2004 floods in Bangladesh

Physical causes of the floods

• Most of the country is the huge flood plain and delta of the rivers Ganges and Brahmaputra
• 70 per cent of the total area is less than 1 metre above sea level
• Rivers, lakes and swamps cover 10 per cent of the land area
• Heavy monsoon rain falls in summer; annual total in Dhaka is high (almost 2000mm)
• Tropical cyclones from the Bay of Bengal bring heavy rain and storm waves in late summer
• Snow melts in the Himalayas in summer and the River Ganges floods

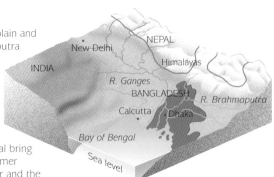

▲ **Figure 1** The natural (physical) causes of flooding in Bangladesh.

There are good **physical reasons** (relief, drainage and climate) why Bangladesh suffers from flooding almost every summer (**Figure 1**). Most of the country is delta and flood plain, landforms formed by regular river floods. The amount of surface water makes them challenging environments for people. At the same time, they are attractive environments because the deep, constantly renewed silt soils are extremely fertile. These support some of the highest agricultural densities of population in the world, especially in a country where padi rice is the main food crop.

Bangladesh is a populous country of 140 million people. When the flooding is more severe than usual, as it was in 2004, considerable loss of life and great suffering result (**Figure 2**). In September 2004, Dhaka had its worst rains for 50 years; on 13 September, 350 milimetres fell in 24 hours.

Possible human causes of the floods

The sources of the Ganges and Brahmaputra rivers are in Nepal and Tibet. In recent years the populations there have grown rapidly, causing the removal of vast areas of forest to provide fuel, timber and grazing land. In Nepal, 50 per cent of the forest cover that existed in the 1950s has been cut down. The forests absorb water from the ground, bind the soil particles and reduce the impact of rain droplets on the ground surface. The removal of the forest cover has increased soil erosion and overland flow. The soil is deposited in the river channels causing the raising of the river beds and reducing the capacity of the rivers. It has been estimated that the river bed of the Brahmaputra is rising by 5 centimetres each year. The building of the Farakka dam in India in 1971 is blamed for raising the river bed of the Hooghly River, a tributary of the Ganges. This increases the risk of flooding.

▲ **Figure 3** Human misery caused by flooding in Bangladesh.

It has been noticed that the interval between big floods is shortening – it used to be between 10 and 15 years, but the 2004 flood came only six years after the major disaster in 1998. This has led to the suggestion that human activities are a

Immediate effects	Later effects
• floods covered over half of Bangladesh • 760 people were killed • 8.5 million people were left homeless • more than 35 million people were affected • rice growing and fish farming were disrupted • roads and bridges were damaged and destroyed	• more than 1 million children suffered from malnutrition and disease in the following months • government rebuilding costs for roads andindustry were estimated at US\$ 2–3 billion • emergency food aid was needed until the following year's harvest

▲ **Figure 2** The effects of flooding in 2004.

Figure 4 Action Plan for Flood Control ▶ in Bangladesh.

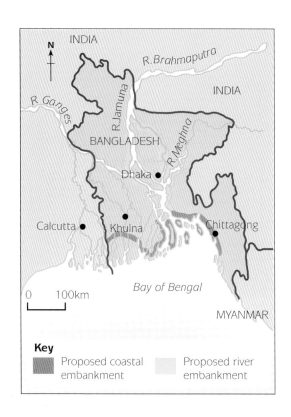

Key
Proposed coastal embankment
Proposed river embankment

contributing factor of increasing importance. If they are, Bangladesh has little control over this. Low-lying, delta countries like Bangladesh are the first to feel the effects from rising sea levels due to global warming.

Flood management

Bangladesh is one of the world's poorest countries, with a GNP of only US$ 330 per head (page 227). In the *short term*, the prime concern is always for the health, survival and suffering of the people affected. A heavy reliance is placed upon emergency aid (food, drinking water, medicines, plastic sheets, boats for rescuing people and animals) from international organizations such as the United Nations, governments in rich countries and charities (pages 246–7). One big problem is distribution, because so much of the country is under water. As the flood waters recede, it becomes easier to set up medical treatment centres, distribute water purification tablets and provide help with repairing houses and restarting economic activities such as farming and fishing.

However, these actions can do nothing to manage the flood problem in the *long term*. In July 1987 the World Bank prepared an Action Plan for Flood Control (**Figure 4**). The plan involved the completion of 3500 kilometres of coastal and river embankments and included seven large dams – partly to stop water from reaching the land and partly to provide up to 15 floodwater storage basins. This is an example of the 'hard engineering' approach to flood management. Millions of dollars of aid were poured into these engineering projects, but the total scheme remains unfinished due to a mixture of corruption and inadequate funding.

From the beginning, some people suggested that what Bangladesh really needed was a mixture of strategies involving flood forecasting and early-warning schemes on the one hand and more flood shelters with supplies for emergency needs on the other. It was argued that these would be both cheaper and more appropriate for farming and fishing communities in rural areas. They would be more in keeping with local knowledge, skills and income levels. They would have the additional advantage of being less likely to damage delicate ecosystems, thereby making a contribution to sustainable development.

Activities

1 Put together the information needed for a 'Case study of flooding in an LEDC – Bangladesh 2004' using the headings:

• Causes of the floods – physical, human and the relative importance of each of these

• Effects of the floods – immediate, later and their overall scale

• Strategies for managing floods – short-term, long-term and relative chances of success.

2

Government Minister
Neighbouring countries are to blame for cutting down trees and blocking the natural flow of the Ganges by building large dams.

Pro-government newspaper
Bangladesh is water and water is Bangladesh. Floods are nothing new to us. People cannot blame the government for this flood.

Anti-government newspaper
Due to bad management and corruption, flood-control projects remain unfinished. Builders bribe politicians to let them build blocks of flats and shopping plazas on waterways used for flood water.

University expert from overseas
Flooding in Bangladesh can never be stopped. Their real problem is population growth of over 2 per cent per year. Every year millions more poor people are being forced to live on flood plains.

(a) Why do people have different views about flooding in Bangladesh?

(b) Do you think that flooding in Bangladesh can be stopped? Give reasons to support your view.

The 1993 Mississippi floods in the USA

▲ **Figure 1** The 1993 Mississippi river floods.

The floods in 1993 were the worst since records began (**Figure 1**). At the peak of the flood the Mississippi river was up to 18 metres deep, 25 kilometres wide and flowing at a speed of 96 kilometres per hour. The floods began when snow melting in the spring was followed by 50 days of very heavy rain and thunderstorms across the American mid-west. The area of the USA affected by the flooding was larger than the whole of the UK (**Figure 2**).

The effects of the flood

- Lives lost: 28.
- Homes lost: 36 000 people.
- Many more people were evacuated.
- Roads and railway lines were under water.
- Electricity lines collapsed, leaving towns without power.
- Six million acres of farmland were flooded, ruining maize crops.
- Millions of tonnes of silt and sand were deposited in the flood zone. This needed to be cleared after the flood.
- Estimated US$ 20 billion were needed to repair the flood damage.

Flood-protection measures

For many years the US Corps of Engineers have tried to reduce the flood risk and prevent serious floods affecting land and property. They have raised and strengthened levées (**Figure 3**), excavated cut-offs to straighten out meanders, dredged the river bed and built revetments (protective walls).

The levées were meant to prevent flood water from spilling out of the river onto the surrounding land. In 1993 the floods were just too great and the water rose over the tops of the levées. Emergency action was taken. In some places, massive quantities of boulders were piled on top of the levées to make them higher, in the hope of keeping the river in its channel. Houses in danger of being flooded were protected by sandbags. In just one town, Sainte Genevieve, 750 000 sandbags were used.

Issues raised by the 1993 floods

After the floods some people began to question the wisdom of flood prevention schemes such as levées. They argued that less damage would have been done if the flood water had been allowed to extend gradually and spill out over farmland rather than the catastrophic flooding which occurred when levées failed. The water would have been absorbed by the land and would have flowed back into the channel once the floods began to recede. Others argue that people have no right to settle on the

Mississippi Basin

▲ **Figure 2** The extent of the flooding in 1993 in the USA.

flood plains and to cover large areas with concrete and tarmac which make them impermeable. They believe that the function of the flood plain is to be a store for water when the river floods.

However, one of the contributory reasons to the severity of the Mississippi floods in 1993 was their occurrence in the upper part of the drainage basin, where levée systems were less well developed. This was because most of the previous big Mississippi floods had been either in the lower course, south of St Louis after the confluences of the big Ohio and Missouri Rivers, or in the delta region, which can receive vast amounts of heavy rain during the hurricane season. Just as the river levels were too high for the channel defences in 1993, the levée systems built to protect New Orleans could not cope with a hurricane as strong and vicious as Katrina in 2005 (page 98) – despite the USA's great wealth.

▲ **Figure 3** A raised and strengthened levée on the Mississippi, an example of hard engineering.

Activities

1 Produce a newspaper article with illustrations for the 1993 Mississippi floods. Your article should contain the following:

- an eyecatching headline
- the dates of the floods
- the causes of the floods
- the long-term and short-term effects of the floods
- the response to the floods.

2 (a) List the similarities and differences in effects between the Mississippi flood in 1993 and the flood in Bangladesh in 2004.

(b) Why do big floods in MEDCs often cost more but lead to less loss of life than in LEDCs?

3 (a) State the advantages and disadvantages of hard engineering schemes for managing river floods.

(b) Give two reasons why some people are promoting soft engineering solutions for managing river floods.

(c) If you were responsible for designing a flood protection strategy in your local area, which schemes would you consider and why?

2000 – a very wet year in the UK

The year 2000 was the wettest for a century in the UK and one of the four wettest years since records began in 1766. April and December were particularly wet months (**Figure 1**); the annual average was 40 per cent above normal. This resulted in frequent and widespread river floods throughout the UK. These reached their peak in autumn, although parts of the north of England were flooded in summer as well; yet in terms of average for the UK, summer was the only consistently dry time of the year. For example, in north-east England up to 75 millimetres of rain fell during the first three days of June, leading to scenes like those shown in **Figure 2**.

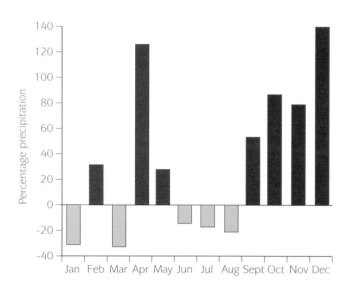

▲ **Figure 1** Percentage monthly precipitation in the UK in 2000 compared with the long-term average.

Exam focus

1 a Study **Figure 1**.

   (i) In how many months was rainfall above average in the UK in 2000? (1 mark)

   (ii) Which was the driest season in 2000? (1 mark)

   (iii) Suggest reasons why flooding was more widespread in autumn than in spring. (3 marks)

b (i) Describe the scene shown in **Figure 2**. (2 marks)

   (ii) The channel of the River Wear is lined by high walls in Durham City. Why did the river still flood? (2 marks)

c Read these two comments about flooding in the UK in 2000:

   A 'In our overcrowded country we are building too close to rivers. We are building too many new housing estates and shopping centres with hard surfaces; heavy rain just bounces off them.'

▲ **Figure 2** The River Wear burst its banks in Durham city centre in June 2000.

   B 'We must learn to accept natural disasters. Nothing can be done to protect us when one month's rain falls in a few hours.'

   (i) Explain each of the comments about the causes of the floods in 2000. (4 marks)

   (ii) With which one do you agree most? Explain your choice. (2 marks)

Chapter 4
Glacial landscapes and processes

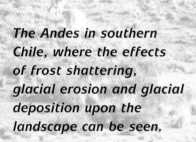

The Andes in southern Chile, where the effects of frost shattering, glacial erosion and glacial deposition upon the landscape can be seen.

Key Ideas

The Earth's crust is modified by glacial processes which result in distinctive landforms:

- glacial processes (abrasion and plucking) combine with freeze–thaw weathering to produce landforms of erosion
- in upland areas valley glaciers form landforms of erosion such as corries, glacial troughs and ribbon lakes
- in lowland areas valley glaciers and ice sheets form landforms of deposition such as moraines and drumlins.

There are interactions between people and glacial environments:

- humans use upland glaciated areas, especially for farming, forestry and tourism
- current issues are related to farming, forestry and tourism
- management strategies are needed for conservation and sustainability.

Valley glaciers and ice sheets

Freshly fallen snow is composed of ice crystals and many air spaces. When you make a snowball, you compress the snow and remove the air spaces. The same happens naturally when snow accumulates; the weight of snow above compresses the air out of the snow below and converts snow into ice. As the ice becomes thicker it will move down the slope by its own weight. When ice moves it is called a **glacier**. Glaciers are of two main types.

1 **Valley glacier** – a moving mass of ice in which the movement is confined within a valley. It begins in an upland area and follows the route of a pre-existing river valley (**Figure 1**). Today most valley glaciers are found near the tops of young fold mountain ranges, such as the Alps, Andes, Rockies and Himalayas. Examples include the Mer de Glace near the ski resort of Chamonix in the Mont Blanc region of south-east France and the Rhône glacier in south-east Switzerland, the source of one of Europe's largest rivers.

◀ **Figure 1**
A valley glacier reaching the sea in southern Chile.

▼ **Figure 2** Antarctica covered by its ice sheet with just enough bare rock for the location of a base for Chilean scientists.

2 **Ice sheet** – a moving mass of ice which covers the whole of the land surface over a wide area. In some cases the ice is sufficiently thick to blanket the entire area of a continent. In Antarctica, where only the peaks of some high mountains stick through the ice, only a tiny strip of bare rock is exposed along a few parts of the coast in summer (**Figure 2**).

Processes of glacial erosion

There are two main processes (or ways) of glacial erosion.

1 **Abrasion** – rocks and rock particles embedded in the bottom of the glacier wear away the rocks over which the glacier passes. These sharp-edged pieces of rock of all sizes are held rigid by the ice above and are used as the tools for abrasion. Smaller rock particles have a sandpaper effect on the rocks over which the ice passes, while the sharp edges of the large rocks make deep grooves, called **striations** (Figure 3).

2 **Plucking** – this is the tearing away of blocks of rock from the bedrock as the glacier moves. These blocks of rock had been frozen to the bottom of the glacier where water had entered joints in the rock and become frozen. The blocks of rock between the joints are pulled away or plucked.

Glaciers would be less effective at eroding the landscape without the help of freeze–thaw weathering (page 24). Before the ice advanced, freeze–thaw left many frost-shattered rocks which were easily removed by the glacier and then used as tools for abrasion. Even when

▲ **Figure 3** Striations on the hard rocks which outcrop in Central Park in New York. They are useful to geologists for working out the direction of ice movement.

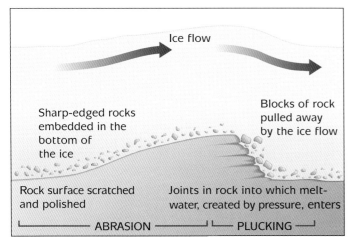

▲ **Figure 4** How the rocks below ice are eroded by glacial processes (abrasion and plucking).

the ice is present, freeze–thaw action affects rocks which outcrop above the surface of the ice because, in a cold climate, there are likely to be many changes of temperature above and below freezing point.

Of the two types of glacier, valley glaciers are considered to be more effective agents of erosion than ice sheets. Confined in a valley, the ice touches both the floor and the sides so that there is more contact between the ice and the rock and therefore more erosion. Also, valley glaciers flow more quickly, partly because of steeper gradients and partly because more meltwater is present to lubricate their flow. There is a plentiful supply of rock fragments from the frost-shattered peaks above, so these glaciers are well supplied with tools for abrasion. However, ice sheets cover and therefore erode a much greater area, so although they erode more slowly, they can still remove a large total amount of rock. As with the other agents of erosion, rocks that are soft or have weaknesses, such as many joints, are eroded more quickly, both by valley glaciers and by ice sheets.

Distribution of landforms of glaciation in the British Isles

If you are sitting north of the line from London to Bristol and take a look out of the window, it must be difficult for you to imagine that just 40000 years ago all the land you can see would have been part of a snow- and ice-covered white wilderness. The British Isles was invaded

ℹ️ Information

Pleistocene Ice Age

- It began 2 million years ago.
- At its peak 30 per cent of the Earth's surface was covered by ice.
- It ended just 10000 years ago.

by ice sheets from Scandinavia during the **Pleistocene Ice Age**, which covered everywhere except for the extreme south of England. In the higher areas, such as the Cairngorms, the Lake District and Snowdonia, heavy snowfall led to the accumulation of snow and ice in hollows on the rocky mountainsides. These were the sources for valley glaciers which flowed down valleys previously eroded by rivers. Over the past 10000 years the world has warmed up. Today no part of the British Isles lies above the **snow line** (the line above which snow and ice remain all year). However, the present-day landscapes of the British Isles show plenty of signs that ice sheets and glaciers once covered the land (**Figure 5**).

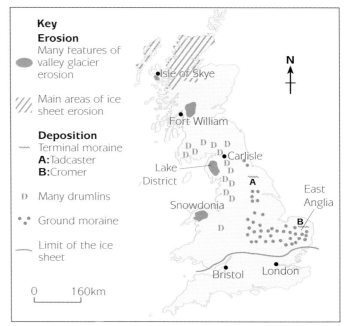

▲ **Figure 5** Distribution of glacial landforms from the Ice Age. The landforms of erosion are explained on pages 56–59 and those of deposition are dealt with on pages 60–61.

Activities

1 State the similarities and differences between glaciers and ice sheets.

2 **(a)** Draw labelled diagrams to show how each of the following processes operates:

  (i) freeze–thaw weathering

  (ii) plucking

  (iii) abrasion.

  (b) Explain why the breakdown and removal of rock is quicker when:

  (i) all three processes operate in the same area

  (ii) rocks have many lines of weakness.

glacial erosion: corries and mountain peaks

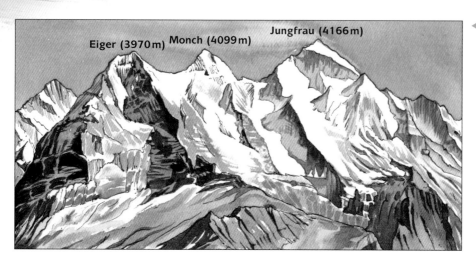

Eiger (3970m) Monch (4099m) Jungfrau (4166m)

◀ **Figure 1** Panoramic view of part of the Swiss Alps taken from a tourist leaflet for the Jungfrau Region. You should be able to identify the corries, arêtes and pyramidal peaks after studying these two pages. How many visitors to the region can do this?

Glaciers modify and enlarge landscape features which existed before the Ice Age. The effects of glacial erosion upon the landscape are greatest in upland areas where glaciers have been present for the longest time and the ice was deeper. In general, slopes are steeper and peaks narrower, especially in areas where processes of glacial erosion are still operating (**Figure 1**).

Corrie (cirque)

The first landform formed by a glacier is the **corrie**. This is a circular rock hollow (hence the alternative name of **cirque** used by some geographers), usually located high on the mountainside, with a steep and rocky backwall

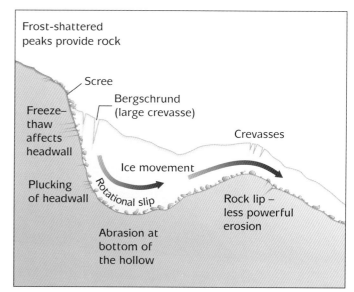

▲ **Figure 3** Formation of a corrie.

up to 200 metres high in the UK, but much higher in the Alps. Although most of the corrie is ringed by steep rocks leading to sharp rocky ridges, the front is open with nothing more than a small rock lip on the surface. The hollow is typically filled with a small round lake, called a **tarn**, after the ice has melted. The information above *describes* the corrie; **Figure 2** shows a corrie which matches this general description.

Corries begin where snowfields (called **névés**), which accumulate below the mountaintops, form ice and grow. As with many landforms, it is necessary to refer to several different processes in order to explain the *formation* of the corrie (**Figure 3**).

▲ **Figure 2** Corrie hollow occupied by Red Tarn on the side of Helvellyn in the Lake District.

- *Freeze–thaw weathering* plays a part in its formation. Frost action on the mountaintops and slopes above supplies loose rocks (scree). Water seeps down the bergschrund crevasse onto the headwall, increasing the amount of freeze–thaw activity, cutting back the headwall and making it steeper.

- Ice sticking to the headwall pulls away blocks of rock by *plucking* as the glacier moves.

- Loose rocks obtained from freeze–thaw and plucking are embedded in the ice and act as tools for scraping out the bottom of the hollow by *abrasion*.

- As a result of the *rotational slip movement* of the ice, there is greater pressure from the ice at the bottom of the headwall and in the base of the hollow than near to the front where the glacier leaves the corrie hollow to flow down the valley; the rock lip forms near the exit as a result of less powerful erosion.

When all the ice has melted the corrie provides an ideal place for a tarn to form. There is a natural ice-carved hollow in which the water can accumulate. The rock lip acts as a natural dam on the one side that is not surrounded by steep slopes. A location in upland areas means that precipitation is likely to be high and there will be a large amount of run-off down the steep sides of the corrie, because the corrie forms a natural catchment area. A tarn fills the floor of most well-developed corries.

▲ **Figure 4** The Matterhorn near Zermatt in Switzerland.

Arête and pyramidal peak

Look at a photograph showing the peaks of any of the world's high mountain ranges, not just the Alps, and you will find pointed mountain peaks and long and narrow knife-edged ridges. An **arête** is a two-sided sharp-edged ridge, whereas the **pyramidal peak**, as its name suggests, is a three-sided slab of rock, of which the most famous example is the Matterhorn (**Figure 4**) with its three near-vertical rock faces. Both landforms are created by the cutting back of the headwalls of corries on the slopes below the peaks by the processes of freeze–thaw weathering and plucking. For an arête, two corries, one on each side of the ridge, cut back until only a narrow piece of rock is left as the ridge top. For a pyramidal peak, three corries cut back. All the peaks continue to be sharpened by frost action.

Activities

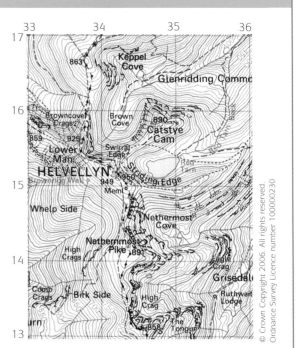

▲ **Figure 5** OS map of Helvellyn at a scale of 1:50000 (2cm = 1km).

1 (a) Using **Figures 2** and **5**, describe fully the features of the corrie occupied by Red Tarn.

 (b) Explain how it may have been formed.

2 From **Figure 5**, draw a labelled sketch map to show how corries and arêtes can be recognized on OS maps.

3 (a) What is the map evidence for visitors to the area?

 (b) Why is the area uninhabited?

Glacial erosion: valley landforms

◀ **Figure 1** The Lauterbrunnen valley in Switzerland, carved out by glacial erosion. The valley is too large to have been eroded by the small stream that flows in it today.

The **glacial trough**, often more simply called a U-shaped valley, is an impressive landscape feature (**Figure 1**). These glaciated valleys can be hundreds of metres deep with vertical rock walls, down which waterfalls cascade from **hanging valleys**. On the top of the valley sides the land often flattens out to form a **high-level bench**, known as an 'alp' in the Alps of Switzerland (the area in the foreground in **Figure 1**). The width and flatness of the valley floor are in marked contrast to the steepness of the sides. These valleys are drained by **misfit streams** which are dwarfed by the size and scale of the new glaciated valley. In some glacial troughs, lakes fill parts or all of the valley floor; these lakes are **ribbon lakes**, so-called because of their shape, which is long and thin. In the lower parts of the valley, examples of landforms of glacial deposition, such as **terminal moraines**, are found. The valley's *long profile* is characterized by its irregular shape (**Figure 2**), providing many hollows for lake formation.

Formation of valley landforms

Everything about a glacial trough shows the power of ice to erode. The former V-shaped river valley is widened, deepened and straightened by the valley glacier into a U-shaped valley. Before the ice, river erosion was confined to the small part of the valley where the river flowed; the glacier, however, fills the whole valley. This means that ice is in contact with all the floor and with both valley sides, so erosion is no longer confined to the centre of the valley. The V-shaped river valley is changed into the U-shaped glacial valley because glacial erosion by abrasion and plucking occurs everywhere in the valley where the ice is in contact with rock. The river moved around obstacles in its path, and its winding course created interlocking spurs. The more powerful glacier cannot flow so freely around corners, and it pushes straight forward, cutting off the edges of interlocking spurs to form **truncated spurs** and straight

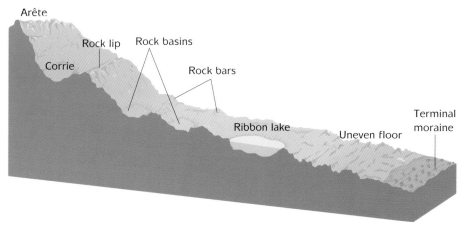

▲ **Figure 2** Long profile of a glacial trough.

valley sides. The ice is thicker in the main valley because it is fed by all the glaciers from tributary valleys. In each tributary valley there was a smaller, less powerful glacier than the main glacier. When only the rivers remained after the ice melted, those in tributary valleys were left hanging well above the level of the main valley floor. The streams from these hanging valleys fall as **waterfalls** into the main valley (**Figure 3**).

As a glacier flows down a valley it picks out weaknesses in rocks, eroding those rocks that are soft and well-jointed more rapidly than those that are hard and resistant. In those places where outcrops of hard and soft rocks alternate, the glacier erodes the soft rock more quickly and more deeply, by abrasion and plucking, forming a **rock basin**. The hard rock is left as a **rock bar**. After the ice melts, the rock basin is left as a hollow on the valley floor between two rock bars, and it is soon filled up by water to form a ribbon lake.

The map (**Figure 4**) shows that the centre of the rock basin which contains Wast Water is over 70 metres deep. Note how the submarine contours indicating water depth show the U-shaped cross-profile of the glacial trough. The steep valley sides continue below the water until the flat valley floor in the centre of the lake is reached.

◀ **Figure 3**
Waterfall from a hanging valley along the side of the Lauterbrunnen valley.

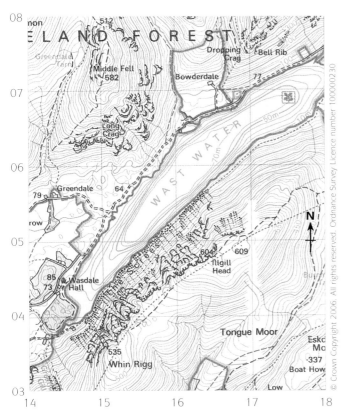

▲ **Figure 4** OS map showing Wast Water, the deepest of the English Lakes, at a scale of 1:50 000 (2cm = 1km).

▲ **Figure 5**   Wast Water and its scree slopes.

Activities

1 Make a sketch of **Figure 1**. Name and label the features of glacial erosion shown.

2 **(a)** From **Figure 4**, draw a sketch cross-section across Wast Water valley.

   (b) Explain how the screes have formed.

Landforms of glacial deposition

Ice behaves in the same way as all the other agents of erosion:

- it wears away the land surface – *erosion*
- it carries away the materials eroded – *transportation*
- it dumps elsewhere the materials it is carrying – *deposition*.

Valley glaciers erode with so much power, and ice sheets erode such great expanses of land, that large amounts of loose rock are available for transport. Glaciers can transport enormous loads. Look at **Figure 1** which shows an **erratic**, the name given to a boulder dropped by ice in an area where it does not belong. This big grit boulder has been dumped on top of the local white limestone rock. This boulder is just one of hundreds that an ice sheet deposited in the same area. Can you imagine a river having the power to transport one of these boulders, never mind hundreds of them? In a river channel a boulder of this size would need to be broken down by corrasion and attrition into small pieces before it could be transported; the ice simply carries it.

All materials transported by glaciers are called **moraine**. Although most are carried in the glacier's base, some are carried on the surface, which show up as dark lines of moraine (**Figure 2**) on the top of the glacier. Piles of material along the sides are called **lateral moraines**; those somewhere in the middle of the glacier, formed after valley glaciers join together, are called **medial moraines**. Two separate lateral moraines unite to form one medial moraine. The material for these moraines is broken off from the rocky peaks above by freeze–thaw weathering and it falls down the valley sides on to the top of the ice.

▲ **Figure 1** Erratic block of grit perched on limestone in the Yorkshire Dales National Park.

▲ **Figure 2** A glacier with lateral and medial moraines on the ice surface.

◀ **Figure 3** Boulder clay.

However, not even a glacier can keep on growing for ever. It reaches a point where the ice loss is greater than the amount of new ice supplied. For example, most valley glaciers begin to melt when they reach lower ground where temperatures are higher. Only a few reach the sea before they have completely melted (see **Figure 1** on page 54). As the ice melts and thins, its carrying capacity is reduced. When the glacier reaches the point of overload (load greater than carrying capacity), it must deposit some or all of its load. Any obstacle along its course encourages deposition.

The general name given to all materials deposited by ice is **boulder clay**. As its name suggests, this is usually clay which contains numerous boulders of many different sizes (**Figure 3**). It is an *unsorted* deposit. This means that large and soft rocks, as well as finer particles, are all mixed together. The boulders it contains are described as *angular*. They have sharp edges, not yet rounded off, as they would have been if they had been transported by rivers. The 'ingredients' of the boulder clay vary greatly according to what the glacier eroded before it reached the area. Sometimes the deposits are more sandy than

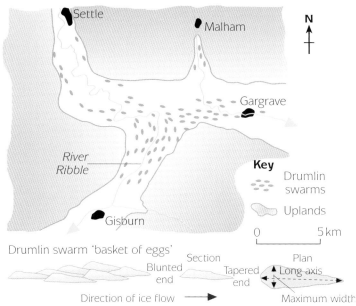

Figure 4 Drumlins in the Ribble Valley.

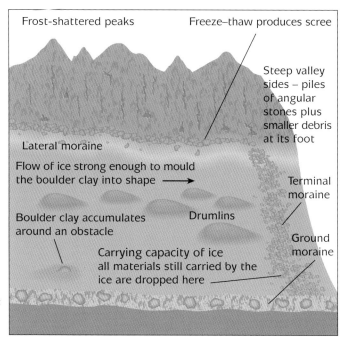

Figure 5 Location and formation of drumlins and terminal moraines.

clayey, which is why some physical geographers prefer to use the term **glacial till** instead of boulder clay to describe all ice-deposited materials. As the glacier continues to push forward, melting more and more all the time, it leaves a trail of boulder clay behind it which forms a hummocky surface of **ground moraine**.

Drumlins

In many of the lowland areas of south-west Scotland and north-west England, glacial deposition has produced a distinctive landscape of many low hills, each one typically about 30–40 metres high and 300–400 metres long. These hills all lie in the same direction and have similar shapes – blunt at one end and tapered at the other; in fact, each hill looks like an egg. Each of these hills is a **drumlin**. Drumlins occur in swarms and are said to form 'basket of eggs' topography, so called because of the appearance of the landscape (**Figure 4**).

Drumlins form when the ice is pushing forward across a lowland area, but it is overloaded and melting. It does not need much to encourage more deposition; any small obstacle, such as a rock outcrop or mound, is sufficient. Most deposition occurs around the upstream end of the obstacle, which forms the drumlin's blunt end. The rest of the boulder clay that is deposited is then moulded into shape around the obstacle by the moving ice to form the tapered end downstream. The drumlin is another landform from which it is possible to detect the direction of ice movement.

Terminal moraine

All the remaining load is dropped and dumped at the glacier's **snout** – the furthest point reached by the ice. This point is marked by a ridge of boulder clay across the valley or lowlands, running parallel to the ice front, and is called a terminal moraine (page 58). Where ice sheets remained

stationary for a long time, such as in central Europe during the main ice advances in the Ice Age, sufficient boulder clay was deposited to form ridges more than 200 metres high. More typically, terminal moraines formed by valley glaciers are between 20 and 40 metres high. Terminal moraines which cross valleys form natural dams behind which river water can pile up and form lakes. These lakes are also long and thin and are called ribbon lakes (page 58). This tells you that landforms with the same appearance can have different methods of formation.

1 a (i) State two features of boulder clay.
(2 marks)

  (ii) Explain why these features show deposition by ice instead of rivers.
(2 marks)

 b (i) Name one example of a terminal moraine. (1 mark)

  (ii) State one difference and one similarity between terminal and lateral moraines.
(2 marks)

  (iii) Explain why terminal moraines are usually larger than lateral moraines.   (2 marks)

 c (i) From **Figure 5** on page 55, describe the distribution of drumlins in the UK.
(2 marks)

  (ii) Explain why drumlins are 'egg-shaped' and are found in 'swarms'.   (4 marks)

Human activities in upland glaciated areas

Difficulties

Many upland areas are naturally unattractive for people to settle in, even without considering the effects of ice on the landscape. Climate deteriorates with height. It becomes colder, which reduces the length of the growing season and narrows the possibilities for farming. There is more precipitation, which also means more cloud and less sunshine, and there is more chance of the precipitation falling as snow, which makes farming more difficult. Many of the changes made by glaciers do not help. Glaciers are such powerful agents of erosion that the land can be scraped bare of all its soil. Valley glaciers increase the steepness of the land and the height of the valley sides, making access more difficult, if not impossible.

Possibilities for settlement and use

Human activity in the Alps was discussed in detail on pages 10–11 in Chapter 1. Although this was in the context of fold mountains, it is often difficult to separate the effects of high mountains from the effects of glaciation upon people and their activities, so most of that material is relevant here as well. Much of the electricity used in Alpine countries such as Switzerland and Austria is supplied from hydro-electric power (HEP). Glacial erosion improves the opportunities to take advantage of the high totals of precipitation in mountainous areas. After glaciation, there are more high waterfalls, and large ribbon lakes provide areas of natural water storage.

In the UK, mountains are less high and not as rugged. The best HEP opportunities are in the Highlands of Scotland. In other glaciated upland areas, the dominant human activities are farming, forestry and tourism.

Farming and forestry

In upland areas, land on the valley floor is precious. Valley glaciers widened valley floors and made them flatter; also it was on the valley floor that the melting glacier left most of its load, giving a greater thickness of soil. Soils formed of boulder clay vary greatly, but some are quite fertile and their high clay content favours grass growth. Land around the farm on the valley floor is used intensively for growing crops and making hay and silage (**Figure 1**), particularly if the aspect (the way the valley slope faces) is south-facing for greater warmth from the sun.

Further away from the farm, the land is only suitable for pastoral farming (keeping livestock). Cattle farming (dairy if possible, beef otherwise) is more likely to be carried out in the lower parts of the valley, while sheep rearing dominates on the steeper slopes and moorlands, where the physical conditions are suitable only for rough grazing. Farming in upland areas is often described as *marginal*, which means that it is difficult for the farmer to make a profit or a good living from farming there. Rocky land around and below the peaks may be useless (**Figure 2**).

In some areas, especially on north-facing slopes and on lower slopes too steep for farming, coniferous trees have been planted. Some of the forests are now mature and used for timber and wood pulp. Planting coniferous trees on the steep slopes of a glaciated valley is one example of *diversification*; diversification means that farmers are creating new and additional sources of income.

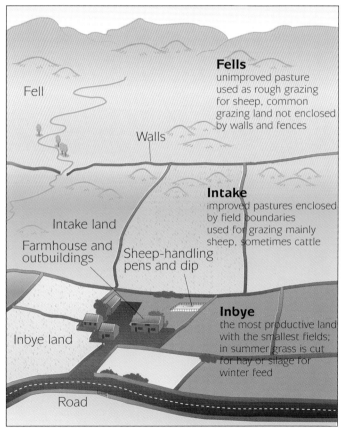

▲ **Figure 1** Layout and land uses on a Lake District farm. A typical farm can be split up into three parts.

▲ **Figure 2** Lake District farm. Notice how much greener the improved pastures are on the valley floor.

Tourism

Without the effects of glaciation, the landscape in the Lake District would be much less attractive to visitors. Glaciation sharpened up the landscape; rounded tops were changed into knife-edged peaks, and valleys were deepened, making the scenery more spectacular for visitors and fell walkers. Glaciation steepened and increased the size of many rock faces, increasing the area's attractiveness to mountaineers and rock climbers. Glaciers formed the large ribbon lakes, without which it would not be 'The Lake District'. Water always attracts tourists. Some come for the easy walks or rambles around the edges of the lakes, or for picnics or boat trips. Others, who are more active or more sporting, principally come to participate in water-based activities such as water-skiing and sailing. Without the effects of glaciation on its landscape, the Lake District would not be the great magnet for visitors that it is today.

Catering for and making money out of tourist visitors is another way many farmers in glaciated upland areas have diversified. Although some money has to be invested in building toilet and shower blocks, much more money can be made in areas with lots of visitors from charging people for camping or for parking a caravan in a field, than from using it for grass. Some rooms in the farmhouse may be used for bed and breakfast; boards advertising this are frequently seen in areas such as the Lake District and Snowdonia. The farm may be some distance from the nearest shop, giving the farmer a captive market for farm produce such as milk and eggs.

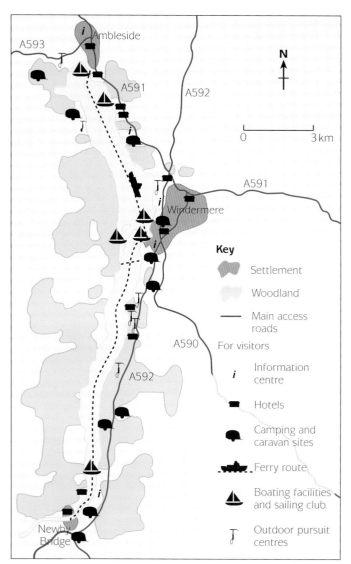

▲ **Figure 3** Windermere – places of interest and facilities for visitors.

Activities

1 **(a)** Using **Figure 2**, draw a labelled cross-section or a sketch to show the three parts into which a typical Lake District farm can be split.

 (b) Identify and explain the physical problems of farming in the Lake District.

2 Lake Windermere as a case study – Part 1.

 (a) Physical: Windermere fills a rock basin and there is a terminal moraine at its southern end. Explain how the lake has been formed by both glacial erosion and deposition.

 (b) Human: describe the tourist uses of Lake Windermere and explain why there are so many visitors to the lake each year. Research other sources of information to support what is given on this page. You could use websites to help you (see Hotlinks, page iv).

Current issues in glacial upland areas and the need for management

Farming

It is becoming more and more difficult for upland farmers to make a living just from farming; increasing numbers are being forced to diversify just to survive. In areas that are popular with visitors, like the Lake District, opportunities for diversification into tourism are plentiful, and are of great economic importance. However, the explosion in visitor numbers, accompanied by the relentless increase in the number of wealthy urban dwellers wanting to buy holiday (or second) homes in scenic rural areas, creates conflicts between visitors and local people (**Figure 1**). Many city people have an idyllic view of life in the countryside; they cannot appreciate how low wages are in rural occupations, and what a struggle it is for farmers in upland areas to make a decent income, even without hassle and damage from visitors.

Conflicts and issues	Management methods
Between farmer and visitor	
• Clambering over and knocking down the old walls between fields	• Making and maintaining stiles
• Letting dogs off the lead – they may worry lambs and sheep	• Educating people about the country code
• Walking through hay fields	• Warning notices telling people to keep to the paths
• Dropping litter	• 'Take your litter home' campaigns
Between rural residents and visitors	
• Visitors buying up houses so that none are available for local people	• Giving planning permission for new houses restricted to locals only
• Visitors pushing up house prices beyond levels affordable by rural people and occupations	• Extending policy of starter homes and affordable housing

▲ **Figure 1** Social and economic conflicts and issues between urban visitors and farmers/other rural residents.

Some farmers are able to benefit from the EU- and UK government-supported programmes for **countryside stewardship**. Under this scheme, farmers receive an annual payment for managing their farmland in ways that benefit the environment and preserve historic landscapes. Of particular benefit to Lake District farmers are grants for repairing dry-stone walls and for planting deciduous trees. Visitors prefer to see slopes covered by native deciduous trees such as oak and ash instead of the regular rows of trees, of uniform appearance, as found in coniferous plantations. Farmers are now being rewarded for farming and conserving the landscape in ways that are sustainable for future generations.

Tourism

Large numbers of visitors anywhere need to be managed, and visitor pressure is never shared out equally across a region. Some areas are more popular than others, either because of ease of access, or the presence of more natural attractions, or the greater availability of services and facilities. The most popular areas are referred to as **honeypots**. It is in these areas that management is most needed. Management has two broad aims:

1 To lessen damage to the environment.

2 To reduce conflicts between different groups of people.

For example, for many years all caravans parked in the Lake District have had to be painted in colours that blend into the landscape, such as green. New developments are required to be eco-friendly, such as building new accommodation out of wood in forest settings.

Footpath erosion

The glaciers left the upland areas with steep slopes and thin soils, which are vulnerable to people pressure. Large numbers of walkers following the same paths cause erosion, which spreads to a larger area as the footpath is widened. The erosion caused by humans can lead to scars on the landscape (**Figure 2**). Locally the damage can be very severe along popular walks on steep slopes. This is a major environmental issue.

Footpath management involves diverting the course of paths and fencing off the old footpaths to allow time for the vegetation to recover. Also, artificial paths are made which are better able to withstand the constant tread of people's feet. Methods include:

• stone pitching – laying hard-wearing stone on the ground

• steps of stone – up and down very steep slopes (see **Figure 2** on page 214 in Chapter 12)

▲ **Figure 2** Footpath erosion on Striding Edge on Helvellyn. This is in square 3415 in Figure 5 on page 57.

- raised wooden boards – above wet surfaces such as marsh and peat.

Footpath repair is essentially about achieving a hard-wearing path that fits in with the landscape. It allows conservation of vegetation and soil on the surrounding land, even in areas under severe pressure from visitors.

Conflicts between different groups of visitors

Even in areas of great scenic beauty, with many and varied opportunities for outdoor activities, the majority of visitors never venture more than 1 kilometre from the car park. For most visitors to the Lake District, the lakes are a more important attraction than the fells. Over the years as visitor numbers increased, conflicts developed on the large lakes such as Windermere and Ullswater between power boat owners and water skiers and all the other lake users and visitors. These conflicting interests were managed by dividing Lake Windermere into zones; speed boats were restricted to certain parts of the lake,

> **Supporter of the ban**
> The speed limit was needed to cut the disruption to other lake users and make the lake safer for anglers and canoeists.

> **Manager of a firm of boat builders and repairers**
> My operation on Windermere is being scaled back. Some of my staff have already left, worried about future prospects.

> **Member of outdoor activities club**
> There is nowhere else in the Lake District to water ski and use power boats. All we wanted to use was 1 per cent of the Lake District.

> **Retired resident**
> It will bring back peace and quiet to the lake and make Windermere the 'jewel in the crown' of Lake District lakes once again.

▲ **Figure 4** Views about the Lake Windermere speed limit.

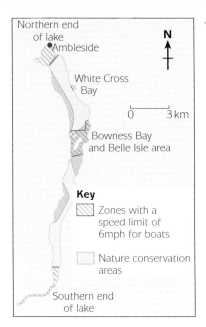

◄ **Figure 3** Lake Windermere: attempts at management to conserve areas that are important for wildlife and to keep parts of the lake free from speed boats.

away from areas of nature conservation and busy places for tourists and sailing boats (**Figure 3**).

In 2005 a lake speed limit of 10mph was imposed on Lake Windermere. Jet skiers and speed boat owners were outraged; the Windermere Action Force (WAF) is campaigning to have the ban overturned. They argue that sports such as water skiing and sailing can go on without conflict. Friends of the Lake District were delighted by the decision and believe that more visitors will come now that peace and quiet have been restored.

Activities

1 (a) Name one *environmental* issue in upland glaciated areas.

(b) Explain the causes.

(c) Describe how it can be managed.

2 (a) State the *economic* benefits of visitors to **(i)** farmers and **(ii)** local people in the Lake District.

(b) Why do social and economic issues arise between **(i)** farmers and visitors and **(ii)** local residents and visitors?

3 Lake Windermere as a case study – Part 2.

(a) Explain why management of visitor activities is needed on Lake Windermere.

(b) Describe the management methods used.

(c) List the arguments for and against the 10mph speed limit on Windermere.

(d) With which one of the views expressed in **Figure 4** do you agree most? Explain why.

1 a Draw a labelled sketch to show three types of moraine. (3 marks)

b Study **Figure 1**.

   (i) Name one type of moraine shown in **Figure 1** and describe its location. (2 marks)

   (ii) State two ways in which the glacier is different between points **A** and **B**. (2 marks)

c Many walkers and skiers visit upland glaciated areas. Explain how these human activities can damage the environment. (2 marks)

d For one upland glaciated area you have studied:

   (i) identify a conflict between people

   (ii) explain why it has arisen

   (iii) describe the management strategies needed to reduce or resolve the conflict. (6 marks)

2 a **(i)** Name the type of lake shown in **Figure 2**. (1 mark)

   (ii) Describe its shape. (1 mark)

   (iii) Explain how the lake may have been formed. (4 marks)

b **(i)** Name another type of lake formed in glaciated areas. (1 mark)

   (ii) State uses of glacial lakes to people. (3 marks)

c Freeze–thaw (frost-shattering) occurs in high mountain areas.

   (i) Explain how it works. (3 marks)

   (ii) Describe how the shape and appearance of the peaks in **Figure 2** shows that they have been affected by freeze–thaw. (2 marks)

▲ **Figure 1** Fox Glacier, New Zealand.

▲ **Figure 2** Aerial view of part of the Alps.

Chapter 5

Coastal landscapes and processes

*On the brink ...
in 1996 a farmer
in Holderness (Yorkshire)
next to her farmhouse
after coastal erosion
had already destroyed a
garage and dairy unit.
Is it still there now?*

Key Ideas

The Earth's crust is modified by coastal processes, resulting in distinctive landforms:

- destructive waves are responsible for coastal erosion
- processes of erosion form distinctive landforms such as cliffs, wave-cut platforms, caves, stacks and arches
- longshore drift transports eroded materials along the coast which are deposited elsewhere by constructive waves
- beaches and spits are examples of landforms of deposition.

The interaction between people and coastal environments causes problems for people and:

- results in the need for coastal management
- requires strategies for solving problems, about which people have different views.

How is the coast eroded?

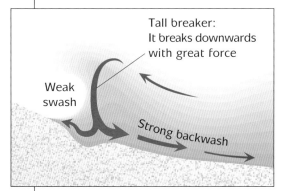

▲ **Figure 1** A destructive wave.

Tall breaker:
It breaks downwards
with great force

Weak
swash

Strong backwash

▲ **Figure 2** Chalk cliffs at Beachy Head, created by destructive waves.

Waves are responsible for most of the erosion along coasts. Wind blowing over a smooth sea surface causes small ripples which grow into waves. When a wave approaches the coast its lower part is slowed by friction with the sea bed, but the upper part continues to move forward. As it is left unsupported, it topples over and breaks forward against the cliff face or surges up the beach. The waves which erode most are called **destructive waves** (Figure 1).

The power of destructive waves

Destructive waves have three main features:

1 They are high in proportion to their length.

2 The backwash is much stronger than the swash so that rocks, pebbles and sand are carried back out to sea.

3 They are frequent waves, breaking at an average rate of between eleven and fifteen per minute.

The height and destructiveness of these waves depend upon the distance over which the waves have travelled and the wind speed. If the waves driven by the wind have crossed over a large area of ocean, they have had time to build up and grow to their full height. A lot of energy is released when they break against the coastline. The length of water over which the wind has blown is called the **fetch**. The greater the fetch and the stronger the wind, the more powerful the wave, especially when driven onshore by storm-force winds. From time to time ideal conditions occur for the formation of huge destructive waves, as they did in Dorset on 3 November 2005 (**Figure 3**). Onshore south-westerly winds were strong, with gusts over 120 kilometres per hour after a long journey over the Atlantic Ocean. Imagine the weight and force of the water crashing against the coastline under these storm conditions.

▲ **Figure 3** 'Waves lash the sea front at Lyme Regis' was one newspaper headline: other newspaper headlines about storms are given on the right.

Insurance companies face £500 million payout for storm and flood damage

We don't like to be beside the quayside

Two flee for their lives as 4x4 ends up in the sea

The newspapers were mainly interested in telling their readers about the damage caused. Can you suggest a caption for **Figure 3** that a geographer would be more likely to write?

Processes of coastal erosion

There are many similarities between the ways in which rivers and waves erode, which is why the names used for the processes of erosion are the same.

A Hydraulic power: This is the sheer weight and impact of the water against the coastline. It is greatest under storm conditions when hundreds of tonnes of water may hit the rock face. Also, air trapped in cracks and caves is suddenly compressed by the breaking waves, which increases the pressure on the rock.

B Corrasion: Another name for this is abrasion. The breaking waves throw sand and pebbles against the rock face. These break off pieces of rock and cause undercutting. In large storms boulders will also be flung against the cliff face causing even greater damage.

C Attrition: Particles carried by the waves are reduced in size as they collide with the rock face and one another. Boulders and pebbles are broken down into sand-sized particles which are easier for the waves to carry away.

D Corrosion: This is the chemical action on rocks by sea water and is most effective on limestone rocks, which are carried away in solution.

The speed of erosion

You should refer to one or more of these processes whenever you are explaining the formation of a landform by coastal erosion. Of the four processes, hydraulic power and corrasion are the most significant, especially under storm conditions, which is when the highest rates of erosion are recorded. The type of rock also affects the speed at which a stretch of coastline is eroded. Rock faces which are riddled with joints, bedding planes or faults will be eroded more quickly than those which are in massive blocks, because there are lines of weakness which the waves can exploit. Cliffs built of a hard rock, such as granite in Cornwall and Devon (page 22), resist erosion longer than those made of soft sediments, such as boulder clay in Yorkshire and Norfolk, which the waves wash away easily (pages 67, 78–79).

In some cliffs the arrangement of the rocks increases the rate of erosion. The clay and sand cliffs at Barton on Sea, east of Bournemouth, are an example. Next to

the sea is a layer of clay which is too weak to withstand the strong destructive waves in the English Channel with their long Atlantic fetch. However, rain water seeps down through the sand which lies above the clay. This saturates the base of the sand layer along the junction with the clay and causes landslides and slumping. This results in erosion at the top of the cliff as well as at the base (**Figure 4**).

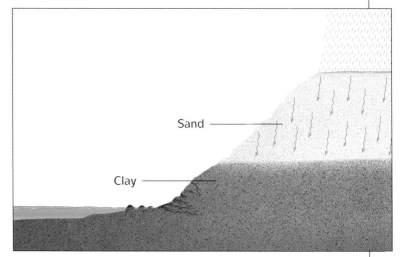

▲ **Figure 4** Cliff erosion at Barton on Sea.

Activities

1 Show that you understand the differences between:

   (a) corrosion and corrasion

   (b) corrasion and hydraulic action.

2 **(a)** Draw a labelled sketch of the cliffs at Beachy Head from **Figure 2**.

   (b) What shows that these cliffs are being actively eroded?

   (c) State two differences between the cliffs at Beachy Head and Barton on Sea.

3 **(a)** Explain why so much coastal erosion takes place during storms.

   (b) Why will a storm cause more erosion along some parts of the coastline than others?

Landforms of coastal erosion

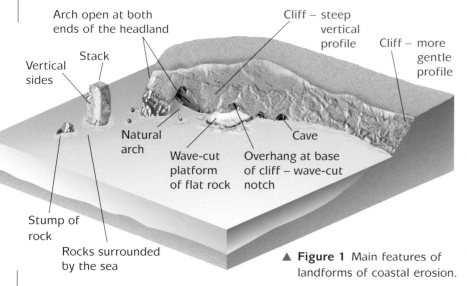

Arch open at both ends of the headland

Stack

Vertical sides

Natural arch

Wave-cut platform of flat rock

Overhang at base of cliff – wave-cut notch

Cliff – steep vertical profile

Cliff – more gentle profile

Cave

Stump of rock

Rocks surrounded by the sea

▲ **Figure 1** Main features of landforms of coastal erosion.

The main landforms of erosion, which can be seen in many places around the British coast, are described in **Figure 1**. Pay particular attention to the labels that are used to *describe* the landforms. No attempt is made on **Figure 1** to *explain* their formation.

Cliffs and wave-cut platforms

The sea **cliff** is the most widespread landform of coastal erosion. Cliffs begin to form when destructive waves attack the bottom of the rock face between the high and low water marks. By the wave processes described on page 69, such as hydraulic power and corrasion, the waves undercut the face forming a wave-cut notch. The rock above hangs over the notch. With continued wave attack, the notch increases in size until the weight of the overhanging rock is so great that it collapses. Once the waves have removed all the loose rocks and stones from the collapsed cliff, they begin to undercut the new rock face which is now exposed to wave attack. Wave erosion, followed by cliff collapse, happens time and time again so that the cliff face and coastline retreat inland. Impressive cliffs are found where rocks resist erosion, such as the 'White Cliffs of Dover' which are built of chalk.

As the cliff retreats a new landform, the **wave-cut platform**, is created at the bottom of the cliff face. This is the gently sloping rocky area between the high and low water marks. It is covered at high tide but exposed as the tide goes out. It is not a smooth platform of rock; rather its surface is broken by ridges and grooves. This is the area of flat rocks that holidaymakers often venture onto when the tide has gone out, looking for crabs, and where they are liable to get trapped as the tide races in! The wave-cut platform is

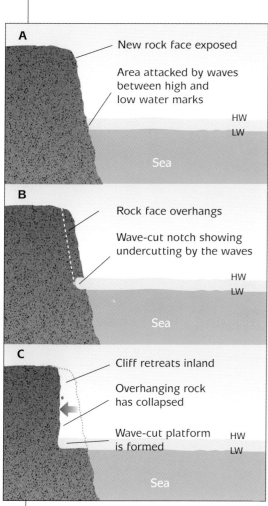

A

New rock face exposed

Area attacked by waves between high and low water marks

HW

LW

Sea

B

Rock face overhangs

Wave-cut notch showing undercutting by the waves

HW

LW

Sea

C

Cliff retreats inland

Overhanging rock has collapsed

Wave-cut platform is formed

HW

LW

Sea

▲ **Figure 2** Formation of cliff and wave-cut platform.

◀ **Figure 3** Marsden Rocks – stacks near Sunderland. Notice in particular the amount of undercutting around the base of the rocks. The smaller stack has since collapsed without warning.

formed where the rock above has been cut away by the waves to form the cliff. Because wave erosion is concentrated where the waves break between the high and low water marks, the rock below is little affected and is left as an area of flat rocks.

Caves, arches and stacks

Waves are particularly good at exploiting any weakness in a rock, such as a joint. By the same processes of erosion, and particularly by hydraulic power and corrasion, any vertical line of weakness may be increased in size into a **cave**. However, the rock needs to be relatively hard or resistant, otherwise it will collapse before the cave is formed. Once a cave has formed, when a wave breaks it blocks off the face of the cave and traps the air within it. This compresses the air trapped inside the cave, which increases the pressure on the roof, back and sides. If the cave forms part of a narrow headland, the pressure from the waves may result in the back of the cave being pushed through to the other side so that it is open at both sides. The cave then becomes a natural **arch**. The base of the arch is attacked by waves, putting more and more pressure on the top of the arch. After continued erosion, and especially if there is a weak point at the top of the arch, the arch collapses and becomes a **stack**. The stack is a piece of rock isolated from the main coastline (**Figure 3**).

You can see that there is a sequence of features formed by wave erosion – notch, cave, arch and stack (**Figure 4**). However, the stack itself is attacked by waves from all sides. It is gradually reduced in size and eventually it collapses so that all the signs of where the coastline used to lie disappear. When you look at a line of cliffs which mark the present-day coastline, as it is shown on maps, you must remember that it may be many kilometres further back than it used to be, as a result of the unceasing energy of destructive waves.

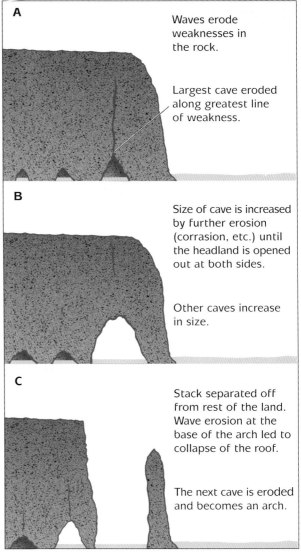

A — Waves erode weaknesses in the rock.

Largest cave eroded along greatest line of weakness.

B — Size of cave is increased by further erosion (corrasion, etc.) until the headland is opened out at both sides.

Other caves increase in size.

C — Stack separated off from rest of the land. Wave erosion at the base of the arch led to collapse of the roof.

The next cave is eroded and becomes an arch.

▲ **Figure 4** Formation of caves, arches and stacks.

▼ **Figure 5** Part of the coastline a few kilometres west of Bournemouth.

Activities

1 **(a)** On a sketch show and label the coastal features on **Figure 5**.

   (b) Explain how they were formed.

2 Why are these cliffs being eroded less quickly than those at Barton on Sea a few kilometres east of Bournemouth (page 69)?

Transport and deposition of material along the coast

Transport

Loose, eroded materials of all sizes are transported by waves and deposited further along the coast. Rivers also carry sediment into the sea, which is picked up and carried away by the waves. The methods of transport are the same as those in the river channel (page 36): large boulders are rolled along the sea bed, smaller boulders are bounced along (saltation), sand grains are carried in suspension, and lime from chalk and limestone rocks dissolves and is carried in solution. The transport of sand and pebbles along the coast by waves is called **longshore drift** (**Figure 1**). Waves often approach a coastline at an angle, but sand grains and pebbles roll back down the slope at right angles to the coastline because this is the steepest gradient. As **Figure 1** shows, a pebble will keep on being pushed up the beach by the waves at an angle, but every time it rolls back down the beach at right angles to the coastline. In this way the pebble is transported along the coastline.

The general direction of longshore drift around the coasts of the British Isles is controlled by the direction of the dominant wind (**Figure 1**). Prevailing south-

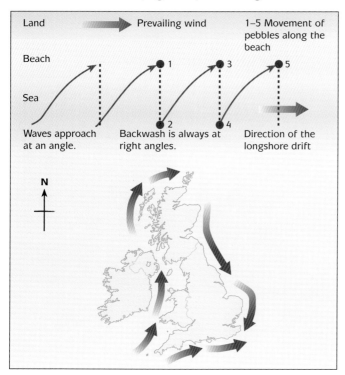

▲ **Figure 1** Direction of longshore drift around the British Isles.

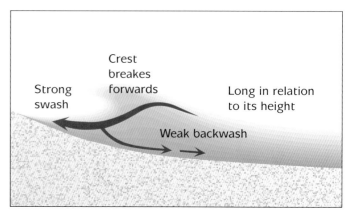

▲ **Figure 2** Features of a constructive wave.

westerly winds cause the drift from west to east along the Channel coast and from south to north along the west coast. The east coast is protected by land from the prevailing south-westerly winds. However, winds from the north cause longshore drift movement from north to south on the east coast. Northerly winds (winds from the north) have crossed a long stretch of open sea so that, although they do not blow as frequently as the westerly winds, they have the greatest influence overall. Longshore drift is important in the formation of all landforms of coastal deposition. Why do Local Authorities need to take the direction of longshore drift into account when planning to protect tourist beaches and construct sea defences?

Deposition

The load of the waves – sand, shingle and pebbles – is deposited by **constructive waves** (**Figure 2**). Such waves add more material than they remove from the coastline. Constructive waves have three main features:

1 They are long in relation to their height.

2 They break gently on the beach so that the **swash** carrying materials up the beach is stronger than the **backwash** carrying them away.

3 They break gently, with between only six and nine waves per minute.

These waves are associated with calm sea conditions when winds are light and are not blowing directly onshore. Therefore they occur more often in summer than in winter. Constructive waves operate most effectively in sheltered coastal locations such as in a bay sheltered by rocky headlands on both sides.

The formation of beaches

Everyone knows what a beach is, but can you describe it in geographical terms? The beach is the gently sloping area of land between the high and low water marks. Most of it is covered by the sea at high tide. Some beaches are straight and may extend for several kilometres. Others, located in bays, are more likely to be curved. The most common beach materials are sand, shingle and pebbles, but you may have seen many other types of materials (both natural and synthetic) washed up on beaches. However, most of these stay on the beach for only a short time before being moved on by longshore drift and the next high tide. The materials from which a beach is formed are carried by the longshore drift. If there is a coastline of weak rocks which has been greatly eroded on the up-drift side, the waves will be heavily laden with material. Where there is a bend in the coastline, deposition by constructive waves is always likely to take place because a more sheltered area has been created. Material accumulates over time and builds up the beach.

How do spits form?

A **spit** is a long and narrow ridge of sand or shingle (**Figure 4**). One end is attached to the land while the other end lies in the open sea. It is really a beach which, instead of hugging the coastline, extends out into the sea. If the spit is formed of sand, sand dunes are usually found at the back of it. Behind the spit there is an area of standing water, some of which may have been colonized by marsh plants. Some spits, particularly those found along the coast of the English Channel such as Hurst Castle spit near Christchurch, have a hooked end (page 82). Others, particularly those found on the east coast, run parallel to the coast, perhaps for several kilometres. Some of these, such as Spurn Point (pages 74–75), extend across estuaries, while others stretch across river mouths, diverting river flow southwards for a time behind the spit.

The formation of a spit begins in the same way as that of a beach. Eroded materials are carried along the coast by longshore drift. Deposition begins at a bend in the coastline. For a spit, however, the deposited materials accumulate away from the coast into the open sea until a long ridge of sand or shingle is built up. Fresh water and sea water are trapped behind this ridge as it forms. As the ridge extends into deeper and more open water, the end of the spit is affected by strong winds. These winds and sea currents help to curve the end of the spit. Do you understand how the direction of the longshore drift along a coastline can be worked out from the form of a spit?

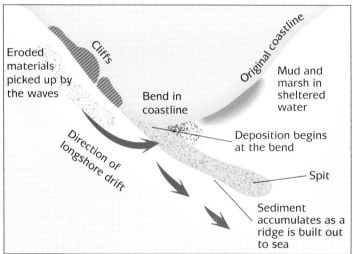

▲ **Figure 3** ▶   **B**
Two different beaches:
A is made of sand;
B is made of shingle.

Eroded materials picked up by the waves

Cliffs

Original coastline

Bend in coastline

Direction of longshore drift

Mud and marsh in sheltered water

Deposition begins at the bend

Spit

Sediment accumulates as a ridge is built out to sea

▲ **Figure 4** Formation of a spit.

1 a (i) State two ways in which constructive waves are different from destructive waves.   (2 marks)

    (ii) Why are constructive waves more likely to deposit than destructive waves? (2 marks)

b (i) What is meant by longshore drift?   (2 marks)

    (ii) Explain why its direction is different along the west coast of the British Isles from that along the east coast.   (2 marks)

c (i) Study **Figure 3**. Describe the differences between beaches **A** and **B**.   (4 marks)

    (ii) Explain how a beach forms.   (3 marks)

Case Study – A stretch of coastline

The Yorkshire Coastline from Scarborough to the Humber estuary

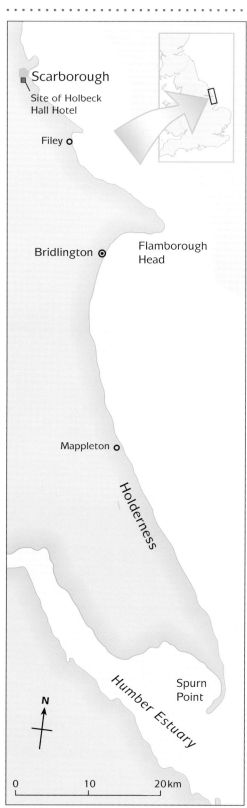

▲ **Figure 1** Map of the Yorkshire coast from Scarborough to the Humber estuary.

◄ **Figure 2** The beach in the South Bay at Scarborough, a landform of deposition.

▲ **Figure 3** Flamborough Head.

▲ **Figure 4** The **spit** at Spurn Point, a large landform of deposition.

Coastal deposition

A: Scarborough – the number one resort of the Yorkshire coast

Sandy beaches are found in the bays on both sides of the castle headland. Sand, carried by the north-to-south longshore drift, is trapped in the bays surrounding the headland. Most tourists go to the beach in the South Bay (**Figure 2**), which is nearer to the town centre. On a hot summer's day the sandy beach is packed with adults sitting in deckchairs, children playing on the beach and people of all ages bathing in the North Sea (despite water temperatures below 14°C). Further south at Filey, the bay is longer and so also is the beach.

B: Spurn Point – Britain's longest spit

The longshore drift of material is from north to south (from top to bottom in **Figure 4**). There is plenty of sand and clay for waves to collect because of the rapid erosion of the boulder clay cliffs along the coast of Holderness from Bridlington southwards (pages 78–79). At the bend in the coastline formed by the Humber estuary, deposition begins. In the past 150 years, a ridge of sand some 8 kilometres long has been built up.

Coastal erosion

A: Dramatic cliff collapses

The collapse in 1993 of the cliff under the five-star Holbeck Hall hotel, located on the southern side of the South Bay in what was thought to be a safe position on the cliff top, came as a shock. It highlighted the serious problem of coastal erosion along the Yorkshire coast (**Figure 5**). It made the local authority think about what needed to be done for coastal protection. Scarborough has since spent over £50 million on sea defences (page 77).

B: Flamborough Head

A thick band of resistant chalk rock outcrops here. The chalk is a much harder rock than the boulder clay which covers Filey Bay to the north and Holderness to the south, and has produced impressive cliffs (**Figure 3**). These are highest along the northern side where they form some of the highest cliffs in England with vertical faces more than 200 metres high. In front of many of the cliffs, rocks are exposed at low tide. The chalk which outcrops on the edge of the headland is well jointed. Wave erosion has been concentrated in the joints and along any other lines of weakness. The bottom of the cliffs is full of caves. Where vertical joints or faults are present, some of the caves have grown and have been cut through to form arches. Small stacks have been left and have names such as the King and Queen rocks or the Adam and Eve pinnacles.

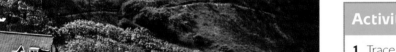

▲ **Figure 5** Holbeck Hall Hotel.

Activities

1 Trace or draw an outline of this stretch of coastline. Mark and name on it one example of each of the following coastal landforms:

- cliff
- cave
- stack
- spit
- wave-cut platform
- natural arch
- beach

2 **(a)** Draw labelled sketches of **Figures 3** and **4** to show the physical features.

   (b) Explain why the coastal landforms shown are so different.

3 Individually or in small groups, put together a leaflet, suitable for issue to visitors by the local tourist office, with the title 'The geographical attractions of the Yorkshire coast'.

Case Study – A stretch of coastline

Coastal management

What is meant by this? **Coastal management** involves controlling development and change in the coastal zone, as well as planning ahead. Within the UK there are two big coastal management issues:

- preventing (or limiting) coastal erosion
- managing the coast for tourist use.

The need for both is increasing.

- Cliffs are crumbling as rates of coastal erosion appear to be increasing. Several reasons are suggested for this, including global warming and rising sea levels, land sinking in south-east England and higher levels of storm activity (page 81).
- People's demands for leisure and tourism, as well as living, along the coast are increasing. Coastal areas have always been popular places to live. Many people dream of living in or retiring to a house next to the sea; sea views, plenty of fresh air and places to exercise the dog are just some of the attractions.

▲ **Figure 1** Coastal management in progress in Scarborough in 2003. The old sea wall is almost hidden behind the great blocks of granite rock, imported from Norway, soon to be used as part of the new defences. What tells you that the castle headland is not as strong as its size would suggest?

Management is most necessary along stretches of coast that are heavily built-up, and in coastal resorts in particular. Today's well-known coastal resorts, from Brighton in the south to Blackpool in the north, grew in Victorian times. As well as attracting holiday-makers and day trippers, today they also house many commuters and retired people. Although concentrated in coastal resorts, tourism has spread along other parts of the coastline. Coastal pathways for walkers and hikers have been created. Is there one close to your home? Some paths have proved so popular that they are being worn away by the pressure of visitor numbers. It is not uncommon to find that the authorities have needed to move the course of a path further inland as a result of cliff erosion. Although waves do most of the work of erosion, visitors can contribute unintentionally by loosening stones and starting rock slides.

How are coasts managed?

In coastal resorts the most common method used to keep the sea out is to build a sea wall, as at Scarborough (**Figure 1**). The wall is often built with a curved lip at the top on the seaward side, to deflect the force of the waves away from the promenade behind. Such walls are expensive to build and costly to maintain. They require a lot of maintenance because they absorb the concentrated energy of the waves, and we have already seen how great that can be (page 68). Their construction can only be justified economically where there are many people and much property to defend.

The beach is the most important natural asset in coastal resorts. To keep the beach, local authorities construct lines of groynes, built out into the sea at right angles to the coastline (**Figure 2**). These trap sediment as it is transported along the coast by longshore drift. They also reduce wave energy, making erosion of the beach and cliffs less likely.

▲ **Figure 2** Groyne at Bournemouth.

Local authorities are only interested in managing their own stretches of coastline; they do not usually consider the consequences further along the coast. Groynes are a simple and generally effective way of keeping a beach, but their construction can have disastrous consequences for beaches on the downdrift

Brushwood-filled gully
Gravel drain
Interlocking steel piles with drainage pipes
Sand
Sea wall
Boulders and tripods
Clay

▲ **Figure 3** Protection for the cliffs at Barton on Sea.

side. Once they are deprived by the groynes of their load of sediment, waves remove sediment from beaches and cliffs further along the coast with renewed vigour. The same groynes, which do a good job of protecting the beach and soft cliffs at Bournemouth, are speeding up the already rapid erosion of the cliffs at Barton on Sea (page 69). Despite having more than £1 million spent on a variety of methods of protecting them, the cliffs are still retreating at Barton on Sea, threatening holiday villages and housing estates.

Natural beaches are dynamic landforms – the sand is constantly deposited and washed away by the waves. Beaches rely upon new supplies of sediment for their survival. Any interruption in supply caused by human features such as groynes, breakwaters, harbour walls and jetties threatens their existence. Management that considers all the relevant factors is required, and a full cost–benefit analysis needs to be done before any money is spent.

▲ **Figure 4** Coastal management completed in Scarborough in 2005.

Coastal management in Scarborough

Why was it needed?
The sand rocks that make up the natural cliffs are easily eroded. The sea wall and other sea defences were inadequate and some were more than 100 years old. During storms, waves smashed over the sea wall and part of the promenade had to be closed – not good for a major seaside resort.

Which management methods were used?
A new sea wall was built, a little further out to sea. Blocks of granite rock, or rock armour, were placed in front of it for protection; each one weighed between 6 and 10 tonnes. Concrete blocks that link together were placed on top of the rocks. Over 4000 of these blocks, called accropodes, were used. The intention is that the new defences will absorb all the energy before the waves reach the coastline.

How much did it cost?
Work began in 2002, when the estimate was £26 million for 2.1 kilometres of coastline along the sea front between the North and South Bays. In 2005 it was announced that the cost of the scheme had doubled to more than £50 million. Would the work have gone ahead if the real cost had been known? Spending such an enormous amount of money on sea defence can only be contemplated in places where many people live and visit, and there is a lot of property to defend.

Activities

1 **(a)** Name three methods of coastal protection.

 (b) For each one:

 (i) describe how it works

 (ii) name and locate a place where it is used

 (iii) explain its advantages and disadvantages for coastal management.

2 **(a)** What was the approximate cost per kilometre of coastal protection in Scarborough?

 (b) Explain why the cost was so high.

 (c) Can costs of this size be justified in other coastal areas?

Coastal management	
Costs	**Benefits**
• What will it cost to complete?	• What are the advantages which justify the cost?
• How much will it cost to maintain?	• For how long will the benefits last?
• Who will be badly affected by it?	• Who will gain from it?
• Which areas will be badly affected by it?	• How large an area will gain from it?
• Will there be environmental damage?	• Will it improve the environment?

3 Investigation – make a study of coastal protection in another coastal resort (for example, the one closest to your home area).

Holderness – problems of coastal management

Being number one in Europe for coastal erosion does not bring much pleasure to the people who live along the coast of Holderness. In a stormy year, the waves remove between 7 and 10 metres of land along unprotected stretches of coast. The coastline is today some 3–4 kilometres further west than it was in Roman times. Twenty-nine villages have been lost to the sea in the past thousand years. All along the Holderness coast farmers keep losing some of their land. Farmhouses are threatened (page 67); caravan sites and holiday homes have already been lost to the sea. The rate of erosion is unlikely to be reduced, because of a combination of land sinking on the eastern side of the British Isles and a possible sea level rise as a result of global warming. Sea levels in Holderness are estimated to be rising by 4 millimetres a year.

Why is the coastline being eroded so quickly? After all, the east coast does not receive as regular a battering from destructive waves as the west coast and exposed parts of the Channel coast. The main factor is rock type. The boulder clay is made up of soft clay and sands which are not consolidated (cemented together). The waves can wash away the clay and sands from between the boulders to leave them unsupported. Also, when it rains, water enters cracks and spaces in the rock; after heavy rains this makes the cliff top unstable and liable to slumping. Most erosion occurs when winds blow from the north or north-east along this coast because the waves cross a long stretch of open sea (a long fetch), which increases wave energy for erosion. There are problems protecting a coastline which stretches for more than 50 kilometres.

Coastal defences at Mappleton

Mappleton is a small village which by 1990 was under real threat of becoming lost village number 30 along the coast of Holderness. The B1242 is the vital road link along this coast and it would have been expensive to find a new route for it. This helped to justify spending almost £2 million upon a coastal protection scheme in 1991 for a village of about 100 people. Blocks of granite were imported from Norway for sea defences at the bottom of the cliff and for the two rock groynes. The purpose of the two rock groynes is to trap beach material, which will protect the rock wall from direct wave attack.

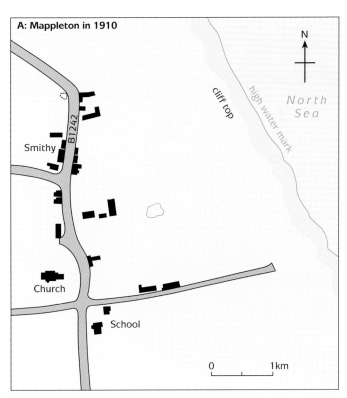

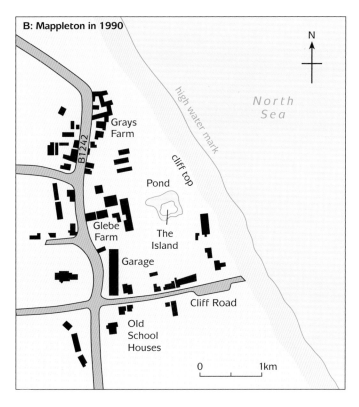

▲ **Figure 1** Mappleton in 1910 and 1990.

▲ **Figure 2** Mappleton and its sea defences. Notice the closeness of the village to the cliff top, the gentle
angle of the soft sand and boulder clay cliffs, the rock wall at the bottom of the cliffs, and the rock groynes.

Some views on coastal protection in Holderness

Holderness Council
We are a small authority with a total annual budget of only £4 million. Spending a large amount of money to protect a village is hard to justify. Many people agree that the village should be allowed to disappear ... but it is terribly difficult to say this.

Ministry of Agriculture
We are moving towards a policy of 'managed retreat'. Although towns, villages and roads would be protected, farmland and even isolated houses would be regarded as dispensable and allowed to disappear. There are food surpluses in the EU so that every bit of farmland is no longer needed.

Dr John Pethick – a top scientist at the University of Hull
Low-lying farmland should be abandoned and cliffs allowed to collapse because they are the main sources of sand and silt that build up to protect other parts of the coast, including towns and cities.

Farmer living just south of Mappleton
My farm is at greater risk from the sea than ever because of the coastal protection works at Mappleton.

Activities

1 Describe the physical features of the coastline of Holderness as shown on **Figure 2** on this page and in the photograph on page 67.

2 **(a)** State three pieces of evidence from this page for rapid erosion along the coast of Holderness.

   (b) Give the physical reasons for this rapid erosion.

   (c) Why is the natural rate of erosion not expected to decrease in the future?

3 **(a)** From **Figure 1** state:

   (i) by how many metres the sea has invaded inland between 1910 and 1990

   (ii) the distance between the B1242 road and the top of the cliffs in 1990.

   (b) With the help of **Figure 2**, describe the methods used to protect Mappleton.

   (c) Explain why the farmer who owns the land south of the cliff road, which can be seen on **Figure 2**, is complaining about the coastal protection works at Mappleton.

4 **(a)** Make two columns and list the arguments for and against protecting the coastline of Holderness.

   (b) What do *you* think would be the best policy? Explain and justify your view.

Case Study – Blackpool

Coastal resorts in the UK – problems and solutions

Blackpool is the UK's premier coastal resort based on number of visitors each year. Its most famous structure is Blackpool Tower, located on the sea front behind the beach (**Figure 1**).

◀ **Figure 1** Blackpool, the UK's largest seaside resort.

The traditional British seaside resorts have been in decline for 40 years, ever since people discovered guaranteed summer sun and warmth in Mediterranean countries. Blackpool was badly affected. Look what happened in the decade from 1990 to 1999:

- visitor numbers per year dropped from 17 million to 11 million
- 1000 hotels ceased trading
- 300 holiday-flat premises closed
- average hotel occupancy rates fell as low as 25 per cent.

Blackpool was not exciting existing visitors enough to make them come back the following year, nor was it attracting sufficient new customers. By 2000 some bed-and-breakfast prices had fallen as low as £10 per night, which left no money for investment in improvements. A downward spiral of decline set in as some parts of town started to look very run-down. This happened despite improvements in road access after the M55 was completed as the motorway link from the M6.

Strategies for solving Blackpool's problems

The local authorities now recognize the need for urgent action to arrest the decline. From 2001 serious efforts have been made to smarten up areas frequented by visitors, by pulling down old buildings and landscaping car parks. Beaches have been cleaned up and beach facilities improved, so that by 2006 three of them were flying EU blue flags. Sand extraction further south along the coast has been reduced. 'Blackpool Illuminations', which are vital for extending the visitor season into the autumn, are being transformed by a £10 million investment after years of 'always being the same'. Other off-season events such as conferences and festivals are being promoted.

Attractions for both summer and winter are essential to offset the effects of the British weather. Some visitors enjoy the thrills of the Pleasure Beach; a new attraction, Water World, opened in 2006, while Blackpool hopes to become one of the first places to be given one of the new 'super-casinos'. There are plans to make more covered ways between the main visitor attractions and around the shops, for greater comfort in bad weather.

Families frightened off by binge-drinking culture of 'stag nights' and 'hen parties'

Beach erosion during winter storms

Beach and sea water pollution

Unemployment out of season

Overcrowding and traffic jams on Bank Holidays

Unreliable summer weather – wet and windy

Cheap package holidays to the Mediterranean taking regular visitors away

▲ **Figure 2** Blackpool's problems.

Activities

1 **(a)** Make a table and rearrange Blackpool's problems under the headings Physical, Environmental, Economic, Social.

   (b) Choose two problems from different headings and explain each of them more fully.

2 Adopt the role of *one* of the people listed below and write a letter to your local paper, outlining your views on how Blackpool should be managed in future;

   - local resident who runs an amusement arcade on the sea front
   - representative of a conservation group
   - member of the local council
   - retired resident of Blackpool.

Coastal issues in the UK

For people living close to the coast, two main issues dominate: coastal erosion and the threat of flooding. Overall the UK is probably not losing any land, since erosion in one place is balanced by deposition in another. However, knowing this does not help people living next to the sea in places where cliffs are crumbling. What is more, for a variety of reasons, rates of coastal erosion and frequency of flooding are expected to increase in future years.

- **The Earth is warming up and sea levels are rising**

In fact, both of these have been happening for 10000 years, since the end of the Ice Age. At that time sea levels were 30 metres lower. Although people debate whether present-day global warming is natural or caused by people, it is a fact that glaciers and ice sheets are still melting into the seas and oceans.

- **Sea levels are rising faster in the east of the UK, where many lowland areas are located**

On average present sea levels are increasing by 1–2 millimetres per year, but the increase is double the average in the east and south of England (**Figure 3**).

▲ **Figure 3** North-west Scotland goes up by 3 millimetres a year, because it is still recovering from the weight of ice sheets on top of it during the Ice Age. South-east England sinks.

- **Strong wind and storm events are increasing**

As the Earth warms up, there is more energy to drive Atlantic depressions. These can be associated with gale- and storm-force winds, leading to the formation of huge destructive waves, capable of causing more erosion in one night than in many years of normal wave activity.

What can be done?

The two extremes are illustrated in **Figure 4**. The balance of opinion within government and local authority circles is moving towards option 1.

Option 1 is more environmentally friendly, costs less and can be considered more realistic in the light of what is happening to sea levels. The London Barrage and the sea defences at Scarborough (page 77) are examples of option 2. Both are vital, but what will their life expectancies be against rising sea levels?

Option 1 Do nothing – managed retreat	Option 2 Total protection – use hard engineering
• Abandon coast to forces of nature	• Maintain and extend existing defences
• No more repairs to old sea defences	• Build hard surfaces such as walls
• Allow the tide to invade low-lying land	• Stop waves from reaching the coastline
• Create salt marshes to absorb wave energy	• Use groynes in bays to trap the sand
• Leave as wildlife habitats for birds	• Keep the beach as wide as possible

▲ **Figure 4** Two options for coastal protection in the UK.

Activities

1 Some of the people likely to have views about coastal protection are:

  A Bird watchers   **B** Holiday camp owners
  C Londoners   **D** Council treasurers in resorts
  E Farmers with land on cliff tops
  F Economic advisers to the government
  G People with coastal retirement homes
  H Families who take summer holidays by the sea

  (a) Draw a line like the one below. Mark the letters **A–H** along the line to show the likely feelings of the people above.

    Option 1   Option 2
    ├──────────────────────────┤
    Do nothing   Total protection

  (b) Briefly explain your choice of position on the line for *four* of the people.

2 **(a)** Visit websites to find out about rates of erosion (e.g. see Hotlinks, page iv) and local views about protection.

  (b) Why do some people have stronger views on coastal protection than others?

OS maps and coastal landscapes

Exam focus

1 Study Figure 1.

a State the:

(i) four-figure grid reference for the square that contains the Old Harry rocks

(ii) direction of Ballard Point from the Old Harry rocks.
(2 marks)

b Draw the different map symbols used for **(i)** Ballard Cliff (west of Ballard Point) and **(ii)** the cliffs between Ballard Point and The Foreland. **(iii)** Suggest one reason why different symbols have been used. (2 marks)

c Describe the map evidence which shows that many people visit the coast at Studland. (3 marks)

d Describe two physical differences between the coastlines north and south of Studland. (4 marks)

e Refer to Swanage Bay.

(i) What are groynes?

(ii) Why were they built here? (4 marks)

2 Study Figure 2.

a (i) Measure the distance in kilometres along the spit, from the southern end of Sturt Pond to the far end of the spit. (1 mark)

(ii) Name two ways in which people can reach the end of the spit in summer. (2 marks)

b Draw a labelled sketch to show the main physical features of Hurst Castle spit. (3 marks)

c (i) Describe the physical features of the area between the spit and the land. (3 marks)

(ii) Suggest two ways in which areas along the coast like this are useful. (2 marks)

d Explain how and why this type of spit was formed. (4 marks)

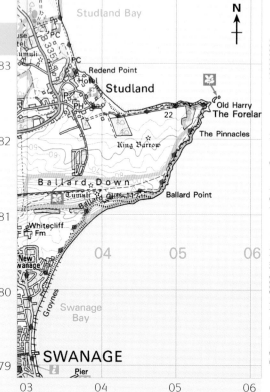

▲ **Figure 1** OS map of Studland and Swanage Bay in Dorset at a scale of 1:50000 (2cm = 1km). The Old Harry rocks and adjacent coastline are shown in Figure 5 on page 71.

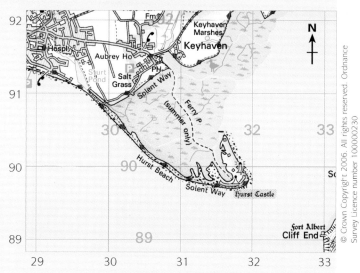

▲ **Figure 2** OS map of Hurst Castle spit in Hampshire at a scale of 1:50000 (2cm = 1km). A reference was made to this in the text on page 73.

Chapter 6

Weather and climate

Winter weather in the Lake District under anticyclonic conditions. Under the settled and calm weather conditions created by an area of high pressure, radiation fog forms in the valleys and frost lingers longer on the shaded slopes.

Key Ideas

Weather and climate are influenced by location:
- the global (world) distribution of climates is affected by factors such as latitude, distance from the sea and prevailing winds
- there are regional differences in climate in the UK based upon factors such as temperature and precipitation
- depressions and anticyclones lead to different and varied weather conditions in the UK.

The interaction between people and the environments and hazards influenced by weather and climate:
- climates affect environments and human activities
- climatic hazards such as drought and tropical storms affect human activities
- people's responses are affected by level of development.

Global distribution of climates

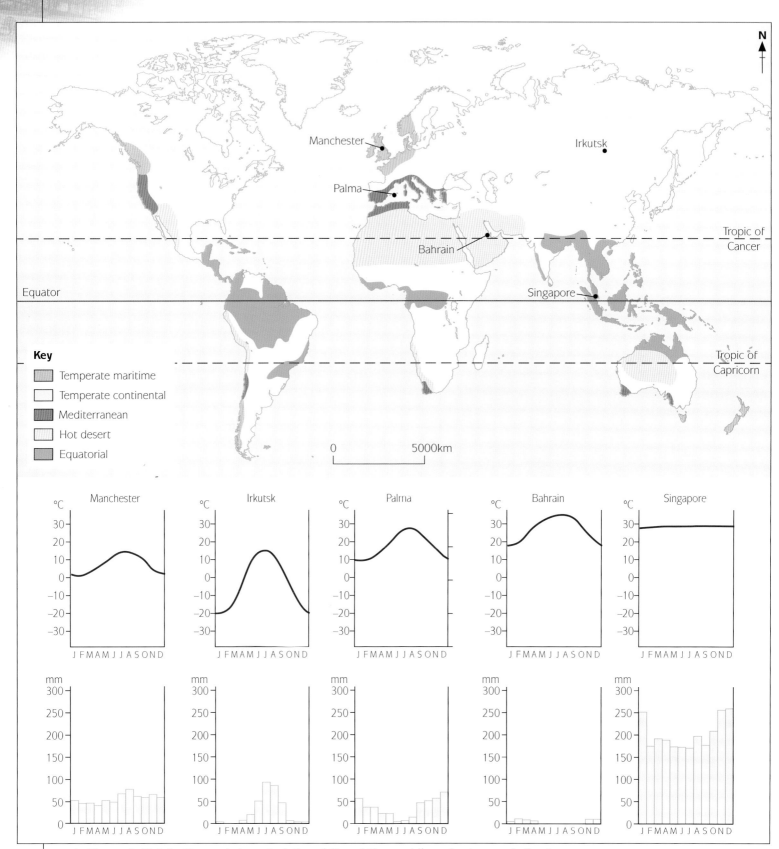

▲ **Figure 1** Distribution and characteristics of five of the world's major types of climate.

What is meant by the climate of a place?

If someone asked you to describe the climate of the UK, what would you say? Wet? Cold? Unpredictable? Unless you had researched the answer, you would be unlikely to give the description used by many geographers, which summarizes the UK's climate as mild (or cool), wet winters and warm, wet summers. This is called a **temperate maritime climate**. *Temperate* means that the UK experiences neither the heat of the tropics nor the coldness of the poles. *Maritime* indicates that the UK feels the influence of the sea. One effect of the sea is to reduce the temperature differences between winter and summer. Another effect is increased precipitation.

Climate is a summary of a place's weather conditions, such as temperature and precipitation, averaged out over a long period of time (usually at least 30 years). It shows the conditions that can normally be expected at a place month by month during the year. If you are going on holiday, for example to Majorca, a study of the climate of its capital city, Palma, would tell you how hot it is likely to be and how much rain you can expect to fall while you are there (**Figure 1**). Note that in **Figure 1** a line graph is used to show average monthly temperatures and a bar graph is used to show precipitation; this is the usual way to show these features.

ⓘ Information

Get to know your latitudes

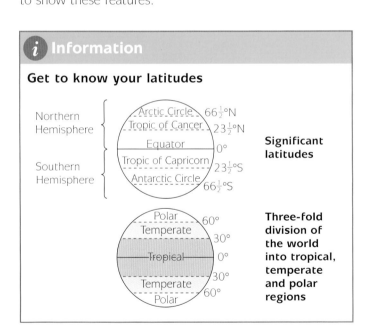

ⓘ Information

How to describe and use climatic graphs

A Temperature graph
1   State the highest (maximum) temperature and name the month.
2   State the lowest (minimum) temperature and name the month.
3   Work out the annual range of temperature by subtracting the lowest from the highest and stating the difference.
4   Describe the shape (gradient) of the temperature line.
5   Check that you have quoted the temperature unit and that your answers are accurate.

B Precipitation graph
1   Name the month with the highest amount of precipitation and state the amount.
2   Name the season or seasons in which most precipitation falls. (Note that in some places the answer could be 'all year'.)
3   Name the month with the lowest amount of precipitation and state the amount.
4   Name the season or seasons in which little precipitation falls.
5   Give some idea of the total amount of precipitation for the year. (Occasionally you may need to calculate the exact total; usually it is sufficient to use descriptions such as 'little' or 'very low' for under 250mm per year and 'high' for over 1500mm per year.)
6   Check that you have quoted the precipitation unit and that your answers are accurate.

C Both graphs
Suggest the name for the type of climate shown (if it has not been given).

Activities

1 Following the instructions in parts A and B of the Information Box above, describe Manchester's climate.

2 Describe the main ways in which Manchester's climate is different from those of **(a)** Singapore, **(b)** Bahrain, **(c)** Palma and **(d)** Irkutsk.

3 From **Figure 1**, describe the distributions of **(a)** equatorial and **(b)** temperate continental climates.

Factors that affect the global (world) distribution of climates

Temperature and precipitation are the measurements used most frequently for describing the characteristics of a particular climate. So what causes temperature and precipitation to vary from one part of the Earth's surface to another?

Factors affecting temperature

1 Latitude

On a global scale the effects of **latitude** are of greatest importance. The highest temperatures are recorded in low latitudes in the tropics between 0° and 23½° north and south of the Equator. Lowest temperatures are recorded in polar latitudes north and south of 66½°. There is a gradual decrease in temperature between the Equator and the Poles because of reduced **insolation** (**Figure 1**). The Sun shines from a high angle in the sky in the tropics all year. As a result, the Sun's light travels directly through the Earth's atmosphere so that less is lost by reflection. Also, because the Sun's rays are (almost) vertical, there is a smaller area of the Earth's surface for each ray to heat up. In contrast, near the Poles the Sun's rays approach the Earth's surface at an oblique angle, which means that each ray has a larger area of surface to heat up. Having had a longer journey through the atmosphere where more of the light was reflected, less sunlight remains to be absorbed by the Earth's surface and used for heating it up.

2 Distance from the sea

Particularly in temperate latitudes, distance from the sea influences temperature. During *summer*, when rates of insolation are highest, the sea heats up less quickly than the land. The Sun's light penetrates below the water surface so that the Sun's rays have more than just the surface to heat up, and constant movement mixes up warm and cool water. If you swim in the Mediterranean Sea in summer the temperature of the water will be up to about 20°C, whereas on land, even in the shade, temperatures between 25°C and 30°C are common. During winter months the opposite happens. The sea retains its store of summer heat longer than the land; on many winter days the temperature of the sea water will be higher than that of the air temperature on land. The result is that places close to the sea have summers that are less hot and winters that are less cold than those a long way inland (**Figure 2**).

3 Prevailing winds

The British Isles and Western Europe have a much warmer *winter* climate than might be expected for their latitude. This is because the **prevailing winds** (the winds that blow most often) are south-westerly and have had a long sea journey across the Atlantic Ocean. The Atlantic Ocean is itself warmer than expected for its latitude because a warm ocean current, the North Atlantic Drift, is present. The south-westerly winds blow over this current and are warmed by it. This shows one way in which the direction of the prevailing winds has an effect on temperatures. Can you understand the two reasons why, if the prevailing winds blew from the east instead of from the south-west, the winter climate of the British Isles would be much colder?

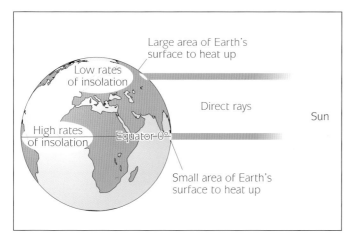

▲ **Figure 1** The effects of latitude on temperature – insolation.

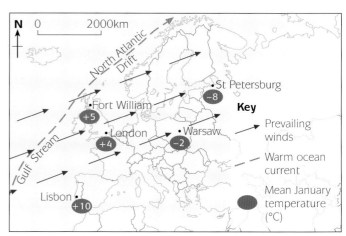

▲ **Figure 2** Why places near to the sea are warmer in winter.

4 Altitude

At the local scale **altitude** (the height of the land) can greatly affect temperature. If you go walking in the hills you will know that it becomes colder the higher up you go. The average rate of temperature loss is about 1°C for every 150 metres you climb. This is because the air becomes thinner so less of the Earth's heat is trapped and more is lost into space. However, when you are in the uplands the loss of temperature may seem greater than the average rate stated above, because the wind chill is higher and cloud is more likely to block out the Sun. However, great variations in temperature occur within upland areas, with sunny south-facing slopes in the northern hemisphere being warmer than slopes that face north; they are also more often in the shade.

Precipitation

Precipitation is the name for all types of moisture from the atmosphere. Usually this means rain (droplets of water), but also other forms such as snow (ice crystals) and hail (bullets of ice). Most precipitation falls after air has been forced to rise to high levels in the atmosphere.

One type of precipitation is **convectional rainfall** (**Figure 3**). The air next to the Earth's surface is heated by contact with the hot surface. This happens all year round near the Equator where rates of insolation are constantly high, and in summer in the middle of the land masses of North America, Europe and Asia. Hot air is lighter and rises up through the atmosphere. As it rises, the air reaches an area of lower air pressure and expands. Expansion causes the air to cool. Continued rising and cooling leads to moisture condensing, forming clouds. The intense heating of the ground drives the hot air (or convection) currents up to such high levels in the atmosphere that tall cumulo-nimbus clouds form. In these clouds water droplets crash against each other and increase in size, until they are too large to be supported by the rising air currents and fall as precipitation. Heavy tropical downpours result, often accompanied by thunder and lightning (**Figure 4**). The other types of rainfall are **relief rainfall** (page 89) and **frontal rainfall** (page 90).

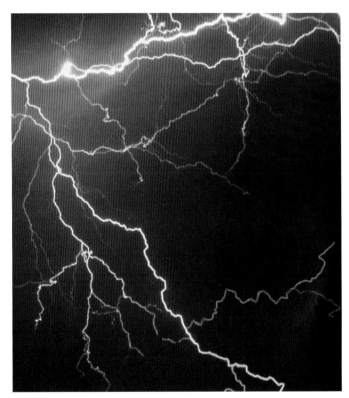

▲ **Figure 4** Tropical thunderstorm in Malaysia.

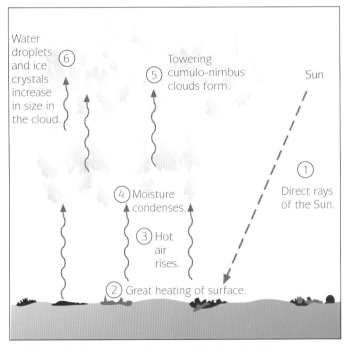

▲ **Figure 3** Formation of convectional rain. The stages in the formation of convectional rainfall are numbered 1–6.

Activities

1 Give reasons for the temperature features stated in each of the following statements **A–D**:

   A Average summer temperatures increase with greater distance from the sea.

   B Average winter temperatures decrease with greater distance from the sea.

   C The British Isles and Western Europe are much warmer than they should be when their latitude is considered.

   D Annual range of temperature increases with distance inland.

The climate of the UK

Mention has already been made of the name of the UK's climate (*temperate maritime*) and of its general characteristics (*mild, or cool, wet winters and warm, wet summers*). However, there are differences in average conditions of temperature and precipitation from one part of the country to another. These are important because they have a great influence on human activities.

Temperature

The temperature pattern changes between summer and winter (**Figure 1**). In *summer* (**Figure 1A**) the south is warmer than the north. The isotherms (lines linking places with the same average temperature) run mainly from west to east. The Sun shines from a higher angle in the sky in the south, which means that rates of insolation are higher here than in the north of Scotland, making for a difference of about 4°C in average temperatures. The built-up area around London, where the effects of the **urban heat island** are felt, is particularly warm. The lines bend southwards near to and over the sea, showing that the sea has a cooling effect in summer.

In *winter* (**Figure 1B**), in some places the isotherms run north to south, and the west of the country is in general warmer than the east. North-west Scotland is warmer than many parts of England. At this time of the year the Sun's influence on temperatures is less because it is at a low angle in the sky and there are fewer hours of daylight. The winter warmth of the sea and the North

Atlantic Drift is of greater importance; their warmth is transferred onshore by prevailing westerly winds, which warm up the western side of the country first.

Influence of temperature on human activities

Farmers are most directly affected by temperatures. Temperature is the main control for the length of the growing season (**Figure 2**). Grass grows for much of the year in coastal regions in Cornwall and Devon, favouring dairy farming. Flowers and vegetables from the same regions are the first to reach markets, fetching higher prices for farmers. Coastal holiday resorts are concentrated along the south coast of England, which has some of the highest summer temperatures. A study of **Figure 1B** should help to explain why so many people also retire there.

Precipitation

Precipitation in the UK is generally highest in the west and lowest in the east (**Figure 3**).

> ### ⓘ Information
>
> **Urban heat island**
>
> Buildings and dark surfaces such as tarmac roads store heat. Further heat comes from car fumes, factories, lights and central heating systems. The larger the built-up area, the greater the effect, which is why it is so noticeable in London (**Figure 1A**), where the centre may be 4–5°C warmer than the suburbs.

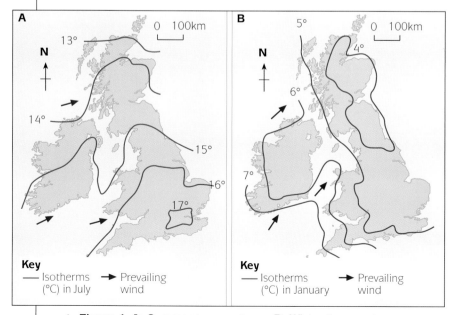

▲ **Figure 1 A**: Summer temperatures. **B**: Winter temperatures.

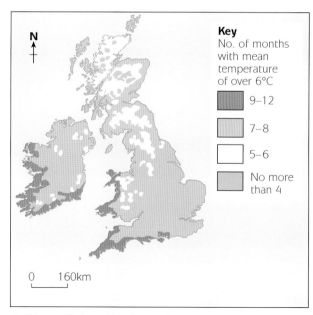

▲ **Figure 2** Length of growing season.

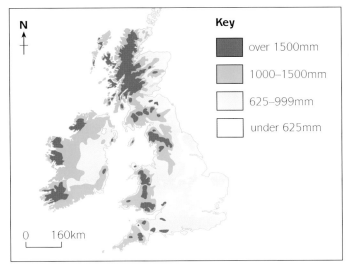

▲ **Figure 3** Average annual precipitation.

Key
- over 1500mm
- 1000–1500mm
- 625–999mm
- under 625mm

N

0   160km

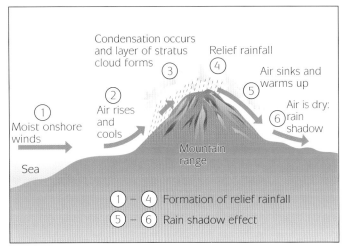

Condensation occurs and layer of stratus cloud forms ③

Relief rainfall ④

Air sinks and warms up ⑤

① Moist onshore winds

② Air rises and cools

Air is dry: ⑥ rain shadow

Sea

Mountain range

① – ④ Formation of relief rainfall

⑤ – ⑥ Rain shadow effect

▲ **Figure 4** Relief rainfall and the rain shadow effect.

The factor most responsible for this is the *direction of the prevailing winds*. The south-westerly winds have had a long sea journey and are laden with moisture when they blow onshore. They release less moisture as they move east. Other factors contribute as well. *Frontal depressions* (areas of low pressure with warm and cold fronts) are also driven from west to east by the westerly circulation and they drop rainfall first in the west. What makes the difference in the amount of precipitation between east and west so large is *relief*. In the west are upland areas; in the highest of these annual precipitation totals exceed 2000 millimetres. The amount of precipitation released increases when winds laden with moisture are cooled even more by being forced to rise over the uplands. In contrast, in some places in eastern England, such as East Anglia, the annual precipitation total barely reaches 500 millimetres. East Anglia is one of the lowest-lying parts of the country; winds and frontal depressions arriving from the west have crossed the widest area of land so that the rain shadow effect is felt more strongly here than elsewhere (**Figure 4**).

Influence of precipitation on human activities

High rainfall washes the minerals (and the goodness) out of the soil. This is called **leaching**. Soils in upland areas become acid and infertile after heavy leaching. In upland areas where the relief is gentle and the rocks are impermeable, such as on many granite moorlands (pages 26–27), the agricultural potential of the land is poor or non-existent. However, the same areas are ideal for water storage even if they are useless for farming. In glaciated uplands, natural lakes may be present or the steep-sided valleys offer suitable sites for creating reservoirs. In western lowlands, where the annual precipitation averages about 1000 millimetres, good grass growth means that dairy farming is the main type

of farming, whereas in eastern parts of England and Scotland crop growing is favoured by annual rainfall totals below 750 millimetres. Cereals such as wheat and barley, and root crops such as potatoes and sugar beet, need water for growth but they do not like too much. Also, wet ground makes the use of machinery difficult.

Activities

1 Study the relief map of part of the north of England (**Figure 5**).

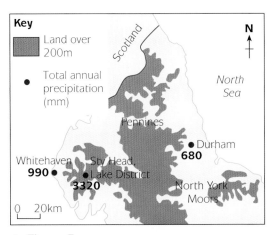

Key
- Land over 200m
- Total annual precipitation (mm)

N

Scotland

North Sea

Pennines

Durham
680

Whitehaven
990

Sty Head
Lake District
3320

North York Moors

0   20km

▲ **Figure 5**

(a) Describe the pattern of precipitation shown.

(b) With the help of a labelled diagram, give reasons for the differences in annual precipitation in northern England.

2 Some of the precipitation falls as snow, especially in upland areas. Suggest ways in which snow (a) hinders and (b) gives opportunities for human activities within the UK.

Weather in the UK – frontal depressions

While climate is the long-term average, **weather** is the day-to-day conditions of temperature, precipitation, cloud, sunshine and wind. A typical weather forecast for the UK on a winter's day might be 'Rain in the morning with strong winds will be followed by sunshine and showers in the afternoon; overnight there will be fog and frost'. A cynic might say that the weather forecaster has no idea what the weather is going to be and has mentioned just about every type of weather that can occur in the UK! That would be a harsh judgement – the British weather is notoriously changeable and usually no two days are alike. There has been a tremendous improvement in the accuracy of weather forecasts, which now rely on satellite data and pictures, radar showing precipitation and computer models for the behaviour of depressions and anticyclones.

Frontal depressions

Depressions are areas of **low pressure**. There is relatively low air pressure at the surface because the air is rising.

Air, carried by surface winds, is drawn into the centre of the depression to replace the rising air. In the northern hemisphere winds blow in an anti-clockwise direction around a depression, spiralling towards its centre. In general, depressions bring spells of unsettled weather with plenty of wind, cloud and rain. Occasional deep depressions (those with a particularly low pressure below 960 millibars (mb)) are responsible for releasing heavy downpours leading to local flooding, or for bringing severe storm-force winds, which cause extensive damage to trees and buildings, such as the famous 'hurricane' which hit south-east England in October 1987.

Formation

Frontal depressions form where warm air from the tropics meets – and is forced to rise above – cold air from the poles. The rising air creates the centre of low pressure. The line which separates the two air masses is called a **front**. As the depression develops and moves east driven by the westerly circulation, two fronts can be recognized.

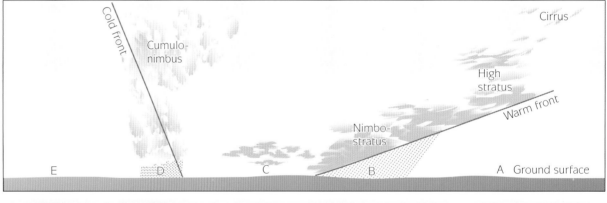

Weather	Mixture of sunshine and showers Temperature falls	Short period of heavy rain, sometimes with thunder and lightning	Rain stops and it becomes warmer	Period of steady rain	Dry but clouding over
Reasons	Main belt of rising air along the front has moved away Pockets of rising air give showers A cold air mass follows the front	Powerful push of cold air Warm air is forced up to high levels in the sky	Warm air is at the surface It is not being forced to rise as strongly as along the front	Moisture in the warm air is now condensing at lower levels	Warm air above warm front is being forced to rise Moisture is condensing to form high cloud

▲ **Figure 1** Cross-section of a frontal depression. Frontal depressions move from west to east across the UK. For the typical pattern of weather associated with the passage of a depression, begin at **A** and work towards **E**.

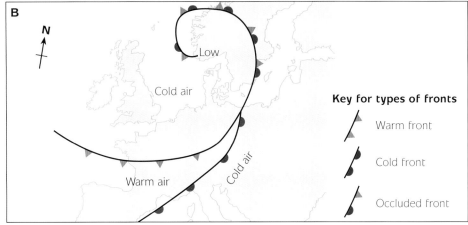

Key for types of fronts

Warm front

Cold front

Occluded front

▲ **Figure 2 A**: Satellite photograph of a frontal depression. **B**: Position of fronts and air masses as shown in **A**.

The leading front is the **warm front**, so called because once the warm front passes over a place, the warm air of the tropical air mass brings warm weather. The front at the rear of the depression is the **cold front**, so called because after it passes the cold air of the polar air mass brings colder weather. In well-developed depressions, like the one in **Figure 2B**, the cold front catches up with the warm front and forms an **occluded front**.

Frontal depressions on satellite photographs and synoptic charts

In **Figure 2A** the centre of the low pressure can be detected by the swirl of cloud. This is where the wind is being sucked into the centre from surrounding areas of high pressure. The swirl is caused by the rotation of the Earth. The fronts show up as trailing lines of cloud on the southern sides of the depression. It is along the fronts that most activity is occurring as air is being pushed upwards leading to condensation of its moisture into cloud. The white speckles of cloud in the area behind the cold front

indicate shower clouds. Cold air is being warmed up by its journey over the sea, encouraging air to rise and to form the cumulo-nimbus shower clouds shown.

1 Study **Figure 3**.

 a State the:

 (i) pressure in the centre of the depression
 (1 mark)

 (ii) wind speed and wind direction at station **A**. (2 marks)

 b (i) Describe the differences in temperatures between England and Ireland. (2 marks)

 (ii) Give reasons for these differences.
 (2 marks)

 c Explain why the western areas of Scotland, Wales and England are being affected by a belt of rain. (5 marks)

 d The weather system is moving eastwards. Describe three ways in which the weather at station **B** is likely to change during the rest of the day. (3 marks)

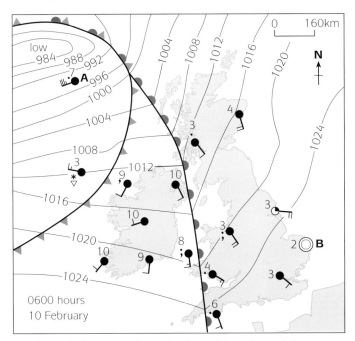

0600 hours
10 February

▲ **Figure 3** Synoptic chart dominated by a frontal depression.

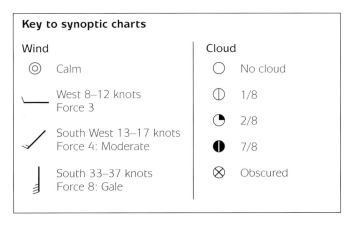

Key to synoptic charts

Wind		Cloud	
◎	Calm	○	No cloud
—	West 8–12 knots Force 3	◐	1/8
╱	South West 13–17 knots Force 4: Moderate	◕	2/8
		●	7/8
⫶	South 33–37 knots Force 8: Gale	⊗	Obscured

Weather in the UK – anticyclones

These are areas of **high pressure** in which the air is descending to the Earth's surface. The air 'piles up' near the surface to create a higher than average air pressure. The air drifts outwards from the centre of the anticyclone and the winds blow in a clockwise direction around the centre in the northern hemisphere. Only light winds are caused by the gentle **pressure gradient** (small differences in pressure over wide areas, which are shown by wide spacing of the isobars on synoptic charts).

A strong anticyclone, above 1030 millibars, may establish itself over the British Isles and stay for several days, or even weeks, blocking out the frontal depressions. At *any time of year*, dry and settled weather with little wind is expected when an anticyclone is dominant. In *summer*, temperatures are likely to be higher than average as strong sunshine from clear cloudless skies heats up the land. This leads to a *heatwave* if the anticyclone does not move and release its grip on the weather. In *winter*, temperatures are lower than average. Sunny but cool days are followed by long cold nights with frost and fog, both of which are most likely on valley floors (page 83). The intensity of the frost can build up night after night leading to a *big freeze*, which brings problems such as burst pipes.

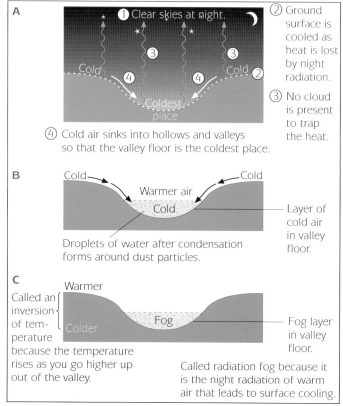

A
① Clear skies at night.
② Ground surface is cooled as heat is lost by night radiation.
③ No cloud is present to trap the heat.
③ Cold ③ Cold
④ Cold ④ Cold
Coldest place
④ Cold air sinks into hollows and valleys so that the valley floor is the coldest place.

B
Cold — Warmer air — Cold
Cold
Layer of cold air in valley floor.
Droplets of water after condensation forms around dust particles.

C
Warmer
Called an inversion of temperature
Colder   Fog
Fog layer in valley floor.
because the temperature rises as you go higher up out of the valley.
Called radiation fog because it is the night radiation of warm air that leads to surface cooling.

▲ **Figure 1** Formation of radiation fog. This type of fog is shown on page 83.

Anticyclonic weather and the reasons for it

Dry weather all year	Little wind at any time	Hot and sunny in summer
• Sinking air is warmed up by increased air pressure near the surface • Warm air can hold more moisture	• Gentle pressure gradient • Little difference in pressure over a wide area	• Greater **insolation** from the sun at a high angle in the sky • Long hours of daylight and cloudless skies for much sunshine
Cold weather in winter	**Frost mainly in winter**	**Fog mainly in winter**
• Sun shines from a low angle in the sky (low **insolation**) • Short hours of daylight from weak winter sun • Long hours of darkness for heat loss from clear skies	• Clear skies allow heat loss from ground surface at night • Ground temperatures can fall below freezing point, especially inland and in valleys • Moisture in contact with the cold ground condenses into ice	• Visibility is less than 1km for fog and 2km for mist • Radiation fog forms on cold nights with little wind (**Figure 1**) • A strong breeze is often needed to clear the fog or it can last all day

▲ **Figure 2** Satellite photograph taken when an anticyclone is dominating the weather over most of the British Isles. Cloud associated with the fronts of the depression centred off the coast of Norway can be seen touching the north of Scotland. Its progress southwards over the rest of the UK has been blocked by the anticyclone, which has given largely cloudless skies to the rest of the UK and much of central Europe.

The effects of fog on human activities

Fog is the most common hazard of anticyclonic weather in the UK. People on the move are most affected. Thick fog, especially if it occurs in patches, can cause road accidents and motorway pile-ups (**Figure 4**). Motorway warning signs telling motorists to slow down, and the addition of lights along those sections of motorway known to be most prone to fog have helped to reduce the numbers of accidents.

Many jet planes are fitted with automatic landing systems, although there are always delays; sometimes the fog is so thick that the pilots cannot see to taxi to the terminal and all aircraft movements have to stop. Flights are cancelled or planes diverted to other airports. Lighthouses have foghorns to warn ships of dangers to navigation, such as rocks, when the light cannot be seen.

▲ **Figure 4** A motorway in fog.

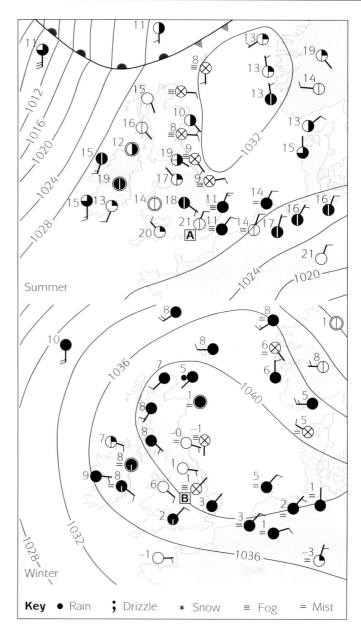

Summer

Winter

Key ● Rain   ⦂ Drizzle   ✳ Snow   ≡ Fog   = Mist

▲ **Figure 3** Synoptic charts for summer (top) and winter.

Activities

1 Make a table to summarize the differences between depressions and anticyclones using these headings: Typical pressure (mb); Wind speed; Precipitation; Weather in summer; Weather in winter.

2 People affected by the weather include airline pilots, cricketers, farmers, fishermen, owners of beach-front shops, skiers. Choose *two* of these people. For each one:

  (a) describe what is good and bad weather

  (b) explain when and why this good and bad weather is likely to occur.

3 **(a)** Describe the differences in weather between stations A and B in **Figure 3**.

  (b) Give reasons for temperature differences between **A** and **B**.

The effects of weather and climate on human activity

This chapter has already mentioned some of the effects of **climate** on human activities such as farming (page 89) and transport (page 93). Although the climate for the UK goes under the general heading of **temperate maritime**, considerable variations in temperature, length of growing season and precipitation exist within the UK (pages 88–89). Climate has contributed to certain distinctive physical and human features of the UK's landscape, such as:

- many surface rivers and reservoirs in the uplands
- natural vegetation dominated by deciduous trees
- good grass growth for livestock farming in the west
- large fields of cereals and root crops in the east.

◄ **Figure 1** Autumn scene in the Severn Valley in Shropshire. What is distinctively British about the scene?

When people from the UK take holidays in countries around the Mediterranean Sea, they are visiting a natural environment quite different from the one back home. The natural vegetation is dominated by evergreen trees, such as the cork oak, although this has been much altered over the thousands of years of human settlement. Mediterranean landscapes today are dominated by a mixture of scrub, bushes with tough leathery leaves and bare rock outcrops (**Figure 2**). Plants must be able to withstand heat and drought. The vine is a typical Mediterranean plant; it uses its deep tap roots to search for moisture in the ground (**Figure 3**).

This contrast in environment is due, in large measure, to the **Mediterranean climate**. What makes it unique among climates is its dry summer (see the graph for Palma in **Figure 1** on page 84). The mainly dry and sunny summers attract millions of visitors who pack the beach resorts (**Figure 4**). This is of great economic importance: without tourism, Spain's great economic development since 1970 would not have been possible. Like the

plants, locals and visitors must adapt to the summer heat. During the 'siesta' hours in the afternoon, cities and resorts in Mediterranean countries are very quiet.

▼ **Figure 2** Mediterranean landscape in southern France, in spring. Many of the flowering plants give off a scent from their oils, which act as a natural protection against the powerful Mediterranean sun.

▲ **Figure 3** Vines planted right up to the village edge. What are the distinctive characteristics of the houses?

▲ **Figure 4** Benidorm in summer. How does this compare with Blackpool (page 80)?

The dry summers are a challenge for all farmers. Citrus fruits, soft fruits such as peaches and salad vegetables grow best with the aid of irrigation. Up to 80 per cent of water supply in Spain is used by farmers. During periods of drought, such as summer 2005, water shortages and heatwaves reduced farmers' incomes by an estimated 35 per cent. Water supply is a big issue in Spain, due to ever-increasing demand from new tourist hotels, villas and retirement homes, as well as swimming pools. Forest fires were a major problem in the drought summer of 2005 in Spain and Portugal.

Fortunately for farmers, winter rainfall is almost as high around the Mediterranean as in the UK. Winters are warmer than elsewhere in Europe, allowing farmers to grow out-of-season fruit and vegetables in open fields, sometimes under plastic covering (**Figure 5**). This produce fetches higher prices in northern markets.

▲ **Figure 5** Growing vegetables in winter in Majorca.

Climatic hazards

From time to time extreme weather conditions exert a much stronger influence than normal over human activities. A short-term event that threatens people and property is known as a **climatic hazard**. The two natural hazards that cause the greatest loss of life are both climatic, namely drought and tropical storms (**Figure 6**).

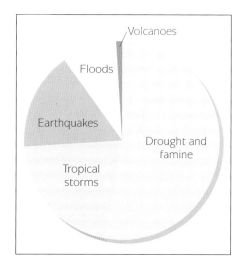

Figure 6 Percentage loss of life from different natural hazards (1970–2000).

Volcanoes
Floods
Earthquakes
Drought and famine
Tropical storms

Drought occurs when dominant wind and pressure patterns differ from the normal, so that expected rains do not fall. In the UK, this happens when a high pressure blocks the free eastward passage of frontal depressions from the Atlantic. In Africa, the cause is high pressure persisting in summer, preventing the arrival of rain-bearing winds from the sea.

> ### ⓘ Information
>
> **Drought**
>
> - This is **not** a hazard for people living in hot deserts; dry weather is expected.
>
> - This **is** a hazard for people living in non-desert areas when expected rains fail to arrive.

Tropical storms are formed when sea water is at its hottest (above 27°C) in late summer and autumn. Air above the sea surface is heated, and warm moist air starts to rise. As it does so, a deep centre of low pressure develops, which sucks up more and more air from the surface. Wind speeds around the centre increase to 150–250 kilometres per hour in a huge circular swirl of cloud. Torrential rain falls from towering cumulo-nimbus clouds, except in the 'eye' in the centre, where the weather is calm and dry. Different names are used – **hurricanes** in the Caribbean, cyclones in the Indian Ocean.

> ### Activities
>
> 1 Make a table with the title 'Summary of contrasts between a temperate maritime environment (e.g. the UK) and a Mediterranean environment (e.g. Spain)'. Use the following headings:
>
>   **(a)** Climatic characteristics (see also **Figure 1** on page 84)
>
>   **(b)** Natural vegetation
>
>   **(c)** Farming (see also Chapter 10 for the UK)
>
>   **(d)** Tourism
>
>   **(e)** Housing styles.
>
> 2 **(a)** Describe two advantages and two disadvantages of the 'dry summers' associated with a Mediterranean climate.
>
>   **(b)** When does a 'dry summer' become a 'drought' in areas with a Mediterranean climate?
>
>   **(c)** Why is water supply an issue of increasing importance in Mediterranean countries such as Spain?

Drought and tropical storms – their impact and human responses

Drought

Water is essential for life. The droughts in various parts of Africa since 1970 have had devastating impacts. Africa is a continent of subsistence farmers: when crops fail or animals die because pastures are burnt off by the heat, people starve. The most vulnerable groups of people – the young, the sick and the elderly – die first. When the drought was at its worst in Ethiopia in 1983, thousands died (page 244). In contrast, in MEDCs (such as Spain and the UK) drought is more likely to cause economic losses than loss of life.

Example of drought in an LEDC – Niger in 2004/5

Niger in West Africa (**Figure 1**) is one of the world's ten poorest countries. It has the highest birth rate (55 per 1000) and the highest fertility rate (an average of eight children per woman); its population growth is very fast. When the rains failed in summer 2004, there were few crops for farmers to harvest in October 2004. Drought in 2004 led to famine in 2005. By July 2005 up to 3 million people (out of a total of 13 million) were desperate for food aid, and an unknown number of children were dying from malnutrition. Worst affected were the Fulani and other nomadic herders: up to 70 per cent of their livestock died because of a lack of fodder.

▲ **Figure 1** The location of Niger.

Food aid was slow to arrive in 2005. It was not until June 2005 that the world's news media became aware of the seriousness of the situation. By mid-August emergency food supplies had still not reached about 1 million people in the villages. Everyone recognises that food aid is no more than a short-term solution to try to alleviate suffering and prevent further deaths. A massive injection of foreign development aid over the long term is the only answer, to enable surpluses to be built up for use in drought years, which will inevitably happen again. Niger lies in the Sahel, an area noted for its unreliable rainfall (page 244).

Example of drought in an MEDC – the UK 1995

During the drought in the UK in 1995, reservoirs throughout England dried up (**Figure 2**). The drought was so severe because of lower than average precipitation in England over a two-year period, combined with a hot and sunny summer dominated by anticyclones. Farmers faced some extra economic costs for water supply and some reductions in crop yields, but, for most people, measures such as a ban on hosepipes were inconvenient, rather than a matter of life and death as in Africa (**Figure 3**).

When will there next be a drought in the UK? South-east England is always the area at greatest risk, because:

- it is the last part of the UK to be reached by Atlantic fronts
- it is mostly lowland, with fewer good natural sites for reservoirs
- underground water stores are being used up faster than they are replenished
- it is experiencing urban growth and there are plans for a lot of new houses
- only between 66 per cent and 80 per cent of average rainfall was received in 2005, which by January 2006 was ringing the alarm bells for future water shortages.

▲ **Figure 2** Haweswater in the Lake District, August 1995. Instead of a reservoir full of water to supply Manchester, this became a tourist attraction.

A

> The sun and heat have come as a welcome fillip to British seaside resorts, with hotels full and seafront stalls doing a roaring trade. Spending in Kent, Surrey and Sussex reached a record £1.5 billion last month, the South East England Tourist Board reported.
>
> By contrast, halfway through the busiest tourist period of the year, more than a million foreign holidays worth £350 million remained unsold this weekend.

B

> Supermarkets announced record sales of soft drinks and ice-cream so far this month. A spokesperson for one of the big chains said 'We listen to the weather forecasts and change or move our stock around accordingly. Hot weather usually means good sales for salad produce and for any kind of drinks, so we extend our shelf space for these items. On the meat counter, people just aren't interested in joints of meat for roasts, but chops, burgers and sausages are selling as soon as we put them out. Everyone must be having barbecues!'

C

> Hot weather has led to bumper cereal crops but dairy farmers are struggling because of dry pastures.

▲ **Figure 3** Cuttings from newspapers in August 1995.

Tropical storms

These are greatly feared by people living in coastal areas in the tropics. Their high winds are capable of flattening everything in their path, while the accompanying torrential downpours are capable of dropping up to 500 millimetres of rain in 24 hours – the same amount that Cambridge receives in a whole year! Storm surges whipped up by the winds can cause a wall of water up to 10 metres high to crash down on coastal areas. There can be extensive flooding in low-lying coastal lands, and landslides of soil, stones and rock are set off on slopes. Devastation can be almost total.

After the storm dies down, people are in a state of shock from *social* losses (deaths of relatives and friends) and *economic* losses (damage to homes, possessions and businesses, loss of crops and animals on farms). Public utilities are badly disrupted: life can be very difficult without access to electricity, telephones, transport and fresh water. Disruption to fresh water supplies, sewage treatment and waste disposal can lead to serious health problems and spread of diseases such as cholera and typhoid. There are often adverse *political* consequences for governments that fail to respond well to the disaster.

Meteorologists can watch tropical storms continuously thanks to weather satellites; warnings can be given to evacuate areas. Police and fire services can practise emergency drills, people can be educated in advance about emergency procedures and emergency shelters can be prepared with rations of food and water. Unfortunately, the behaviour of tropical storms is notoriously unpredictable; they can suddenly become stronger and change direction. Some people are always unwilling to leave their homes and property unattended. Others are too poor or do not have the means of transport to move inland out of the storm's reach. Theory and reality can be two different things, with the most severe impact often falling on poor people living in LEDCs (**Figure 4**).

	A Rich world: Hurricane Floyd, North Carolina, USA (September 1999)	B Poor world: Hurricane Mitch, Honduras and Nicaragua, Central America (October 1998)
Wind speed	up to 250kph	up to 260kph
People affected	3 million	4 million
Loss of life	7	18000 dead or missing
Damage	scores of houses destroyed	over 1 million homeless
	roads washed away	landslides washed away whole villages
Estimated losses	US$ 16 billion	US$ 7 billion
Percentage of losses insured	75%	2%

▲ **Figure 4** The impact of two tropical storms compared.

Activities

1 (a) Refer to pages 233 and 244.

    (i) Draw a simple sketch map to show the location of the Sahel.

    (ii) State the climatic causes of drought in this part of Africa.

(b) Describe how human factors made the impact of drought worse in Niger in 2004/5.

2 (a) Describe how **Figures 3** and **4** show that the poor suffer more than the rich in climatic disasters.

(b) Explain why strategies for preventing loss of life in cyclones are usually more successful in MEDCs than in LEDCs.

2005 – A record-breaking year for tropical storms in the Caribbean

◀ **Figure 1** New Orleans was full of water after it took a direct hit from Hurricane Katrina at the end of August 2005.

Autumn 2005 was the busiest and most expensive hurricane season on record (see the Information Box). For the first time ever, US meteorologists exhausted the alphabetical list of names (alternate male and female) for storms, and had to resort to using letters from the Greek alphabet. Higher than usual sea water temperatures supplied the energy that caused the storms and maintained their strength. With the world warming up, meteorologists issued a warning that the next few years could prove to be just as stormy. Some are convinced that human activity is helping to fuel the monster hurricanes.

A trail of death and destruction was left across the Caribbean, Central America and Gulf States of the USA in 2005. The biggest loss in the USA was caused by Hurricane Katrina, which killed an estimated 1321 people and left hundreds of thousands homeless when it devastated New Orleans and other low-lying parts of Louisiana and Mississippi. Although production from oil and gas fields in the Gulf of Mexico was disrupted, and the prices of oil and petrol rose sharply, the overall loss to the American economy was negligible because the two Gulf States only contribute 2 per cent of the USA's wealth. Those who died were mainly poorer people who did not own cars, and did not have the money to escape inland before the full fury of the hurricane hit New Orleans. To keep a sense of perspective, how does this compare with Hurricane Mitch's death toll in LEDCs in Central America (page 97)?

ℹ Information

The 2005 hurricane season	In 2005	Previous record	Annual average (over 150 years)
Named storms	26	21 (in 1933)	10
Storms of hurricane strength	14	12 (in 1969)	6
Hurricane hits in the USA	4	3	1–2
Category 5 hurricanes (top strength)	3 (Katrina, Rita and Wilma)	2	1
Cost of insurance claims	US$50bn	US$25bn	–
Largest insurance claim for a single hurricane	US$30bn+ (Katrina)	US$22bn+ (Andrew, 1992)	–

28 August New Orleans takes a direct hit from Hurricane Katrina, one of the strongest hurricanes ever to reach the USA.

23 September Millions of Texans from large cities such as Houston attempt to head northwards to escape the threat of Hurricane Rita, causing 150km+ traffic jams.

5–10 October Torrential rains from Hurricane Stan set off mudslides and avalanches in El Salvador and Guatemala, killing up to 1000 people as homes and villages are wiped out.

27 October Mexico counts the cost in lost tourist dollars as Hurricane Wilma (number 21 in 2005) destroys beaches and hotels, and drives thousands of visitors home.

November–early December Tropical storms Alpha to Epsilon make it a late finish to the hurricane season.

▲ **Figure 2** Diary of the 2005 hurricane season.

Activities

1 (a) Give three pieces of evidence that show that the 2005 hurricane season was more active than usual.

(b) Why might this increased activity not occur again?

2 Explain why the poor suffered most from Katrina.

3 Investigate two of the hurricanes named on this page by visiting websites (see Hotlinks, page iv). Summarise the impact, effects and severity of each one.

Chapter 7
Ecosystems

A young boy in Nigeria takes the measure of an old and once mighty ironwood tree, which has just been felled by clearance of the tropical rainforest.

Key Ideas

Globally different ecosystems can be recognized:
- the main factor controlling the world distribution of ecosystems is climate
- three ecosystems which extend over great areas are coniferous woodlands, tropical rainforests and savanna grasslands
- the nature of the soils associated with these ecosystems is influenced by both climate and vegetation.

There are interactions between people and ecosystems:
- tropical rainforests and savanna grasslands are under increasing pressure as human uses increase
- both environmental problems and human conflicts are increasing
- there is an increasing need for global management and development that is sustainable.

Global distribution of ecosystems

An **ecosystem** is a living community of plants and animals which also includes elements of the natural environment, such as climate and soil. Each element in the system, whether living or natural, depends upon, and influences, others. This is why the diagram summarizing the ecosystem (**Figure 1**) shows some two-way relationships.

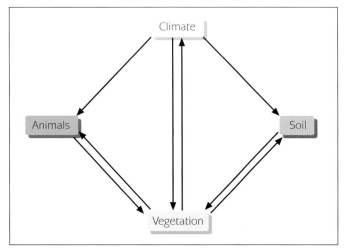

▲ **Figure 1** Ecosystem – systems diagram.

On a world scale, *climate* is the main factor which determines the nature and extent of the natural vegetation cover. **Figure 3** shows the distribution of three of the world's main ecosystems. Look back at **Figure 1** on page 84 which shows the world's main climatic regions. Is the distribution of the ecosystems similar to that of the climatic regions?

Within the tropics, areas covered by **tropical rainforest** closely coincide with areas experiencing the hot and wet Equatorial climate. These areas are close to the Equator and down the wet eastern sides of the continents. Also largely within the tropics are the areas with a hot desert climate. In these areas there is plenty of sunlight and heat for plant growth, but water is in short supply. The effects of lack of water on vegetation are shown in **Figure 4**. In hot deserts there are large areas of bare ground between the plants; plant roots below the surface are often 20 times or more longer than the height of the plant above the surface, which reflects the desperate need to search for underground supplies of water.

Between the rainforests and the hot deserts is the **savanna** or tropical grassland. In this zone, temperatures remain high for most of the year and rainfall varies from about 500 millimetres in the drier areas to over 1000 millimetres closer to the rainforests. Savanna areas have marked wet and dry seasons, so most of the vegetation is drought-resistant. The vegetation cover is mostly grassland but with scattered trees and shrubs where moisture is available. During the dry season the grasses die back and many trees lose their leaves. Some even drop their branches in order to survive the drought.

In temperate latitudes, temperature becomes a more important factor for controlling the nature of the vegetation cover. With increasing distance from the Equator, lack of warmth and shortness of the growing season restrict plant size and variety until no trees will grow in tundra lands (**Figure 4**).

◀▼ **Figure 2 A:** Savanna grassland, Zimbabwe.
B: A baobab tree in Zambia in the dry season.

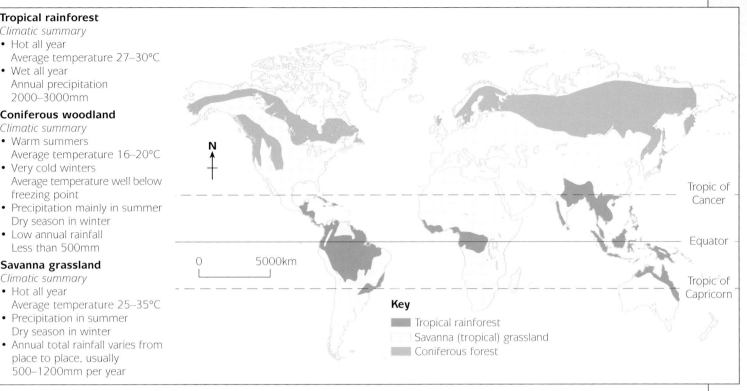

Tropical rainforest
Climatic summary
- Hot all year
  Average temperature 27–30°C
- Wet all year
  Annual precipitation
  2000–3000mm

Coniferous woodland
Climatic summary
- Warm summers
  Average temperature 16–20°C
- Very cold winters
  Average temperature well below
  freezing point
- Precipitation mainly in summer
  Dry season in winter
- Low annual rainfall
  Less than 500mm

Savanna grassland
Climatic summary
- Hot all year
  Average temperature 25–35°C
- Precipitation in summer
  Dry season in winter
- Annual total rainfall varies from
  place to place, usually
  500–1200mm per year

Tropic of Cancer

Equator

Tropic of Capricorn

0   5000km

Key
- Tropical rainforest
- Savanna (tropical) grassland
- Coniferous forest

▲ **Figure 3** World distribution of three ecosystems.

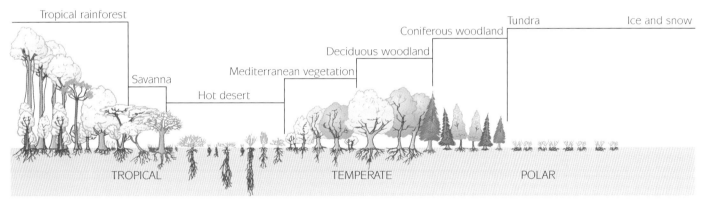

Tropical rainforest

Savanna

Hot desert

Mediterranean vegetation

Deciduous woodland

Coniferous woodland

Tundra

Ice and snow

TROPICAL   TEMPERATE   POLAR

▲ **Figure 4** The changing pattern of vegetation cover from the Equator to the Polar regions in the northern hemisphere.

Activities

1 Use an atlas and **Figure 3**.

(a) (i) Name one country covered by large areas of tropical rainforest in each of these continents: South America, Africa and Asia.

(ii) Name the continent with the largest continuous area covered by tropical rainforest.

(iii) Describe the distribution of tropical rainforest in Africa.

(b) (i) Name one country covered by large areas of coniferous woodland in each of the following continents: North America, Europe and Asia.

(ii) Name the continent with the largest continuous area covered by coniferous woodland.

(iii) Describe the distribution of coniferous woodlands in North America.

(c) (i) Describe the distribution of savanna grasslands in Africa.

(ii) In what way is their distribution different in South America?

2 (a) From **Figure 4**, describe the ways in which vegetation changes within the tropics as the distance from the Equator increases.

(b) (i) Name the main climatic factor responsible for these changes.

(ii) Explain why this factor is important.

Coniferous woodlands

Characteristic features

The *woodlands* are evergreen. The coniferous trees of which they are composed, such as fir, pine and spruce, are remarkably uniform in shape, height and size. There may be only two or three species of tree in a square kilometre of woodland, which increases the uniform appearance of the woodlands over wide areas of land. There tends to be only one layer of vegetation – the tree layer. In the forest gloom, caused by the trees growing so closely together, little else grows. The forest floor is covered by a thick mat of dead needles (**Figure 1**). The dark woods do not attract bird life. Little food is provided and only a few animals, such as deer, can feed by browsing the trees.

The *trees* reproduce from cones, which protect their seeds. The trees have a conical shape with branches sloping downwards along the whole length of the trunk. They are softwoods. Their leaves are small needles which give the trees their evergreen appearance. They have thick bark which contains resin. Tree roots are shallow, spreading out only small distances near the surface.

Adaptations to climate

Each of the vegetation characteristics mentioned above has a purpose. The trees have adaptations which allow them to survive in areas with a cold continental climate. In winter this is one of the most challenging climates for any kind of plant or animal life because of intense cold, snow-covered surfaces and strong, cold winds. Summers are short and not particularly warm, and water is often only readily available in early summer. The climate of Irkutsk is an example (page 84). Some of the adaptations of trees to climate are explained in **Figure 2**.

▲ **Figure 1** Coniferous woodland.

Tree characteristic	Adaptation to the climate
Conical shape	They are flexible and bend in the strong winds.
Downward sloping branches	Snow slides off them more quickly.
Evergreen	Leaves are always present so that trees can begin to grow as soon as it is warm enough in spring/ early summer; necessary because of the shortness of the growing season.
Needle leaves	They reduce water loss by transpiration; necessary when water is not available (e.g. when ground is frozen in winter and when little rain falls in summer).
Thick bark	This protects trunk from extreme winter cold.

▲ **Figure 2** Adaptations of trees to climate.

Soils

A very distinctive soil – the **podsol** – forms under coniferous woodlands (**Figures 3 and 4**).

Look carefully at the colours of the different layers of soil in **Figure 3**. Working downwards from the surface the following layers can be recognized.

1 A narrow layer near the surface which is dark, and can be almost black. This is the **humus layer** next to the surface; the humus gives it the dark colour.

2 A wider layer, of a much lighter colour, almost grey. This is called the **A horizon**; minerals and organic material, which give soils a dark colour, have been washed out from this layer (a process called **leaching**), which is why it is so light in colour.

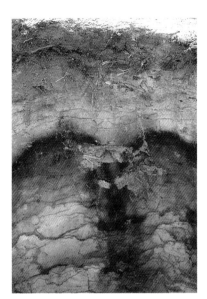

▲ Figure 3 Podsol soil.

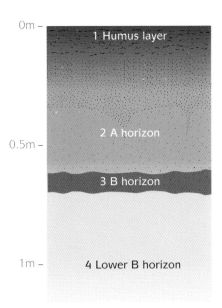

```
0m  —     1 Humus layer


          2 A horizon
0.5m —
          3 B horizon


1m  —     4 Lower B horizon
```

▲ Figure 4 Profile of a podsol.

3 A reddish-brown layer that looks quite different from the layer above it. This is the top of the **B horizon**; the dark-coloured organic material and the dark red oxides of iron washed out from the A horizon are re-deposited here. They are compressed together which forms a **hard pan** just over half a metre below the surface.

4 A wider layer that is orange-brown or orange-yellow in colour. This is the **lower B horizon** which gets its colour from deposited clays.

Management of coniferous woodlands

Although the soils below coniferous trees are of little use for farmers, their softwood is very valuable as timber for construction and for making pulp and paper. Large coniferous plantations cover many steep valley slopes in upland areas, such as the Lake District, northern Pennines and Highlands of Scotland. Not only are the forests useful in reducing run-off and soil erosion, they are also a profitable use of land otherwise marginal or uneconomic to farm. Some plantations are also used for recreation: in Kielder Forest in Northumberland, forest trails, picnic and camping sites and accommodation in log cabins have been provided for visitors. New trees are planted after clearance, thereby providing for the economic needs of future generations, an example of sustainable development.

Soil is the name for the loose material above the solid rock. Climate and vegetation are important factors affecting the formation of podsol soils to create the distinctive layers. The most significant feature of the *climate* is that precipitation is greater than evapotranspiration. This means that water drains downwards through the soil carrying organic material and minerals such as iron and clay with it. The most significant feature of the *vegetation* is that the pine needles decay slowly and what little humus they produce is acidic. There is little in the soil to hold on to the minerals and to stop them being washed out by leaching. Also, the soil is too acidic for much earthworm activity, which is why there is little mixing up of the different layers. Podsols are infertile soils, of little use for cultivation.

Information

Soil

Soil has four constituents:

1 Mineral matter – mainly from weathering of the parent rock below the soil

2 Organic matter – humus is formed from the decomposition of plant remains by organisms in the soil

3 Air   **4** Water.

Activities

1 From the graph for Irkutsk in **Figure 1** on page 84, describe the main features of the cold continental climate.

2 (a) Draw a coniferous tree and add labels to describe its characteristic features.

(b) Choose three of the features labelled in **(a)** and explain how each is an adaptation to the climate.

3 Make a copy of the podsol profile in **Figures 3** and **4**.

(a) Colour it in or clearly label the different layers.

(b) Label the positions of the humus layer and the A and B horizons.

(c) Mark L for leaching in a place where it occurs.

4 (a) Define 'leaching'.

(b) Explain why it occurs.

(c) Explain how leaching influences the characteristics of a podsol soil.

(d) Why does it reduce the usefulness of the soil to farmers?

5 (a) Draw a spider diagram to show the physical and human advantages of coniferous plantations in upland areas of the UK.

(b) Why can coniferous plantations be used as an example of sustainable land use?

(c) Suggest reasons why some people object to increasing the area used for coniferous plantations.

Tropical rainforests

Metres

A Discontinuous canopy of tree crowns of the tallest trees (called emergents)

B Continuous layer of the main canopy formed by the crowns of the many tall trees

C Discontinuous under-canopy of trees between 10m and 20m high

D Layer of shrubs and young trees

E Herb layer with ferns 6m or more high

▲ **Figure 1** The five layers in a tropical rainforest.

Characteristic features

The vegetation mass is much greater in tropical rainforests than in other ecosystems. Despite the sheer amount of vegetation present and the way in which climbing plants and creepers run from tree to tree in a chaotic manner, it is possible to recognize five distinctive forest layers (**Figure 1**). The canopy provides a habitat for monkeys and numerous birds such as macaws; on the ground floor are some larger animals such as tapirs and anteaters. There are countless insects everywhere. However, it is the plant life that is really abundant.

These are evergreen forests. The tall trees are deciduous, but they shed their leaves at different times and for only six to eight weeks each year, so that the forests always look green. The tallest trees, by reaching up to 50 metres high, stand head and shoulders above the forest canopy; these are hardwoods and include types such as mahogany and ironwood. They have long trunks without any branches until their rounded crowns extend out over the canopy. Their leaves are oval in shape with extended points known as drip tips, and they have dark green and leathery upper surfaces. The smooth bark is thin. Their shallow roots, which mainly extend sideways below the ground surface, extend above the ground as buttress roots.

Adaptations to climate

The tropical rainforest's biodiversity is a response to climate. There are constant high temperatures, with a mean monthly average above 27°C, accompanied by high solar light intensity. Rainfall is regular and high, with above 2000 millimetres falling during the year, which creates humid conditions. There is no more favourable climate on Earth than this for plant growth.

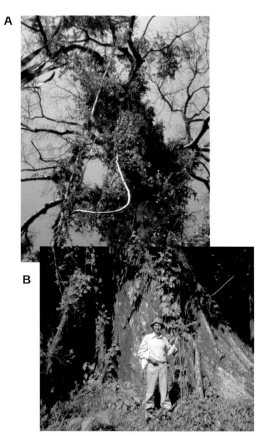

▲ **Figure 2** Tropical rainforest kapok tree. **A:** Top above the canopy. **B:** Bottom on the forest floor.

▲ **Figure 3** A latosol.

Plant communities are fiercely competitive. There is survival of the tallest as the tall trees are drawn upwards by the heat and light, which is why leaf growth is concentrated in the canopy. The leathery upper surfaces of the trees' leaves are necessary to withstand the great power of the Sun's rays. The drip tips help the leaves to shed water during the heavy rains.

In the lower layers of the forest, sunlight is in short supply. Ferns are adapted to life on the forest floor by having leaves which intercept a high proportion of the light that reaches them. The shrub layer is sparse because of lack of light, although shrubs quickly take advantage of any gap in the forest canopy.

Soils

Figure 3 shows a **latosol**, which is the name given to soils which form under tropical rainforest. They are red or yellowish-red in colour throughout; you cannot pick out distinct horizons like you could in **Figure 3** on page 103 for the podsol. Latosols are deep soils, often 20–30 metres deep, whereas podsols are 1–2 metres deep. The layer of black humus at the top forms a narrow horizon. The red and yellow colours below this come from the oxides of iron and aluminium, which remain in the soil after other minerals have been washed out by leaching.

Looking at the density and diversity of the vegetation cover you might think that it is growing from the world's most fertile soil. Nothing could be further from the truth. Precipitation is much greater than evapotranspiration; even with the protection given by the forest canopy, there is the downward movement of rain water through the soil. Leaching washes organic material and silica downwards and then out of the soil. The most important activity is concentrated at the top of the soil profile. This is where the rapid recycling of forest nutrients upon which the life of the forest depends takes place (**Figure 4**). Leaf fall and falling branches provide a continuous supply of litter to the forest floor. The high temperatures and rainfall encourage intensive biotic activity which leads to rapid decomposition of the organic material.

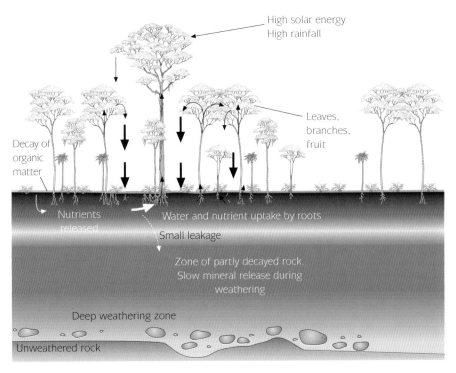

▲ **Figure 4** Nutrient recycling of tropical rainforest.

Human use of tropical rainforests and current issues

▲ **Figure 1** Banana plants in a native clearing. How great is the risk to the environment?

▶ **Figure 2** Tropical rainforest biodiversity in Ecuador. Species of flowering plants are even found growing in the crowns of the emergent trees 40 metres above the ground.

ⓘ Information

Ecuador is one of the smaller South American countries. Although it is 30 times smaller than Europe, it has more plant species. A typical 5-hectare patch of rainforest contains:

- 750 species of trees
- 1500 species of flowering plants
- 400 species of birds
- 150 species of butterflies
- 100 species of reptiles
- 40 000+ species of insects.

Human settlement in rainforests is long established, but often of low density. Indigenous (native) peoples, living in tribes or groups, either collected and hunted to use the food naturally available in the forest, or practised slash and burn and grew crops such as manioc, or did both. With slash and burn, the clearing was small and the group cultivated only for as long as the soil retained its fertility, probably two or three years. They then moved to another part of the forest, hence the alternative label of shifting cultivation. Indigenous tribes, low both in numbers and levels of technology, barely left a mark on the forest, which re-invaded within a few years as if there had been no human presence. This is an example of human use of rainforests that is **sustainable**. Another example of sustainability is rubber tapping from wild rubber trees dotted around the Amazon forest. Provided each rubber tapper has a large patch of forest in which to work, the wild rubber trees will have sufficient time to recover between each collection.

The importance of tropical rainforests

The tropical rainforest is unique among ecosystems because of its **biodiversity**. This term refers to the great number and variety of living species, plants and animals; it is estimated that 50 per cent of the world's 10 million species live in the tropical rainforests. A 5-hectare block of coniferous woodland may contain only two or three different species of tree; in the same area of tropical rainforest there may be 750 species (**Figure 2**). The Earth's genes, species and ecosystems have evolved through 3000 million years and form the basis for human survival. Wild varieties of plants are the basis for new seeds for farmers to use; many of today's drugs, such as aspirin, are derived from plants. When new combinations of genes are sought in the future for new food crops or cures for diseases, without the rainforests and their biodiversity, the number of potentially useful species from which to choose will be reduced. Some environmentalists emphasize the importance of tropical rainforests as suppliers of oxygen (the 'lungs of the world') and carbon dioxide stores (against the enhanced 'greenhouse effect').

Tropical rainforest destruction

Activity from outsiders, unlike that of the indigenous peoples, has in nearly all cases been destructive. Governments, companies and wealthy businessmen realize that rainforest regions are rich in natural resources. The first forest clearances are usually associated with road building. Roads attract farmers, loggers and miners, enabling them to open up wider areas of forest away from the roadsides. Farming by outsiders is no longer slash and burn; cattle ranching is a more likely activity (**Figure 4**). The ranchers are interested in only one thing – to replace forest with pastures. Often they do not even save and sell the valuable hardwood timber and it is just burned (**Figure 3**).

Logging companies are only interested in certain types of tree, but one of the characteristics of rainforests is that individual species of tree are widely dispersed. In order to reach the limited number of trees considered to be commercially useful, all the other trees are felled and cleared. Oil, gas, iron ore, bauxite (for aluminium), nickel and gold are just a few of the natural resources that have attracted mining companies to rainforest regions. Despite all the problems with access, once a mineral deposit of commercial size is discovered, then roads, railways and pipelines are built with little or no thought for the forests or their inhabitants. In remote locations, without government supervision or environmental controls, mining operations, disposal of untreated waste, leaks and spillages cause land and water pollution.

Local effects of rainforest clearance

These are both human and environmental. The fate of indigenous peoples is directly linked to that of the rainforests. They are pushed back into smaller and smaller areas of forest by the advances of loggers, miners and farmers, each group supported by superior modern technology. Direct contact results in the spread of diseases to which they have no resistance, and to a destruction of their traditional culture and ways of life. The natural forest is a closed system with little leakage of nutrients. However, once the forest is cleared, the rich nutrient cycle shown in **Figure 4** on page 105 is broken and nutrients are soon washed out with nothing to replace them (**Figure 5**). This has a devastating effect on soils and hydrology.

▲ **Figure 3** Timber from rainforest clearances piled up and burned in Brazil. How great is the risk to the environment?

▲ **Figure 4** Farming on land previously covered by rainforest in Costa Rica (Central America). How great is the risk to the environment?

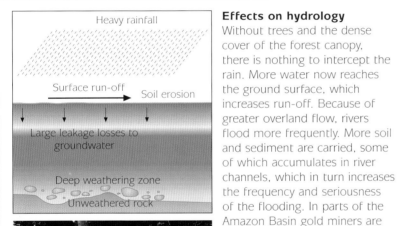

Soil erosion in a clearing

Effects on hydrology

Without trees and the dense cover of the forest canopy, there is nothing to intercept the rain. More water now reaches the ground surface, which increases run-off. Because of greater overland flow, rivers flood more frequently. More soil and sediment are carried, some of which accumulates in river channels, which in turn increases the frequency and seriousness of the flooding. In parts of the Amazon Basin gold miners are creating a potential hydrological catastrophe; an estimated 1000 tonnes of mercury, used to amalgamate gold, has entered the Amazon river system with unknown consequences for human and animal health. Little treatment of waste takes place from any of Amazonia's many mining operations before it is fed into the rivers.

Effects on soils

After a large area is cleared, the surface is exposed to the full force of the often daily, and always heavy, tropical downpours. These erode the topsoil and wash it away into rivers. There is less organic matter in the soil to hold back the water as it seeps downwards, so leaching increases. Most of the soil's minerals are quickly removed.

▲ **Figure 5** Destruction of the nutrient cycle after forest clearance and its consequences.

Activities

1 (a) Define 'biodiversity'.

(b) Give two examples to show that tropical rainforests have the greatest biodiversity on earth.

(c) Why is biodiversity used as an argument for preservation of rainforests?

2 Study **Figure 4**.

(a) Describe the land uses shown.

(b) How great is the risk to the environment? Explain your answer.

Tropical rainforest in Brazil

While the natural deciduous forests have been cleared almost to extinction in Europe and North America, large areas of tropical rainforest remain in other parts of the world, mainly because of obstacles such as difficult access and infertile soils. Rainforests are mainly located in LEDCs (**Figure 1**). However, many are now under threat as governments view their untapped resources as a passport to economic development. The country with the largest area of untouched rainforest is Brazil, the world's fifth largest country by area. Tropical rainforest is the natural vegetation cover in 60 per cent of Brazil; despite clearances since 1960, rainforest still covers almost half of its total area. Even in areas close to Manaus, the largest town in the Amazon Basin with over 2 million people, large areas of untouched forest remain (**Figure 2**).

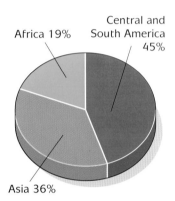

Central and South America 45%
Africa 19%
Asia 36%

▲ **Figure 1** Areas of rainforest in 2005.

▼ **Figure 2** Satellite photograph of part of the Amazon Basin near Manaus. Below Manaus the Negro River joins up with the main channel of the Amazon.

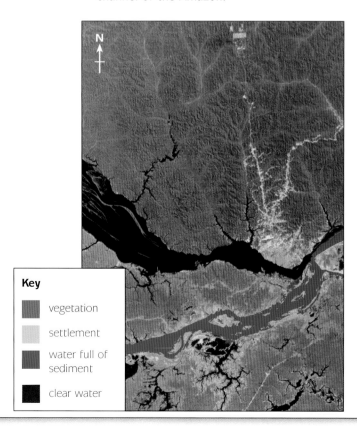

Key
- vegetation
- settlement
- water full of sediment
- clear water

Forest destruction and the reasons for it

What alarms environmentalists and others living outside Brazil is the persistent and continuous forest destruction, which in some years has seen an area the size of Wales (20 754 square kilometres) disappear (**Figure 3**). Although there is still a lot of rainforest left in Brazil, environmental groups are dismayed by the figures showing an increase in the rate of clearance since the low point in 1997. They argue that it cannot keep on increasing for ever. However, the figures suggest that powerful economic, social and political reasons lie behind the forest losses. These are summarized in the Information Box.

Year	km²
1996	30000
1997	12600
1998	16200
1999	16700
2000	19200
2001	17600
2002	24600
2003	23400
2004	25200

▲ **Figure 3** Estimated rainforest losses in the Amazon Basin of Brazil 1996–2004.

ℹ Information

Development of the Amazon Basin

The following are some reasons why Brazilian governments have encouraged and supported development in the Amazon region since the 1960s.

1 Economic
- to export minerals, gain foreign exchange and pay off international debts
- to use the minerals and other raw materials in growing industries
- to extend areas of agriculture and export beef and soybeans; soybeans were Brazil's leading export crop in 2004 (up 50 per cent from 2001)
- to become a more economically developed country.

2 Social
- to relieve population pressure along the coast
- to give landless peasants the chance to own land for the first time.

3 Political
- since 1960 it has been government policy to open up the interior ('march to the west')
- to take people's minds off problems such as poverty and landlessness.

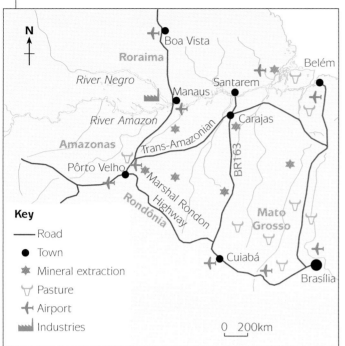

◀ **Figure 4** The Amazon Basin in Brazil.

Under pressure from the agri-business lobby, paving BR163 was planned to begin in 2005.

- Roads open up areas over which the government, police and army have little control.

- Land is invaded on either side to clear trees and raise cattle.

- Loggers build unofficial networks of roads through the forest from the main roads.

- The prospects of paving a road lead to land rushes by soybean farmers.

- More and more of the natural forest will be cleared and lost.

▲ **Figure 5** Arguments against building and paving roads in the Amazon Basin.

Exam focus

1 a (i) Use the values from **Figure 3** to draw a bar graph of forest losses in Brazil.
(3 marks)

(ii) Describe the pattern of forest losses from 1996 to 2004. (3 marks)

b Refer to **Figure 6**.

Large landowner
Nearly half the area of Brazil is still covered by tropical rainforest

Government minister
Brazil spends 40 per cent of its annual income repaying loans

Company director
More land is needed for agri-business, Brazil's number 1 export earner

Farm worker
I am one of 400 000 landless peasants in Brazil. Why can't I have land of my own to farm?

▲ **Figure 6** Comments about rainforest destruction in Brazil.

(i) Identify one comment which is social and another one which is economic. (2 marks)

(ii) Choose two of the four comments and explain their importance for rainforest destruction in Brazil. (4 marks)

c Name one group of people likely to be against paving the BR163. Explain their objections.
(3 marks)

The roads shown in **Figure 4** provide the framework for access into the Amazon rainforest. They are in the process of being improved and extended, although not without controversy (**Figure 5**). Brazil's big share of the Amazon Basin is rich in natural resources. The largest concentration of mineral resources is at Carajas, where there are major deposits of iron ore, bauxite, nickel, copper and manganese. Most of the cleared land is used for farming. Cattle ranching is everywhere on the cleared land. Much of the recent advance of agriculture westwards and northwards from Brasilia, has been by farmers growing soybeans. This crop offers farmers large profits. Most of the crop is exported to Europe, where soybeans are in high demand as animal feed. Good market prices for soybeans and for raw materials in general are boosting Brazil's export earnings, allowing the government to pay off some of the country's massive international debts.

The government of Brazil has designed sensible policies to combat deforestation including:

- increasing protected areas

- hiring more forest patrols

- supporting forest-based activities.

However, in practice their effectiveness is reduced by shortage of resources, corruption, red tape and opposition from powerful business interests.

Savanna (tropical) grasslands

Characteristic features

Moving away from the Equator, the rainforests become lower and less dense until they are replaced by savanna grasslands. Although savanna vegetation is dominated by tall tufted grasses, it is mixed with trees such as umbrella-shaped acacias and bushes (see **Figure 2** on page 100). In drier areas baobab trees and scrub are also found. There are fewer trees than in the rainforests and only two main layers of vegetation.

Savannas occupy a larger area in Africa than in any other continent, where they are renowned for their animal life. Herds of zebra, gazelles, wildebeest and buffalo roam the grasslands looking for the best grazing and water holes; elephants and giraffes graze from bushes and trees as well. They are followed by the carnivores – lions, leopards, cheetahs and hyenas.

Adaptations to climate

The grasses and scattered trees are a response to the savanna climate (**Figure 1**). In drier areas there may also be areas of bare ground between the clumps of grasses. The major feature of the climate is the wet season in summer and the long dry season in the winter months. The total rainfall may vary between 500 and 1200 millimetres per year. Temperatures are always high, between 25 and 35°C. This means evaporation is high.

The trees are drought resistant. They are usually deciduous and lose their leaves during the long dry season. Some trees only keep their leaves for a few weeks during a year, so the trees only grow very slowly. Some trees, such as the baobab or 'upside down' tree, store water in a soft, spongy layer in their large trunks, up to 9 metres in diameter. This protects the tree from fires caused by lightning or people, and provides a reserve water supply during the dry season. Many of the trees grow their leaves a few weeks before the rainy season starts. This enables the trees to take full advantage of the first rains that fall. Many trees, such as acacias in Africa and eucalyptus in Australia, have branches quite low to the ground that spread out into a huge umbrella-shaped crown. The large canopy shades the ground below the tree, reducing evaporation from the soil. The trees also often have small waxy leaves to reduce transpiration. Tree roots are often long tap roots to reach underground sources of water.

The grasses grow quickly during the wet season, often to over 3 metres high (elephant grass). They produce their seeds towards the end of the wet season, then the grasses above the ground die back, turn brown and add large amounts of organic matter to the soil.

Soils

Soils are in general more fertile than in wetter regions nearer the Equator. During the dry season, high temperatures encourage high evaporation rates. As a result there is an upward movement of water through the soil. The water carries with it dissolved minerals such as calcium. The calcium is deposited in the upper layers of the soil, making the soil calcium rich. However, the heavy rains in the wet season may also cause some leaching of the soils. The litter layer is large, especially at the start of the dry season when the grasses die off. Decomposition is quite rapid in the wet season but much slower in the dry season.

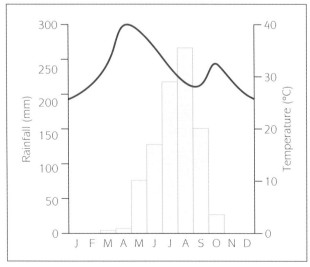

▲ **Figure 1** Climate graph for Kano in Nigeria.

▲ **Figure 2** Giraffes are adapted to grazing leaves from trees and bushes.

Human uses of savanna grasslands

Many parts of the African savannas have had a long history of settlement by tribes of livestock farmers (such as the Masai, page 223) and groups of cultivators. During the last century, populations increased rapidly. This increase has been matched by an equally rapid increase in environmental problems, notably **soil erosion** and **desertification**; population pressure has made a major contribution. The arrival of large numbers of tourists on safari to see big numbers of wild animals in their natural habitats is quite recent; it has led to conflicts of interest in parts of east and southern Africa (page 221).

A study of **Figure 3** suggests that population growth is the underlying cause of desertification, increasing the demand for life's essentials (food and fuel for cooking). Herds of livestock are increased in size and larger areas are planted with crops. In wet years few adverse environmental effects are noticed, but in dry years it is very different. **Overgrazing** and **overcultivation** eventually lead to removal of the vegetation cover leaving bare ground. The next heavy rains wash away the soil and cause huge gulleys (**Figure 4**). There is also a non-human contribution to desertification; drought conditions, possibly linked to global climatic change, are a factor, but most believe that human activity which overuses the environment is the main cause.

Once the stage shown in **Figure 4** is reached, the process of desertification seems irreversible, at least in the short term. Strategies to halt and reverse the process are better targeted at the causes as shown in **Figure 5**, especially when accompanied by a programme of sustainable environmental management at the local level (page 112).

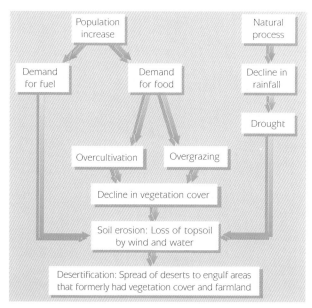

▲ **Figure 3** The process of desertification.

▲ **Figure 4** Overgrazed and gulleyed savanna, Kenya.

Cause	Method
Deforestation	• tree-planting schemes • alternatives to fuel wood e.g. biogas
Overgrazing	• reduce numbers by using higher-yielding breeds • rotate grazing land
Overcultivation	• drought resistant and higher-yielding seeds • crop rotation
Population pressure	• policies for reducing birth rates • alternative employment e.g. tourism, craft industries

▲ **Figure 5** Tackling the causes of desertification.

Activities

1 **(a)** Describe the main characteristics of savanna vegetation.

   (b) Explain how **(i)** grasses and **(ii)** trees are adapted to the climate.

2 **(a)** What is desertification?

   (b) Explain the human causes of desertification.

   (c) How important are natural processes in causing desertification and increasing its extent?

3 Study the information about desertification.

   (a) (i) Describe what is shown in **Figure 4**.

       (ii) What would this area have looked like previously?

   (b) Is the process of desertification reversible? Explain how the answer can be both 'yes' and 'no'.

   (c) Why is a package of measures to control the spread of desertification likely to be more successful than just one?

Case Study – Forest management in Malaysia

Sustainable use of ecosystems

The word **sustainable** is applied to human activities which have a long future because people are working with nature and the environment upon which they depend. It has been estimated that the cash benefits from the value of forest products, tourism, flood control and soil protection from the conservation of forests are three times greater than from timber and farming the cleared land.

Sustainable forestry in Malaysia

Malaysia is one of the better economically developed countries in Asia. About half of its area remains under natural or semi-natural forest. The Forest Management Plan was developed in an attempt to achieve sustainable forestry instead of more predatory logging and destructive clearances.

Sustainable forestry can be achieved by:

A Dividing the forests into two *groups*:

  • Protection and conservation forests: these include National Parks, wildlife and bird sanctuaries.

  • Production forests: in these forests logging takes place but it is carefully planned and controlled.

B Making a *survey* of the area to be logged and its resources.

C Using *selective logging*. Only between seven and twelve mature and fully grown trees per hectare are cut down in each logging cycle. This allows the logged area to regain full maturity after 30–50 years. The forest recovers because the younger trees and saplings are given more space and sunlight to grow.

D *Monitoring* what happens. At all stages it is necessary to check that the work being done conforms to the plan. Illegal activities and clearances are easier to detect now that aerial photography and remote sensing are available. They still occur, which is why the number of staff enforcing the rules has increased, as have the size of the fines and length of prison sentences.

Exam focus

1 Study **Figure 1**.

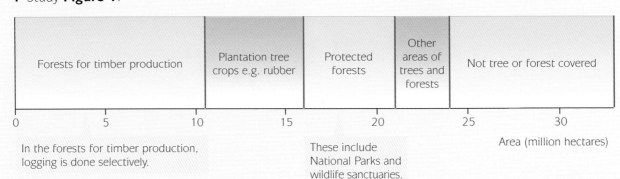

▲ **Figure 1** Land uses in Malaysia, 2004.

a (i) Give the area of protected forests in Malaysia. (1 mark)

(ii) Why can tourism help to protect natural ecosystems? (2 marks)

(iii) Explain what needs to be done to make tourism sustainable in protected areas. (3 marks)

b Approximately what percentage of the total area of Malaysia has some kind of forest and tree cover on it? Show your working. (2 marks)

c Explain the ways in which the Malaysian government is trying to ensure that the forests for timber production are being exploited in a sustainable way. (3 marks)

d Many people believe that plantation tree crop farming is more sustainable than other types of farming in the tropics such as cattle ranching. Give reasons for this. (4 marks)

Chapter 8

Population

Places where there are very few people ... places where there are many people ... why are some parts of the world crowded with people while others remain empty?

Key Ideas

The global distribution of population is uneven:
• physical factors and human factors help to explain why some areas are crowded and others have few people.

Population change depends upon birth rate, death rate and migration and presents challenges to human populations:
• high birth rates and low death rates have led to rapid world population growth
• there are contrasting population problems between LEDCs and MEDCs
• some countries need to reduce population growth, others are having to cope with ageing populations
• migration may be forced or voluntary and there are several different types
• migration has advantages and disadvantages for both losing and receiving countries and regions.

The world distribution of population

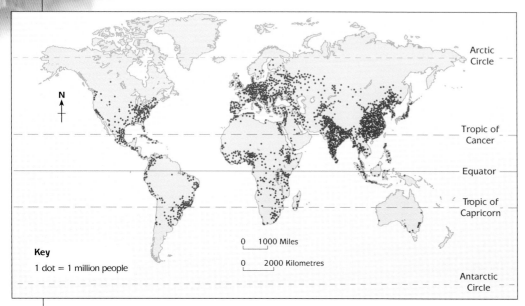

▲ **Figure 1** Dot map showing world population distribution.

Population means people; there are over six billion people living in the world today. They are unevenly spread across the Earth's surface. Some parts of the world are very densely populated while others are almost totally deserted. The *distribution* of the population is the way people are spread out across the surface of the Earth. This is shown in **Figure 1** as a **dot map**.

The *density* of population is the number of people who live in an area, measured in people per square kilometre. **Figure 2** is a **choropleth** (shading) **map** showing the world's population density.

The world's population distribution can be divided into three main categories:

1   Areas with high densities of people (over 50 per square kilometre): South and East Asia, Europe and the north-east of North America

2   Areas of medium population density (10–50 per square kilometre): California, the coast of Brazil, the Nile valley in Egypt and south-east Australia

3   Areas with low population densities (less than 10 per square kilometre): Sahara Desert in North Africa, northern Russia and Canada.

Explaining the population distribution

The world's population distribution is affected by both physical factors and human factors (**Figure 3**). Physical factors such as relief, soils and climate can encourage high densities of population. **Densely populated** areas are likely where there are areas of low, flat land with fertile soils and a temperate climate. However, where the climate is very hot or very cold, where there are high mountains or deserts, the environment is hostile to people and these areas are often only **sparsely populated** (**Figure 4**).

The majority of people live in the northern hemisphere between 20 and 60 degrees. Asia and Europe house about 85 per cent of the world's population, partly because these areas have been settled for such a long time. In some textbooks they are referred to as the 'Old World'.

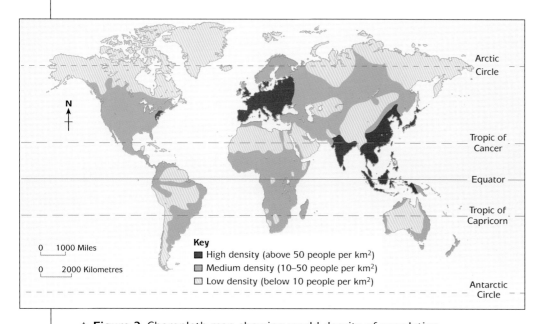

Key
■ High density (above 50 people per km²)
▨ Medium density (10–50 people per km²)
☐ Low density (below 10 people per km²)

▲ **Figure 2** Choropleth map showing world density of population.

PHYSICAL FACTORS	Reasons for areas of high population density	Reasons for areas of low population density
Relief	Lowland which is flat or gently sloping such as river flood plains, e.g. Ganges valley in India	Mountainous areas with high altitude and steep slopes, e.g. Alps, Andes
Climate	Moderate climates with no extremes. Enough rain and warm temperatures to allow crop growth, e.g. UK, Japan	Extreme climates: very cold, very hot and too dry, e.g. tundra, Sahara
Soil	Thick fertile soils such as loam and alluvium, e.g. south-east England	Thin, rocky and acid soils, e.g. hot deserts, mountains
Vegetation	Areas of open woodland and grassland, e.g. Pampas in Argentina	Very dense jungle which is difficult to penetrate, and swamps, e.g. Amazon Basin
Accessibility	Coastal areas with easy access, e.g. coasts in South America, UK	Interior areas with poor access, e.g. central South America
Resources	Plenty of water, timber, minerals such as coal, oil, copper; opportunity for fishing, e.g. coalfields in Western Europe	Few economic resources, e.g. southern Chile, Sahel

HUMAN FACTORS	Reasons for areas of high population density	Reasons for areas of low population density
Economic	Intensive farming using much hand labour, e.g. rice growing in river valleys of Asia, irrigated farming in Nile Valley	Extensive farming using machines instead of people, e.g. Great Plains of USA and Prairies of Canada
	Skilled and varied labour force, e.g. in the large towns and cities, especially in the MEDCs	Limited job opportunities in farming and industry, e.g. Amazon Basin, Sahara, tundra, Alaska
	Good infrastructure (roads, railways, services) and access to imports and exports, e.g. Japan, Europe, USA	Poor infrastructure (roads, railways, services) and limited access, e.g. Sahara, Amazon Basin, interior Australia
	Large, rich markets for trade	Poor trading links and markets
Social	Some groups of people prefer to live together for security and companionship, e.g. Japanese, Americans	Some groups of people prefer to be more isolated, e.g. Scandinavians
Political	Stable government, e.g. Singapore, Taiwan	Unstable governments and civil war, e.g. Afghanistan

▲ **Figure 3** Reasons for different densities of population.

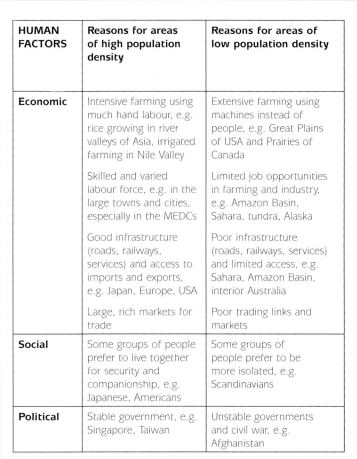

y
Too cold
Too dry
Too hot and wet
Too high
Temperate (favourable)

0   1000 Miles

0   2000 Kilometres

Arctic Circle

Tropic of Cancer

Equator

Tropic of Capricorn

Antarctic Circle

▲ **Figure 4** Limitations on population density.

Activities

1 Study **Figures 1** and **2**.

   (a) Describe what they show about the distribution and density of population in **(i)** Europe and **(ii)** Asia.

   (b) Describe the similarities and differences in population between South America and Africa.

2 **(a)** Identify and name from three different continents three areas of:
   (i) high population density
   (ii) low population density.

   (b) Choose one area of high and one of low density. Explain the factors responsible for the differences in density.

3 'Most of the world's people live within 300 kilometres of the coast.'

   (a) Write down evidence for and against this statement.

   (b) In your view, how true is it as an overall summary of world population distribution?

Small-scale variations in population distribution and density

A case study of the UK

The UK is part of Western Europe, one of the areas of the world with a very high population density. However, as **Figure 1** shows, the population density is not high everywhere in the UK. The population distribution in the UK is very uneven. There are great differences from one area to another, from densities of under 10 per km² in the remote upland areas to densities of over 1000 per km² in the cities.

The sparsely populated areas in the UK are mainly upland areas (**Figure 2**) such as the Pennines, Dartmoor and the Scottish Highlands. The upland areas have a harsher climate than the lowlands. The terrain is rugged and the steep slopes make the building of communications and settlements difficult. The steep slopes, thin soils and short growing season prevent crop growth. The areas are often remote, with poor access and few services. In the uplands there are limited employment opportunities. The main occupations are extensive sheep farming, forestry, quarrying, the water industry (reservoirs and HEP) and tourism. Larger concentrations of people are only found where a resource has been developed, such as at Aviemore for winter sports or around the slate quarries in Snowdonia.

The high-density areas in the UK are the conurbations, the large cities and the industrial areas, e.g. Tyneside, Merseyside and Greater London (**Figure 2**). About

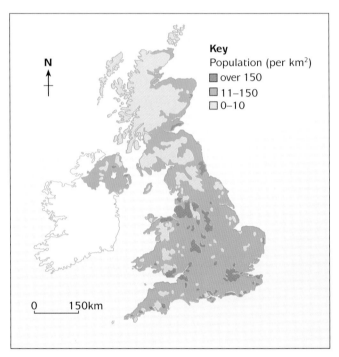

▲ **Figure 1** UK population density in 2004.

80 per cent of the UK population lives in towns and cities and over 50 per cent in the seven largest city areas, the conurbations. Most of the conurbations in the UK, except Greater London, are former coalmining areas which were developed during the Industrial Revolution. London has very high densities of population because of its importance as the capital city, and as an important port. The areas of high density are located in the lowlands where the relief is gentle and the climate more pleasant. There are dense networks of communications including roads, railways, airports and ports.

Large areas in the UK have medium densities of population (20–200 people per square kilometre). These are mainly fertile farming areas with market towns and villages, such as East Anglia, or places with small industrial towns and villages, such as County Durham with its landscape of former mining villages and small towns.

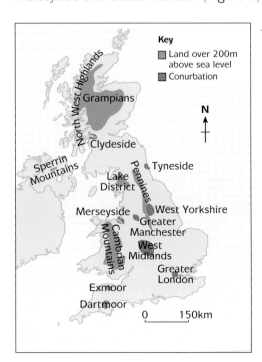

◀ **Figure 2** Upland areas and conurbations in the UK.

ℹ Information

A conurbation is a huge urban area created by the growth of one or more cities which merge with each other and engulf smaller towns and villages.

Case study of Brazil

Population distribution in Brazil is even more uneven than in the UK. The majority of Brazilians live within 300 kilometres of the coast, while the interior in the Amazon Basin is almost empty. This historical pattern is little changed by recent movements into the interior.

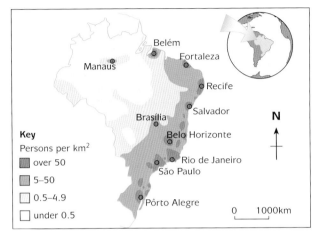

▲ **Figure 3** Brazil population density in 2004.

The coast

The highest densities of all are within the triangular-shaped area between Rio de Janeiro (the old capital city on the coast and main port), São Paulo (the industrial centre and with about 18 million people one of the world's three largest cities) and Belo Horizonte (an industrial centre based upon mining in the area around it). This is the economic heart of Brazil, a big country of over 170 million people that has developed into a medium-income economy (page 227). The 'Triangulo' is the centre for industry, business and trade. The big cities are a magnet for people from all parts of Brazil; the population keeps increasing due to a combination of in-migration and high birth rates.

▲ **Figure 4** View over Rio de Janeiro with the Sugar Loaf Mountain in the background. Very high densities of population are found between the rock outcrops.

Advantages of the coast over the interior include:

- easy access by sea to other parts of Brazil and for trade overseas

- a reliably wet, but not excessively hot, tropical climate

- fertile red soils and some of the best farmland in Brazil – for home food supplies (vegetables, fruits) and for export (coffee, soybeans).

The interior

The main factors causing the low population densities are **physical factors** and the **remote location**.

- Equatorial climate is hot (28°C) and wet (over 2000 milimetres) all year with high humidity.

- Density of the jungle vegetation makes it difficult to penetrate and to clear, and communications are poor.

- Soils are infertile and easily leached and eroded once the trees are removed (page 107).

- Diseases spread quickly, e.g. Yellow Fever.

There are pockets of higher density around well-established Amazon river ports such as Manaus, which prospered in the late nineteenth century because of the rubber trade. Since 1960 the policy pursued by most Brazilian governments has been to open up the interior to develop its timber and mineral resources and encourage people to move away from the overcrowded coastal cities. There are fingers of settlement along the new roads such as the BR163 (page 109). Despite all the publicity that clearance of the Amazon forest has received, the overall population density remains very low, with under 10 per cent of Brazilians living in the interior. Some of the new activities are predatory, such as logging and mining; others, such as cattle ranching, are extensive, needing large areas and few people. New settlers face many physical problems.

Activities

1 Using **Figures 1** and **2**, compare the distribution of
   (a) low density of population and upland areas and
   (b) high density of population and conurbations in the UK.

2 Explain (a) physical and (b) human reasons for the general increase in population density from north west to south east in the UK.

3 (a) Describe the pattern of population density in Brazil from **Figure 3**.

   (b) State the historical and physical reasons for this pattern.

   (c) How and why is the pattern changing (see also pages 108–9)?

Population change

A population may increase or decrease over time. How a population changes depends on the birth rate, the death rate and migration (see the Information Box). A population grows if the birth rate is higher than the death rate, i.e. there is a natural increase, but a few countries have a natural decrease where the death rate is greater than the birth rate. In a few countries, such as the USA, migration can have a large impact on population size.

World population growth

The growth of world population over the last 200 years has been spectacular (**Figure 1**) and it has not stopped yet. From 1950 there was a population explosion and the total of 6 billion people on Earth was reached in 1999. Exponential growth is the term used to describe such a rapid increase. Although there is some evidence that the rate of growth is at last beginning to slow down, the world's population continues to grow because a great majority of countries have higher birth rates than death rates, leading to a natural increase.

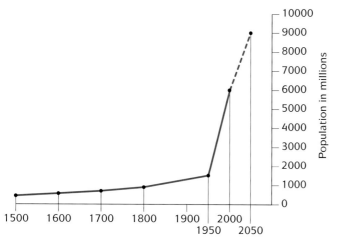

▲ **Figure 1** Growth of world population from 1500.

	Country	Highest		Country	Lowest
1	Liberia	55.5	1	Latvia	7.8
2	Niger	55.2	2	Bulgaria	7.9
3	Somalia	51.8	3	Ukraine	8.1
4	Angola	51.3	4	Germany	8.2
5	Uganda	50.6	5	Slovenia	8.2
6	Mali	49.6	6	Sweden	8.2
7	Sierra Leone	49.1	7	Austria	8.3

▲ **Figure 2** Countries with the world's lowest and highest birth rates per 1000 in 2003.

ⓘ Information

Understanding population terms

Crude birth rate – the number of live births per 1000 population per year.

Crude death rate – the number of deaths per 1000 population per year.

Natural increase – birth rate higher than death rate: birth rate minus death rate.

Natural decrease – death rate higher than birth rate: death rate minus birth rate.

Annual population change – the birth rate minus the death rate plus or minus migration.

Migration – the movement of people either into or out of an area.

Low birth rates in Europe	
Social	• family planning is practised by nearly all couples • women are well educated and career orientated • one or two children is accepted as a normal family size.
Economic	• children are unlikely to contribute to the family income • the average cost of bringing up a child in the UK in 2005 was over £60 000.
Political	• governments support and finance family planning.

High birth rates in Africa and the Middle East	
Social	• family planning is not widely used, especially among the poor • many women receive little formal education and marry young • socially, families of five or more children are considered to be quite normal.
Economic	• children are expected to work to supplement the family income • one child who does well may lift the family out of poverty.
Political	• some governments and religions do not approve of birth control • family-planning clinics are not always available, especially in rural areas.

▲ **Figure 3** Reasons for low birth rates in Europe and high birth rates in Africa and the Middle East.

Birth rates

The average birth rate in the rich, industrialized countries (MEDCs) is around 12–13 per 1000; in poorer developing countries (LEDCs) it is about 26 or 27 per 1000. There tends to be a general relationship between birth rate and level of economic development – the more economically developed the country, the lower its birth rate (look at **Figure 2** on page 228). **Figure 2** lists the countries with extremes of high and low birth rates in 2003. It is dominated by countries from only two continents – which are they? All the countries in column 1 have average incomes per head per year under US$ 500, compared with over US$ 20 000 in the three richest countries in the lists (Germany, Sweden and Austria). In fact, birth rates in all European countries are low, not just in those countries named in **Figure 2**. In contrast, many countries in Africa and the Middle East have birth rates well over 40 per 1000. The social, economic and political reasons for these are given in **Figure 3**.

Death rates

Unlike birth rates, death rates are similar between MEDCs and LEDCs; the world average for both is between 9 and 10. During the second half of the twentieth century, death rates fell everywhere, due to the spread of medical knowledge and improvements in primary and secondary healthcare. Primary healthcare is preventing disease, by immunization for example; secondary healthcare is treatment of illnesses by doctors and nurses. Countries with death rates above 20 per 1000 are now quite exceptional (**Figure 4**). The countries named in **Figure 4** are all from one continent: Africa. All are poor. Included are war-torn countries with high levels of disease, such as Sierra Leone, and countries badly affected by the spread of HIV/AIDS, such as Zimbabwe. Here the trend towards lower death rates has been reversed.

Natural increase and decrease

The wider the gap between high birth rates and low death rates, the greater the size of the natural increase. Angola features in both **Figures 3** and **4**, which makes it possible to calculate the natural increase. At 27.7 per 1000 (or 2.77 per cent) it is still high, despite Angola's higher than average death rate. LEDCs, with their high birth rates and low death rates, are overwhelmingly responsible for continuing world population growth.

A higher death rate than birth rate, resulting in a natural decrease of population, is something new. However, it is a trend that is on the rise in European countries (**Figure 5**). Birth rates are falling as social trends change in some countries in northern and western Europe. People are marrying at an older age and women are delaying starting a family until they have established themselves in a career. Other couples are foregoing children in favour of a higher standard of living. In the

formerly communist eastern European countries, many people are pessimistic about the new economies and their future circumstances.

Death rates are already low, and cannot be expected to fall much further. You need to remember that the death rate is a ratio per 1000 of the population. As a result of an increasing proportion of elderly people within the population, there are many people per 1000 reaching the natural age for death. This is despite the arrival of new drugs for illnesses previously considered untreatable.

	Country	Death rate
1	Sierra Leone	29.3
2	Zambia	28.0
3	Zimbabwe	27.0
4	Lesotho	25.7
5	Swaziland	25.4
6	Malawi	24.1
7	Angola	23.6
8	Mozambique	23.5
9	Central African Republic	22.1
10	Rwanda	21.8

▲ **Figure 4** Countries with the world's highest death rates per 1000 in 2003.

Country	Birth rate (per 1000)	Death rate (per 1000)	Natural change (per 1000)	%
Sweden	8.2	10.6	-2.4	-0.24
Germany	8.2	10.2	-2.0	-0.2
Austria	8.3	9.4	-1.1	-0.11

▲ **Figure 5** Examples of countries with a natural decrease in population.

Activities

1 Total world population (billions): 1999 6.0. Estimates for the future: 2010 7.0; 2023 8.0; 2050 9.0

Describe how the data show that the rate of world population growth is expected to slow down by 2050.

2 Data for four countries in 2003:

	Birth rate (per 1000)	Death rate (per 1000)	Income per head (US$)
UK	11	10	23 900
Hungary	9	13	5 200
China	14	7	900
Nigeria	40	14	350

(a) For each country, calculate the rate of annual population change.

(b) Why do rates of population change vary so greatly between countries?

The Demographic Transition Model and population structure

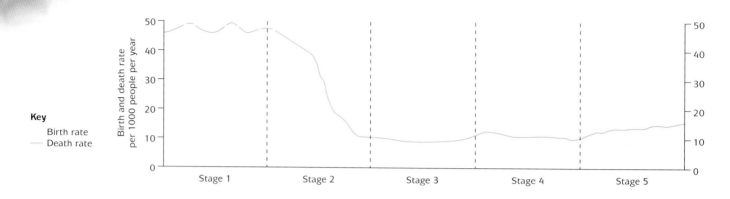

Stage	1	2	3	4	5
Birth rate	High	High	Decreasing	Low	Low
Death rate	High	Decreasing	Low	Low	Low
Natural increase	Low	Becoming high	High becoming low	Low	Natural decrease
Countries	None – all countries have progressed into another stage.	LEDCs with low levels of economic development, many of them in Africa.	LEDCs with improving levels of economic development, many in south and east Asia, Mexico, Brazil.	MEDCs in North America, Australasia, Japan and many European countries.	A few MEDCs with very low birth rates, mainly in Europe.

▲ **Figure 1** The Demographic Transitional Model.

The Demographic Transition Model (DTM) in **Figure 1** is used to show how countries pass through different phases of population growth over time. For many years there were only four stages, but the recent emergence of countries with a natural decrease of population created the need for a stage 5. Close relationships between birth and death rates are visible in stages 1, 4 and 5, but not in stages 2 and 3, where birth rates are noticeably higher than death rates. The majority of countries are currently passing through these two stages, which explains the large growth in world population.

In many ways the critical stage is stage 2. Death rates take a great fall due mainly to improved medical care, but also as a result of cleaner water supplies and increased food output. Birth rates remain high because the economic, social and political reasons for high birth rates referred to on pages 118–19 still apply – more children means more workers in the fields, children can look after their parents in old age and some governments either cannot afford or are unwilling for religious or other reasons to support family planning. The adoption of improved medical care is always faster than that of contraception.

When birth rates start to fall in stage 3 it is a major change. The underlying reason is the introduction of

family planning and birth-control policies, but economic growth is a great help for financing the policies, improving education and changing social attitudes towards large families. Economic development is accompanied by more people living in cities, where costs of bringing up children are higher; on farms children are workers. Improved educational opportunities for women make a big difference as well; more women entering higher education, marrying later and pursuing careers greatly reduces average family sizes.

Population structure

Population structure is the composition of a country's population by age and sex. Data for these, usually collected by a **census**, is shown in a **population pyramid**. In the pyramids in **Figure 2**, horizontal bars are used to show the numbers of males and females in each five-year age group (0–4, 5–9 etc.) on different sides of the central dividing line. Sometimes the actual numbers of people are used for constructing pyramids instead of percentages. Pyramids give a good visual impression of the differences in population make-up by age between countries; striking differences are shown here between Ethiopia (an LEDC) and the UK (a typical MEDC).

The pyramid for Ethiopia displays many of the characteristic features of a less economically developed

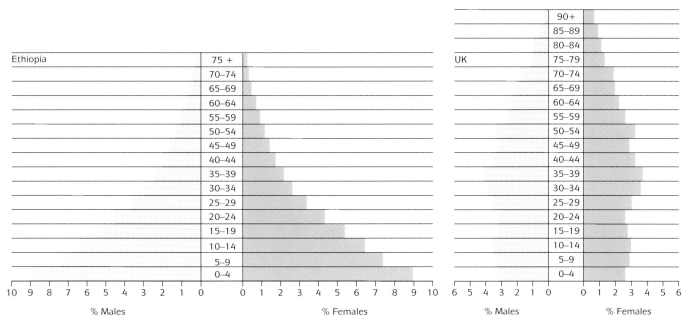

▲ **Figure 2** Population pyramids for Ethiopia and the UK.

country, particularly the wide base showing a population structure dominated by young people, due to high birth rates. The graph has an almost perfect pyramid shape, progressively tapering towards a narrow top with few people above the age of 65. The pyramid for the UK is taller and the top is more pronounced, showing significant numbers above the age of 65. The UK's birth rate is low and the narrow base shows this. It is the middle-aged groups that are dominant in pyramids for MEDCs. From their shapes it is possible to suggest stages in the Demographic Transition Model – Ethiopia is definitely in stage 2, while the UK has reached stage 4. In summary, population pyramids for LEDCs are wider at the base, narrower at the top and less tall than those for MEDCs.

| Stage 1 | Stage 2 | Stage 3 | Stage 4 | Stage 5 |

▲ **Figure 3** Pyramid shapes associated with different stages in the DTM.

Activities

1 (a) Draw a diagram of the Demographic Transition Model to show all five stages.

(b) Use different shading for the areas that show natural increase and natural decrease on your diagram.

2 (a) What is the key change between stages 1 and 2 in the model and why does it happen?

(b) What is the key change between stages 2 and 3 in the model and why does it happen?

3 (a) Describe what the population pyramids show for **(i)** Ethiopia and **(ii)** the UK.

(b) Draw simple summary sketches of the two pyramids and label the differences between them.

◀ **Figure 4** Dependants in the UK.

It is customary to subdivide the structure of a country's population into three age groups, namely young (0–14), middle-aged (15–64) and old (65 and above). The middle-aged are distinguished from the other two as the working or independent population; they are the group in society that works, earns money, contributes to pensions and pays income taxes. Young and old have in common that they are dependants; although some of them work, the majority depend upon services such as education and healthcare, paid for by taxes collected from the working population. The dependency ratio, the ratio between the dependent and independent populations, can be worked out from a population pyramid.

Population problems and issues in LEDCs and MEDCs

Population problems are very different in LEDCs and MEDCs; in fact, they are opposites. The first group of countries is trying to provide for young and growing populations resulting from many years of rapid and persistent population increase, while the second group is trying to come to terms with **ageing populations** (an increasing proportion of old people) and a reduction in numbers of people of working age.

LEDCs – more and more people

Crippling rates of population growth have led to many problems – economic, social and environmental (**Figure 1**). The result is low standards of living, poor quality of life and poverty; all increase vulnerability to natural and human disasters (pages 232–3).

Many of the *environmental* problems are most noticeable in rural areas. As population growth has increased demand for food and natural resources such as fuel wood and water, farmers have responded by using the land more intensively, in some places to the point where the land and soil have been destroyed. The general term for this damage is **environmental degradation**. One example is **desertification**; a summary of its causes is given in **Figure 2**. This occurs where deserts spread and engulf areas that formerly carried surface vegetation cover and farming settlements. It is a process whereby land is turned into a desert as a result of human activities, although climatic hazards, such as low rainfall and drought, can make it worse. Overgrazing, overcultivation and overuse of irrigation water leading to salinization (soils becoming more salty) create land surfaces unable to support vegetation.

Areas most at risk are semi-arid regions where rainfall is concentrated in one season and the amount that arrives is very variable from year to year. When two or three dry years occur together, human pressures on the natural vegetation increase. One of the worst affected areas is the Sahel on

In rural areas:
- overgrazing and overcultivation
- water, land and air pollution
- deforestation, land degradation, soil erosion and desertification.

In urban areas:
- overcrowding and the growth of shanty towns
- water, land and air pollution
- traffic congestion.

In the country as a whole:
- shortages of resources, food and raw materials
- unemployment and under-employment
- lack of money for basic health care and schooling
- rising crime, political coups, huge debts
- low living standards.

▲ **Figure 1** Problems of population growth in LEDCs.

the southern edge of the Sahara desert. Countries located here have some of the highest birth rates in the world (for example, Niger 55, Somalia 52, Mali 50 and Chad 49). The relentless rise in numbers means that subsistence farmers can no longer accumulate a food surplus in wet years to see them through the dry years. In dry years, overcultivation, overgrazing and further deforestation for fuel wood start the train of events that ends with desertification.

The poverty in rural areas is transferred to the urban areas by migration. It is in urban areas that *socio-economic* problems such as overcrowding, poor housing, inadequate public services, and unemployment are most concentrated. The *political* problem for governments and city authorities is how to plan and pay for public services (health, education), public utilities (clean water supply, sanitation, electricity) and housing. The *environmental* consequences

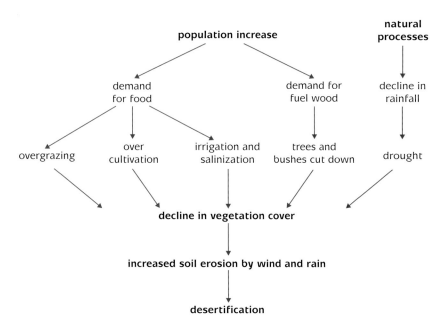

▲ **Figure 2** Causes of desertification.

▲ **Figure 3** Self-help housing in Mumbai (Bombay), the type of housing in which more than half of the city's 16 million inhabitants live.

of population growth in urban areas without effective pollution controls are alarming. Air pollution is at dangerous levels for human health in many big cities, from traffic congestion on roads and streets full of old trucks and cars and unsupervised factory emissions. Water pollution is widespread because rivers are used for waste and litter disposal. Sprawling shanty towns lead to removal of vegetation on the urban edges, while inside cities they climb up hillsides, increasing the risk of landslides.

MEDCs – ageing populations

In MEDCs like the UK, life expectancy is increasing due to continued improvements in medical knowledge and the discovery of new drugs. The over-85s are the fastest-growing demographic group; there were more than 1.1 million people over 85 in the UK in 2001 compared with only 200 000 in 1951. The elderly require special services and increasing levels of health care as they get older. Already there is more need of nursing and care homes, warden-assisted flats and bungalows, and home services such as daily care and meals on wheels. In addition for the government there are the direct economic costs of pensions and housing benefits.

Most European countries have generous state pension systems, which are unsustainable if the forecasts for 2050 shown in **Figure 4** turn out to be correct. Pensions and health services are paid for by taxes, most of which come from the working population. At the same time as more money is needed for the elderly, the economically active working population is shrinking. European

governments have just begun to wake up to the problem and are increasingly worried about their continued ability to finance state pensions. As long as birth rates remain low, there will not be enough workers to generate all the money needed. Needless to say, telling today's workers that they must pay more in taxes and into pension funds to support the elderly is not popular.

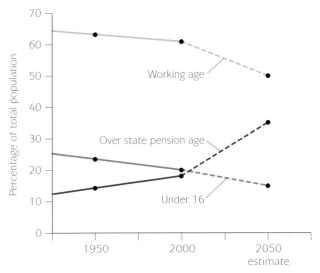

▲ **Figure 4** Demographic trends in the UK.

Activities

1 Make a table entitled 'Problems of high population growth', using the headings Economic, Social and Environmental. Fill it using **Figure 1** and information from the text.

2 (a) Explain the physical and human causes of desertification.

(b) Which one do you consider to be the more important cause of desertification – physical or human? Explain your answer.

3 (a) (i) Make a frame and draw a labelled sketch of the housing area in **Figure 3**.

   (ii) Add labels for the urban problems likely to exist here.

(b) Explain how population growth leads to urban problems like these.

4 (a) Define the term 'ageing population'.

(b) What is the evidence from **Figure 4** that the UK has an ageing population?

(c) Explain the economic problems that result.

Strategies for coping with population problems in LEDCs and MEDCs

The LEDCs have two difficult tasks: (i) to control the size and growth of their present populations and (ii) to cope with the consequences of past, and likely future high rates of population growth.

Controlling population growth

Many governments in LEDCs now recognize high birth rates as a problem, and have family planning and population policies in place (**Figure 1**). The policies vary from persuasion and incentives, as in Sri Lanka, to passing strict laws reinforced by severe punishment, as in China. Some countries, such as Iraq and Saudi Arabia, have shown little interest in controlling their populations. In most cases this is due to religious beliefs. In Muslim and Catholic countries religious teaching opposes any form of contraception.

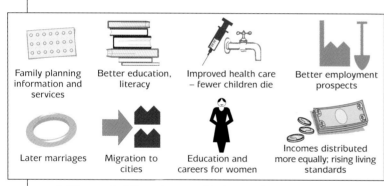

Family planning information and services — Better education, literacy — Improved health care – fewer children die — Better employment prospects

Later marriages — Migration to cities — Education and careers for women — Incomes distributed more equally; rising living standards

▲ **Figure 1** What changes birth rates?

China's 'one child' policy

In the 1980s China's population was already over 1000 million. The government decided that the existing 'two child' policy was not sufficient to reduce population growth. China's population would continue to grow at least until 2025, by which time there would be 1.8 billion Chinese. The government introduced the 'one child' policy in the hope that the population would stabilize at about 1.2 billion early in the twenty-first century. **Figure 2** shows an example of the advertising information used by the Chinese government.

The policy of 'one couple, one child' is very strict. It is virtually illegal to have more than one child and families are criticized and fined; forced abortions and sterilizations have also been reported. The policy has been quite successful in the cities, but 75 per cent of China's population are peasant farmers living in the rural areas.

WHY HAVE ONLY ONE CHILD?

For you with one child:
Free education for your one child.
Family allowances, priority housing and pension benefits.

For those with two children:
No free education, no allowances and no pension benefits.
Payment of a fine to the state from earnings.

To help you
Women must be 20 years old before they marry.
Men must be 22 years old before they marry.
Couples must have permission to marry and have a child.
Family planning help is available at work.

REMEMBER: One child means happiness.

▲ **Figure 2** Part of the Chinese government's advertising campaign.

Chinese couples in the countryside want large families to help with work in the fields and to look after them in old age. Chinese culture has always held boys in higher esteem than girls, and there have been reports of infanticide where girl babies have been killed by couples who want sons. There are also concerns that the nation is breeding a society of spoilt children, mostly boys, who will have difficulty later finding partners because of the shortage of women. Such actions and concerns have forced the government to relax the rules in rural areas to allow up to two children.

Policies in other Asian LEDCs

Countries without elected governments, like China, can impose an unpopular policy and make it work, but it is different in a democracy. An earlier Indian government forced through a massively unpopular programme of compulsory sterilization, and was voted out at the next election. The present voluntary programme does not work well in backward rural states where social and

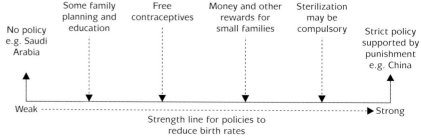

No policy e.g. Saudi Arabia — Some family planning and education — Free contraceptives — Money and other rewards for small families — Sterilization may be compulsory — Strict policy supported by punishment e.g. China

Weak ----- Strong

Strength line for policies to reduce birth rates

▲ **Figure 3** Strength line for national population policies.

	Total population 2003 (millions)	Birth rate per 1000 in 2003	Death rate per 1000 in 2003	Average annual population growth 2000–2005 (%)	Estimated population in 2050 (millions)
China	1285	14.3	7.0	0.73	1395
India	1025	23.8	8.5	1.51	1531
Pakistan	145	36.3	9.6	2.44	348
Thailand	64	17.8	7.1	1.01	77

▲ **Figure 4** Population data for four Asian countries.

religious traditions remain strong. Between now and 2050 India is expected to overtake China as the world's most populous country (**Figure 4**).

Thailand has a successful and effective family-planning programme, which uses a mixture of media, economic incentives and community involvement to increase contraceptive use. The immediate benefits of smaller families are stressed; the key message links family planning and low family size to high standards of living. To get everyone involved, even in remote rural areas, fun events such as birth-control carnivals are organized. In Thailand the message is carried everywhere, not just left in the cities. Farmers who have registered for family planning are given financial benefits, such as above-market prices for their crops and reduced transport costs to market. There is a big difference in Islamic Pakistan, which does not have a well-promoted population policy and where there is constant political rivalry with India, its much more populous neighbour.

Coping with ageing populations in MEDCs

Retirement ages are rising everywhere. Some have already gone up, for example to 70 in Iceland and 67 in Norway and Denmark. The rise in the female retirement age in the UK from 60 to 65 is being phased in over a number of years, and it is likely that the state pension age for all will eventually rise to 67 or 68 in association with reforms to the state pension system. In response to Europe's plunging birth rates, Germany in 2006 became the latest of the EU countries to offer incentives to couples to have more children, including tax breaks and up to €1800 a month for parents to take time off work.

However, schemes to encourage couples to have more children have not worked well in the past, and most politicians accept that there are limits to what a government can do to influence highly personal decisions like having a child. The only way for Europe to maintain its working population in the future will be through immigration, but this is very unpopular with the public. Therefore, increasing the retirement age and forcing people to make higher pension contributions during their working lives seem to be the only options for governments, though neither of these is exactly popular. There are going to be bumpy times ahead for European governments as they grapple with the pension problem, particularly because the 'grey vote' is significant at election time.

Are there any benefits of young and old populations?

There is always a positive side. Young people are dynamic, have more energy to work and are more likely to adopt new ideas – in other words, they can power economic development. Some pensioners in MEDCs (but by no means all) are well off and determined to 'spend the kid's inheritance' (dubbed SKIHERs). They pour money into the leisure sector at off-peak times of the year and mid-week, while some manufacturing companies have created successful businesses by targeting the elderly for the bulk of their sales, such as the makers of stair lifts and mobility vehicles.

Activities

1 **(a)** Name one country that has a population policy for reducing birth rates.

   (b) Describe the methods used.

   (c) How successful has it been?

2 **(a)** For the four countries in **Figure 4**:

   (i) calculate rates of natural increase in 2003

   (ii) draw a graph to show total populations in 2003 and estimated populations in 2050.

   (b) Explain why rates of population growth vary greatly between one LEDC and another. Support your answer with examples.

3 **(a)** Name two government strategies for coping with an ageing population.

   (b) Explain the problems of implementing them.

Migration

Country	Birth rate	Death rate	Natural increase	Actual annual rate of growth	Migration difference
In-migration					
USA	13.1	8.3	4.8	10.3	+5.5
Australia	12.7	7.4	5.3	9.6	+4.3
Out-migration					
Mexico	22.2	5.0	17.2	14.5	-2.7
Jamaica	24.2	6.1	18.1	9.0	-9.1

▲ **Figure 1** Population growth and migration (all values are per 1000 people).

Migration is the movement of people to live in a different place, either within the same country (**internal migration**) or to another country (**international migration**). Internal migration can have a big influence on regional population growth or decline. International migration can mean that the growth rate of population in the country is significantly different from the rate of natural increase indicated by statistics for birth and death rates (**Figure 1**). More people have **emigrated** (moved away) from Jamaica and Mexico to other countries than have moved in, whereas more have **immigrated** into the USA and Australia than have moved out. Both Jamaica and Mexico have migration losses; from both countries there is a long history of out-migration to the USA, their much wealthier neighbour.

What makes people move?

In most decisions more than one factor is involved. Often these factors are a mixture of push and pull (**Figure 2**). **Push factors** are people's dislikes about where they live – the disadvantages of living there. There may be no work, or the work that exists may be badly paid, or available for only part of the year. This is an example of an economic factor. Public services and utilities such as schools, hospitals, electricity and clean water supply may not be widely available. These are more social factors affecting people's quality of life.

A natural disaster, such as an earthquake or flood, might force people to leave immediately: environmental factors like these can be so strong that people do not even think about staying.

Pull factors are attractions of the place people are moving to – the advantages of moving there. The place may offer physical advantages such as a wetter climate, more reliable rainfall and better soils. Or it could offer economic attractions, such as job opportunities with good prospects of improved standards of living. Attractions can also be predominantly social, like moving closer to family and friends.

Although in general there have been great improvements in communications and transport in the world, there are still **obstacles** to the free movement of people. Between countries there are border controls: visas and entry permits are often needed. Within countries there may be roadblocks and political or military controls between different regions, especially during times of civil strife and war. For some people, social obstacles to migration are strong, such as leaving the family behind and fear of the unknown; however, the strength and importance of these varies greatly from person to person.

Types of migration

A simple classification is between forced and voluntary migrations (**Figure 3**). **Voluntary migration** is when a person makes the decision to move. The decision is usually made after weighing up the advantages and disadvantages of moving. It is likely that both push and pull factors will be involved, but the relative strength of these varies greatly. The most common type of international migration in the world today is **economic migration**; basically this is movement for work, usually from poor to rich countries. Economic migrants are seeking the better life that a higher income brings. The push factors for this are no work and low standards of

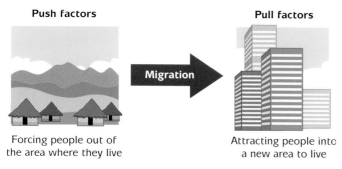

Forcing people out of the area where they live

Attracting people into a new area to live

▲ **Figure 2** Push–pull model of migration.

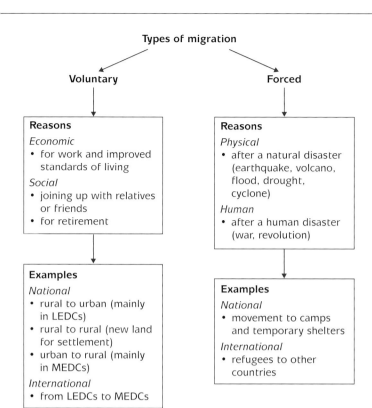

▲ **Figure 3** Types of migration.

▲ **Figure 4** African refugees.

living; the powerful pull factor is the amount of money that can be earned by working in an MEDC compared with the home country. Even if a person is in work in an LEDC, the higher wages in MEDCs are still an attraction. This suggests that the pull factors are often stronger than the push factors for economic migrants, such as Mexicans crossing the border for work in the USA (pages 128–9).

Forced migration is compulsory migration and people have little choice about moving. Forced migrants who move to another country are **refugees**. Both physical and human factors can cause this movement. The major volcanic eruption in the Caribbean island of Montserrat (pages 16–17) forced out over half this island country's population to neighbouring islands, the UK and USA. War and persecution of people of different ethnic groups are the most important human factors in forced migration. Africa is the continent that had the most refugees in 2005; upheavals caused by armed conflicts and civil wars have been made worse by physical problems such as drought, which has led to widespread famine and mass movements of people. The scene shown in **Figure 4** has become depressingly frequent in recent years.

Effects of migration

Movements of people can arouse strong feelings – among families and friends, in the national population of countries receiving large numbers of immigrants and among politicians and city officials trying to provide for large numbers of new arrivals. As with most things

involving people, there are both positive and negative impacts from migration. Advantages and disadvantages for both losing countries (from which people migrate) and the receiving countries (into which people move) are included in the summary in **Figure 5** for migrations that involve cross-border movement.

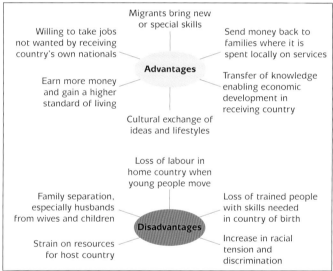

▲ **Figure 5** Examples of the advantages and disadvantages of international migrations.

Activities

1 Briefly state the differences between the following pairs:

   (a) international and internal migrations

   (b) immigration and emigration

   (c) push and pull factors for migration

   (d) voluntary and forced migrations.

2 Use **Figure 5**. Rearrange the advantages and disadvantages in two lists: **(a)** for the losing country and **(b)** for the receiving country.

3 Draw two spider diagrams to show advantages and disadvantages of rural to urban migration within LEDCs.

International migrations

Although international migrations receive a lot of publicity in the media, these are relatively less important than in previous centuries. In the nineteenth century, large numbers moved voluntarily from Europe to live in new countries, such as the USA, Canada and Australia, where there was plenty of space and few people. These were seen as lands of opportunity for people hoping to begin new and better lives. At that time, some migrations were also forced, notably African slaves to the southern USA, Caribbean islands and Brazil. Today, two types of international migration dominate: voluntary movements of economic migrants and forced movements of refugees.

Economic migrants

This migration from poor to rich countries is on the increase. In 2005, about 200 million people lived in countries other than their country of birth, double the number compared with 1980. Although this is a worldwide movement to MEDCs from LEDCs, there are regional patterns:

* Hispanics (Spanish-speaking people) from South and Central America and the Caribbean migrate in the greatest numbers to the USA

* most African migrants go to Europe

* migrants from South Asia (India, Pakistan, Bangladesh) seek work in the oil-rich countries around the Gulf, as well as in Europe and Australia.

The changes reported in **Figure 1** for El Salvador, a poor country of only 4 million people in Central America, are typical of what happens in other LEDCs that have many nationals working overseas. Most economic migrants share the same wish – to fund the family home and eventually to return there, where their comparative wealth will support a good lifestyle and improved status. This leads to increases in the inequalities in wealth between families with and without workers abroad.

Case study – Mexican workers in the USA

The over 1000 kilometre land border between the USA and Mexico is one of the greatest wealth divides in the world (see **Figure 2** on page 226). A rich neighbour, so easy to reach, is too big an attraction. Each year, millions of Mexicans are picked up by police and army as they try to cross the border and are sent back, only to try again. At least 1 million a year make it. There are plenty of employers in the USA willing to hire them, even without the correct papers. One businessman described them as the 'life-line of the American economy, particularly agriculture'. Who would pick the lettuces, grapes and tomatoes at harvest time without the Mexicans? They fill shortages in the labour market and work hard because they have families to fund.

In 2006 illegal immigration became a big political issue in the USA. During the first five years of the Bush administration their numbers have gone up by 6 million. Complaints against these 'illegals' made by US politicians and the public include:

* high costs of healthcare and education for their families (estimate US$ 11 billion per year)

* excessive Hispanic cultural influence; many speak only Spanish

* wage rates are depressed for jobs done by the poorest native-born Americans

* threats to national security along an unsecured border.

GREAT AT HOME
– THANKS TO THE MONEY FROM OVERSEAS

Remittances from nationals working abroad, mainly in the USA, are the lifeblood of the economy of El Salvador. The US$ 2 billion plus they send back each year makes up 15 per cent of the country's GNP.

Go to a small town in El Salvador from which a high percentage have migrated, even in the middle of nowhere, and everyone seems to live in a good house with wrought-iron gates, own a pick-up up truck and walk around in name-branded clothes and trainers. There seem to be an awful lot of banks for a small town.

A billboard in the park gives a clue to this apparent wealth. It advertises an American airline that carries migrant workers to the USA. As everywhere, money from family working overseas is spent first on the house and possessions, and on fast food! In towns where the money has been flowing for twenty or more years from workers overseas, improvements go further – paved streets, a water-treatment plant, a new secondary school and football ground, even drinking-water supplies to surrounding farming villages.

What the town hasn't got is work. Migration is painful for the families; many would like more work here. Realistically, even with work in El Salvador, they could never achieve the quality of life they have become accustomed to with remittances from family members working overseas.

◀ **Figure 1** Newspaper report from El Salvador, March 2004.

Information

Illegal immigrants in the USA

- estimated number in 2006: 12 million, half of them Mexican

- California has the most, followed by Texas and Florida

- 40 per cent work in agriculture

- others mainly do unskilled service work: home helps, cleaners, kitchen staff, etc.

- remittances to Mexico from the USA rose massively from US$ 6.6 billion in 2002 to US$ 20.2 billion in 2005

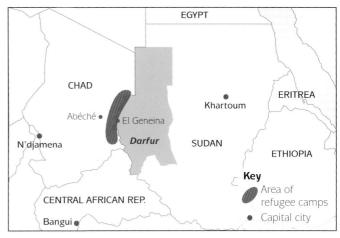

▲ **Figure 2** Darfur emergency refugee camps in Chad.

The main counter-argument is the estimated economic contribution from illegal immigrants of US$ 21 billion per year. All previous plans to seal the land border with Mexico have failed and few are confident that new plans announced in 2006 will be any different. As long as migrants can earn in an hour in the USA what they can earn in a day back home in Mexico, they are going to keep on coming.

Refugees

Strictly speaking, refugees are people who have been forced out of their home country, although the word is also used by some people to refer to people displaced within their country and forced to flee to another region. The main reasons for fleeing are usually human rather than physical; at the top of the list of causes are civil wars and persecution for racial or religious reasons. Their numbers are increasing, for example from 1.5 million in 1960 to 21 million in 2005.

The UNHCR (United Nations High Commission for Refugees) was set up in 1950 to safeguard the rights and look after the well-being of refugees. Refugees living in other countries often live in extreme poverty, lacking food, clothing, shelter and medical care. They have no citizenship and few, if any, rights; without any prospect of improvement, they are dependent upon the UNHCR, aid and charities. Visit the UNHCR website (see Hotlinks, page iv) for up-to-date information about world refugees.

Refugee crisis in the Darfur region of Sudan

Between 2003 and 2005, 200 000 refugees fled from Darfur to neighbouring Chad; at least a million more people were displaced within Darfur and an estimated 200 000 were killed. Darfur is an arid, poverty-stricken region of black farmers. In 2003 a rebel group in Darfur began attacking government targets, accusing the Arab-dominated Islamist government in Khartoum of oppressing black Africans. Government-organized and private Arab militias were accused of trying to 'cleanse' black Africans from large areas of disputed farmlands. Refugees from Darfur painted a horrifying picture of militias riding into villages on horses and camels, slaughtering men, raping women and stealing everything. Some sought safety in neighbouring Chad, itself a poor African country, where they camped along the border, only to be exposed to cross-border Muslim militia attacks. UNHCR runs ten camps in Chad where food, clean water and other relief supplies are provided for refugees. Such are the cultural, racial and religious differences and hatred between the two groups that, despite massive international pressure, no ceasefire has

Activities

1 State the similarities and differences between economic migrants and refugees.

2 Draw a large table to show the economic, social and political advantages and disadvantages of:

   (a) immigration of Hispanics into the USA

   (b) out-migration of workers from El Salvador and Mexico.

3 Put together a case study of refugees (from Darfur or another example). Use the following framework:
   - sketch map showing locations
   - causes
   - effects
   - attempts to reduce the effects.

Internal migrations

Rural areas	Urban areas
Push factors	*Pull factors*
• poverty	• better-paid jobs
• work only in farming	• work in factories, offices and shops
• land shortages, overuse of farmland and drought causing food shortages and famine	• reliable food supplies
• lack of services, shortage of clean water	• schools, hospitals, safe water supply and electricity
• remoteness – dirt track links only	• focus of roads, mainly paved roads
• little hope of change and improvement; old and traditional ways of life	• always changing; new skyscrapers and road systems; proper shops; dynamic feel to the place

| Farming village in the mountains of Peru | Business zone in Lima, the capital city |

▲ **Figure 1** Push and pull factors for rural to urban migration.

Internal migration is more common than international migration because distances are shorter and obstacles to movement are fewer. In LEDCs one type dominates – rural to urban migration. The underlying reason is development – as countries develop economically, the number of people engaged in agriculture (a rural activity) declines in favour of those in industry and services (largely urban activities). Present-day urban growth is concentrated in LEDCs (pages 150–1).

Rural to urban migration

In LEDCs, people began to move from the countryside to the towns in the early twentieth century. The rate of movement accelerated so that by the end of the century

about 40 per cent of people in LEDCs were classified as urban (compared with nearer to 80 per cent in MEDCs). Nigeria in West Africa is a typical case: just 5 per cent of its population was urban in 1921; by 1991 this had risen to 30 per cent. In 2005 it was 45 per cent. Lagos is the magnet for people from all regions of Nigeria; it is the second-largest city in Africa (after Cairo) with a population officially estimated at 9 million, although it is admitted that no one knows the real numbers. What is known is that it is growing by least 5 per cent per year.

This movement is age- and sex-selective. Typically it is the young people, in some countries mainly males, who move to find work in the big cities and mines. Although the number one attraction is job opportunities, the chance of a better education for the children is also important, as are the dynamic and modern images of the big cities as shown in films and TV programmes. The movement to the cities is speeding up as rural people become more aware of the size of the development gap between rural and urban areas, as suggested by a study of the two photographs in **Figure 1**. More and more people have sons or relatives already based in the city, which makes it easier for them to make the decision to move. The decision becomes even easier to make during times of difficulty such as drought or wars, when migrants flood into the cities. As a result of the influx of young workers, the population structure of large cities is different from that of the country as a whole (**Figure 2**).

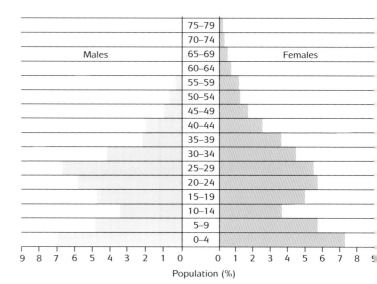

▲ **Figure 2** Population structure of an African city.

Urban to rural migration

The movement to the cities began earlier in MEDCs and has run its course. Now it is in reverse: movement out of urban areas into the countryside is a stronger movement today. In the UK there has been a flight from the big cities in particular (**Figure 3**), where the contrasts between living in the middle of a large urban mass and in a village or small market town in the countryside are at their greatest. The air is cleaner, the pace of life is gentler, there is little traffic congestion and noise, and they are surrounded by green fields; they are considered good places for weekend leisure activities and bringing up families. Commuting to the cities was made possible by new roads and modern cars; these, more than anything else, created the choice to live in rural areas. Unlike the developing world, in MEDCs there is little difference between rural and urban areas in the provision of essential public services.

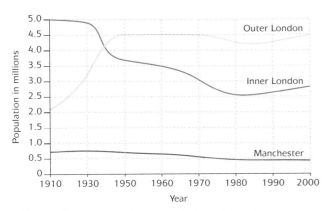

▲ **Figure 3** The changing population of London and Manchester.

Environmental	Socio-economic
Old terraced housing	Declining industries
Air and noise pollution	Higher rates of unemployment
Derelict land and buildings	Low incomes
Lack of open space	Ethnic minority groups
Vandalism and graffiti	High crime rates
Inner-city Birmingham	Village in nearby Shropshire

▲ **Figure 4** Push factors from the inner cities in the UK.

Good and bad effects of these migrations

In LEDCs, elderly people, mothers and young children are the ones left behind in the villages. After the most dynamic element in the population has migrated, rural areas are left as economic and social backwaters with even less chance of investment, change and improvement. The money and goods from family workers in the cities can make the difference between starvation and survival in bad times; some of the pressure for increased food output is relieved by their departure. The cities are full of people with potential; this large pool of cheap labour attracts investment from overseas companies. Their factories, offices and call centres are almost always based in big cities, widening the wealth and development gap between urban and rural areas. However, many cities are already full of unskilled and uneducated workers and the arrival of more only worsens the multitude of existing urban problems (see pages 152–3).

In MEDCs, urban commuters moving into rural areas may not be immediately welcomed by local people. They reduce the number of houses available and increase house prices to levels that rural workers cannot afford; they are also less likely to use local services and support community events. Adding new housing estates can 'urbanize' rural villages, devouring green fields and destroying their peace and quiet. Within cities, as the better-off have moved out, they have been replaced by poorer and less fortunate groups of people. This is most apparent in the inner cities and inner suburbs, some of which are dominated by pensioners, ethnic minorities, the unemployed and students. This has been accompanied by increases in social problems and crime levels.

Activities

1 Describe how **(a)** the pyramid in **Figure 2** suggests rural to urban migration and **(b)** the graph in **Figure 3** suggests urban to rural migration.

2 **(a)** Describe the differences in **(i)** provision of services and **(ii)** economic activities between rural and urban areas in LEDCs.

   (b) Why is the wealth gap between rural and urban areas in LEDCs so wide?

3 **(a)** Make two lists for push and pull factors for urban to rural migration (as in **Figure 1**).

   (b) Why is the gap in living standards between rural and urban areas in MEDCs less wide?

Case Study – Migration in the UK

Migration in the UK

Figure 1 shows how population changed in the regions of the UK between 1981 and 2001. It gives evidence for the continued drift of British people from north to south and highlights the way in which population growth was most concentrated in the south of the country, below the line from the Wash to the Severn Estuary.

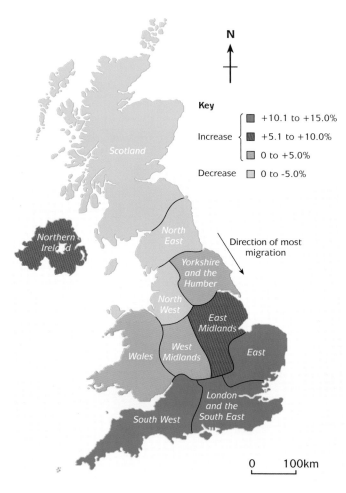

N

Key

Increase
- +10.1 to +15.0%
- +5.1 to +10.0%
- 0 to +5.0%

Decrease
- 0 to -5.0%

Scotland

Northern Ireland

North East

Direction of most migration

Yorkshire and the Humber

North West

East Midlands

West Midlands

Wales

East

London and the South East

South West

0   100km

▲ **Figure 1** Population change in the UK, 1981–2001.

Out-migration from the North

The North is the part of the country with the harshest physical conditions, where the climate is colder and a higher proportion of the land is upland. During the industrial era, which was dominated by coalmining and heavy industries (steel, shipbuilding, engineering and large-scale textile manufacturing), it was well placed. However, since the 1960s the decline of mining and heavy industries as major employers has been spectacular, and in some cases almost total. Half a million coal miners in 1950 had fallen to 10000 in 2000. New manufacturing and services industries attracted to the north of

England and Scotland, partly with the help of financial incentives supported by the government and EU, have not been on the scale needed to soak up the thousands of jobs lost in the traditional industries such as coalmining, shipbuilding, metal smelting, cotton and woollen spinning and weaving. For many people, the young in particular, the only option was migration to where the jobs were – and they were overwhelmingly concentrated in London and the South East. Much service-sector work in government, finance and company administration is still London based. Modern light and high-tech industries prefer a South East location because of its proximity to the largest and wealthiest markets in the UK, the larger EU market and the motorway and airport hubs for the best national and international communications (see **Figure 1** on page 188).

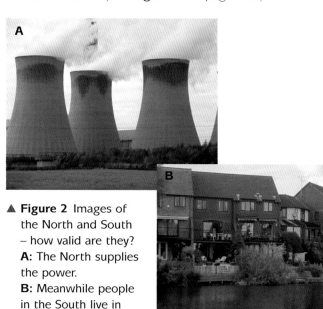

▲ **Figure 2** Images of the North and South – how valid are they? **A:** The North supplies the power. **B:** Meanwhile people in the South live in comfort on the banks of the River Thames.

Migration within the South

Although London itself has the greatest concentration of jobs, many people prefer to live outside London and commute by road or rail. For several decades the population of London has been falling (see **Figure 3** on page 131), as it became possible to commute longer distances – an example of urban to rural migration. A strong London effect is felt up to 150 kilometres away (or more), with many regular London commuters living as far north as Peterborough and Cambridge and as far west as Oxford and Bournemouth – and beyond. Coastal resorts, university towns and a host of smaller market towns and villages are preferred places to live because of their higher quality of life and lower property prices. Population growth has been very marked north and east of London, for example along the M11 corridor towards Cambridge. South West England is part of a high population growth zone in **Figure 1** because it remains popular as a place for retirement, with the UK's highest average temperatures, great coastal and inland scenery and housing more reasonably priced than in the South East.

The issue of future growth in the South East

The Labour government announced in 2003 that over a million new homes were needed in the South East by 2021. Although these were to be built wherever land was available, four growth areas were identified (**Figure 3**). Some were on brownfield sites – land that had previously been built on and used. Thames Gateway is an example. Here new housing is seen as a means of improving the appearance of a derelict zone along the north bank of the Thames which includes the site of an old power station, abandoned docks and piers on the river side and polluted areas used for illegal dumping. At present it is not a pretty sight! Others will need to be built on greenfield sites – countryside only ever used for rural land uses. Increasing the size of Ashford in Kent (sometimes called 'The Garden of England') and extending fingers of development further into East Anglia are more controversial because countryside will be lost to be replaced by housing estates.

Indeed, the whole programme is controversial – over a million new homes is a huge number for an area that is already overcrowded, short of water and with estuarine areas among the first to be affected by rising sea levels. However, the suggestion is rooted in need. It reflects an economic reality – many businesses want a location in the South East and many people are happy to work there; it is also driven by social factors – a real preference for living in a house with a garden and a garage instead of a tower block, and a real need for more homes as increasing numbers of families break up and split into two households.

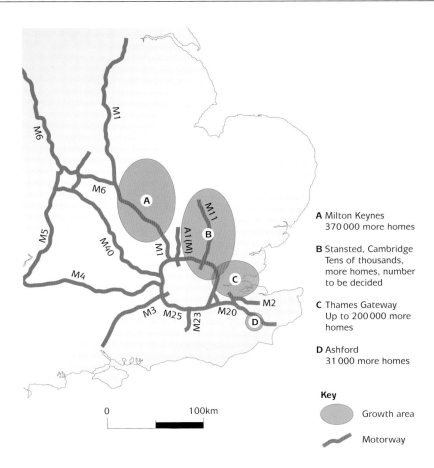

A Milton Keynes
370 000 more homes

B Stansted, Cambridge
Tens of thousands, more homes, number to be decided

C Thames Gateway
Up to 200 000 more homes

D Ashford
31 000 more homes

Key

Growth area

Motorway

▲ **Figure 3** Growth areas in the South East proposed in 2003.

Activities

1 (a) Describe how **Figure 1** shows that North to South migration has taken place within the UK.

(b) Give the economic push and pull factors for this migration.

(c) Why is this direction of migration difficult to reverse?

2 (a) Describe the main plans for future growth in the South East.

(b) State and explain one environmental, one economic and one social argument against the plans.

(c) Are there any alternatives to further growth in the South East? Answer as fully as you can.

Europe – ageing populations and immigration

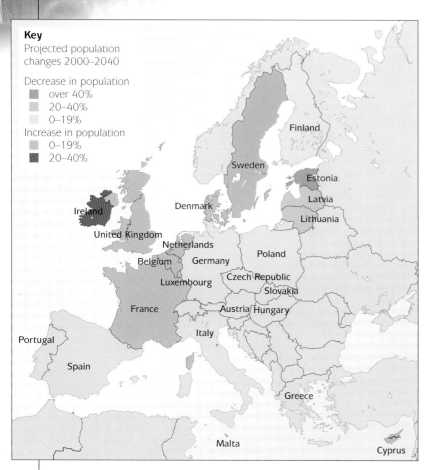

Key

Projected population changes 2000–2040

Decrease in population
- over 40%
- 20–40%
- 0–19%

Increase in population
- 0–19%
- 20–40%

▲ **Figure 1** Population trends in the EU 25 (2000–40).

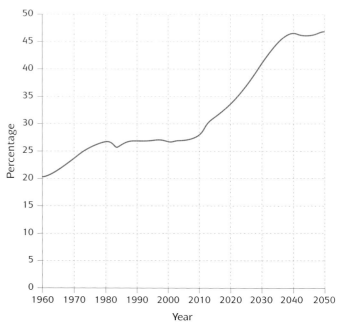

▲ **Figure 2** Pensioners (65 and over) as a percentage of the working adult population (aged 20–64) in the UK, 1960–2050. Source: *Daily Telegraph* 2006

	Country	Number of immigrants	Net migration
1	Spain	610 000	+1.44%
2	Italy	558 000	+0.96%
3	UK	223 000	+0.37%
4	France	105 000	+0.17%
5	Germany	82 000	+0.10%
	'The Mediterranean is now Europe's immigration frontier.'		

◀ **Figure 3** Net migration into the EU in 2004 – the top five countries.

Exam focus

1 a Study **Figure 1**.

   (i) State the difference between Estonia and Ireland. (1 mark)

   (ii) Describe the general pattern of population increases and decreases within the EU. (3 marks)

   (iii) Give two reasons why populations are declining in some EU countries. (2 marks)

b Study **Figure 2**.

   (i) Describe what the graph shows. (2 marks)

   (ii) Explain why it is likely to lead to economic problems for UK governments. (4 marks)

c (i) What is meant by the comment at the bottom of **Figure 3**? (1 mark)

   (ii) Suggest where most of the immigrants into EU are likely to be coming from. (1 mark)

   (iii) Explain why economic migrants are attracted to EU countries. (3 marks)

d Explain why immigration is one strategy for reducing the economic disadvantages of ageing populations. (3 marks)

e Outline the problems that economic migration can cause in:
   (i) MEDCs receiving migrants
   (ii) LEDCs losing migrants. (5 marks)

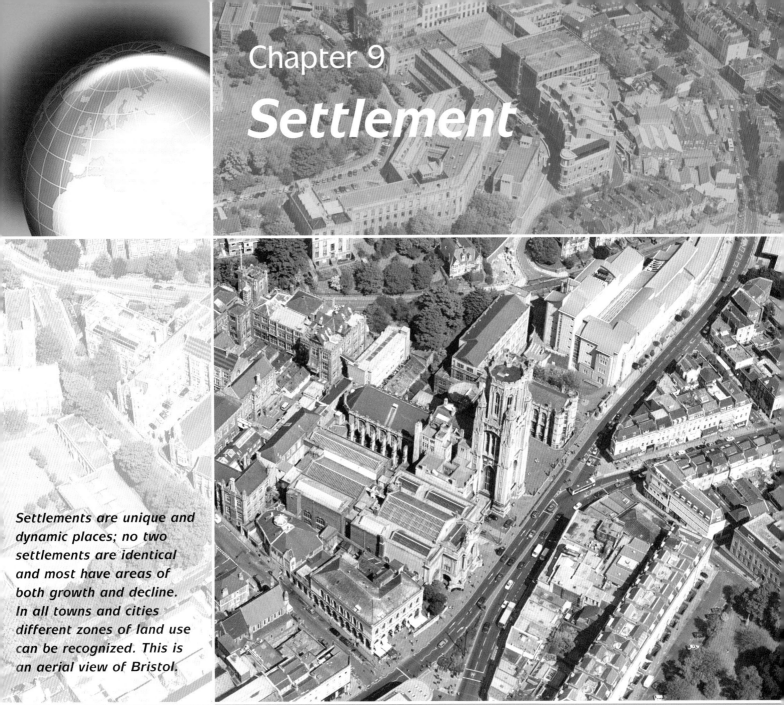

Chapter 9
Settlement

Settlements are unique and dynamic places; no two settlements are identical and most have areas of both growth and decline. In all towns and cities different zones of land use can be recognized. This is an aerial view of Bristol.

Key Ideas

Settlements vary in site, size, structure and function:
- different siting factors affect settlement location
- there is a hierarchy of settlements
- settlement function may change over time
- cities can be divided into urban zones.

Urbanization is a global phenomenon and presents challenges to human populations:
- in MEDCs there are redevelopment issues in CBDs and inner cities, and conflicts about how best to develop or protect the rural–urban fringe
- in LEDCs rates of urbanization are greater, leading to the widespread growth of shanty towns
- some attempts have been made to solve the social, economic and environmental urban problems that result, but the scale of the problem is formidable.

Settlement: site, situation and growth

Settlements are places where people live and work. In most countries there is a whole range of different-sized settlements, from individual farms and hamlets through to huge sprawling cities.

The physical land on which a settlement is built is called the **site**. Early settlements were located with great care. People had to grow their own food, find their own water supply, use the resources of the local area and even defend themselves against hostile neighbours. Think about how different life is today, especially in developed countries. Most families buy their food in huge supermarkets, there is electricity, piped gas and water supplies. Defence is no longer a factor.

Over time some settlements grew, others remained small, while some disappeared from the landscape. **Figure 1** shows the remains of an abandoned settlement on the Isle of Arran in Scotland. Settlements which grew were those with a good **situation**. The situation of a settlement is its location in relation to the surrounding area. A good situation is an area with the potential for settlement growth. Some settlements were at a route focus or bridging point. These encouraged trade which led to growth. Others were in rich farming areas and became market towns into which local farmers brought their produce for sale.

Site factors
Water supply
Water is essential for life. In arid or dry areas settlements locate near rivers, streams, wells or springs. These are called **wet-point sites**. In the chalk and limestone areas

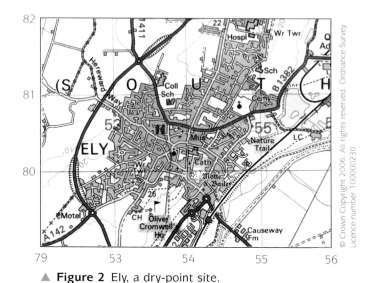

▲ **Figure 2** Ely, a dry-point site.

of the UK there is often a line of spring-line settlements at the base of escarpments (page 31).

In some locations the land is marshy and prone to flooding. Settlement in these locations tends to locate on valley sides, gravel terraces or small hills above the flood plain. The Fenland area of the UK was once a marshy lowland and many of the settlements, such as Ely in Cambridgeshire (**Figure 2**), are sited on small mounds. These are **dry-point sites**.

Aspect and shelter
In the northern hemisphere the south-facing (*adrêt*) slopes are warmer than the cooler north-facing (*ubac*) slopes (**Figure 3**). The south-facing slopes are also sheltered from cold north or north-easterly winds. More settlements are sited on the south-facing slopes in the northern hemisphere and also on the lower slopes where there is even more shelter.

▲ **Figure 1** An abandoned settlement on the Isle of Arran, Scotland.

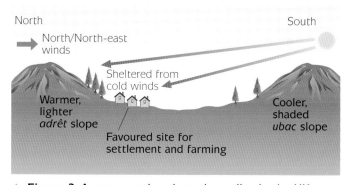

▲ **Figure 3** A cross-section through a valley in the UK running from west to east.

Defence

There are large numbers of settlements today which have the remains of castles, forts or town walls. This indicates just how important selecting a defensive site was in earlier times. Corfe Castle is sited where it controls a gap through a ridge of chalk (**Figure 4**). Later growth when defence became a less important factor was to the south, where the relief was less steep.

Resources

It was essential that early settlers had access to the resources they needed. A food supply was vital, so the areas with the most fertile soils often supported more settlements as more people could be fed. There is a higher density of rural settlement on the fertile loams and boulder clay soils in East Anglia than in the Pennines or the Chilterns where soils are thinner and less fertile.

Early settlements often relied upon timber for fuel as well as for building materials, so nearby woodland was an advantage. In other areas settlements were sited near a quarry or mine for building materials and mineral resources. County Durham has many examples of villages which only developed because of the local coal seams which could be mined.

Communications

Settlements often grew at **bridging points** or fords, around a crossroads, in gaps through hills or at a junction of valleys. Good communications often gave the settlement an advantage over others so that it

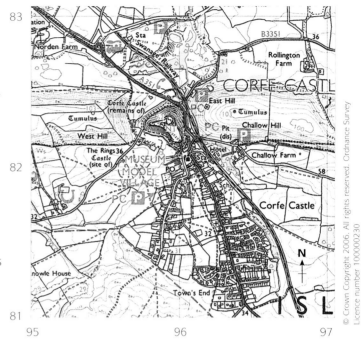

▲ **Figure 4** Corfe Castle: potential for growth?

grew as a **route focus** and attracted trade from other local settlements. Further encouragement to growth was given if the settlement had a castle, cathedral or monastery. The settlement grew as more people were needed to provide services for the soldiers or monks. The sketch map in **Figure 5** shows the site of Berwick-upon-Tweed in Northumberland and the reasons why it has grown into a town.

Activities

1 What is the difference between **(a)** site and situation of a settlement and **(b)** dry- and wet-point sites?

2 From **Figure 2**, draw a labelled sketch map to show how Ely has a dry-point site.

3 **(a)** Describe and give examples of two physical factors that favour the growth of a settlement as a route focus.

(b) Why are settlements located on sites with good communications more likely to grow today than those with good defensive sites?

4 Refer to an OS map of your home area.

(a) Describe the site and situation of the settlement in which you live.

(b) Has it grown in the last 20 years? If it has grown, explain how and why: if it has not grown, suggest reasons why not.

▼ **Figure 5** Sketch map of Berwick-upon-Tweed.

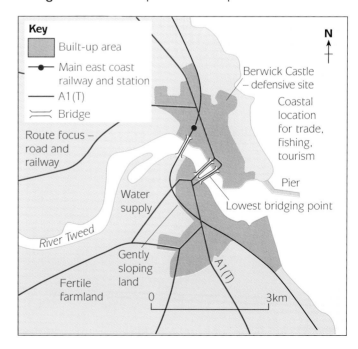

Hierarchy of settlement

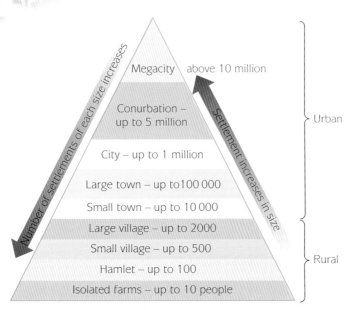

▲ **Figure 1** A hierarchy of settlements according to population size.

A settlement hierarchy arranges settlements in order of importance, with individual farms and hamlets at the bottom and the single largest city, which is usually the capital city, at the top. Three different measures are often used:

- the size of the settlement in terms of its population
- the range and number of services
- the sphere of influence or the size of the area served by the settlement.

Settlement size

Figure 1 shows a hierarchy of settlements according to population size. There is no agreement about the size of a hamlet, a village or a small town. In the UK today, many

villages have grown enormously with the addition of modern housing estates. Their populations have grown to over 2000 people, yet they would still be called villages because they do not fulfil all the requirements of a town, such as the wider range of services and employment opportunities.

Services

The shops and services in a settlement provide the local population with its needs. The larger a settlement, the more services are needed to provide for the population. **Figure 2** suggests a hierarchy based upon the services in settlements of different sizes.

Sphere of influence

The sphere of influence is the area served by a settlement, sometimes called its **catchment area** or **hinterland**. The larger the settlement, the greater the number and variety of shops and services and the wider the area from which people will travel to use the centre. London's sphere of influence is the whole country. Outside London towns such as Plymouth, Newcastle, Leeds and Norwich serve local regions. Market towns serve smaller villages and farms in the area. A village only serves itself and some surrounding farms.

Smaller settlements tend to have fewer shops and services than larger settlements. The shops, such as a general store, newsagent, small supermarket and chemist, tend to provide low-order or **convenience goods** such as newspapers, bread and milk. In larger settlements there are more shops and services. They include shops selling convenience goods but there are also department stores and national chain shops selling jewellery, sports equipment and furniture. These are called high-order or **comparison goods**. The types of goods and services in a settlement are linked to the following.

1 **The threshold population** – the minimum number of people required to support a service so that it remains profitable. In the UK this is about 300 for a village shop, 500 for a primary school, 25 000 for a shoe shop, 50 000 for a medium-sized store and 100 000 for a large one.

2 **The range of a good** – the maximum distance people are prepared to travel to use a shop or service. Most people do not travel great distances to buy a newspaper or do their shopping but they are prepared to travel further to purchase clothes, jewellery or furniture, which are more costly and bought less often.

Major shopping centre.
Several central covered centres.
Several suburban and edge-of-city centres.

One covered area
in city centre.
Many shopping streets.
Several edge-of-city centres.

Several shopping streets.
One or two edge-of-city centres.

One main shopping street and market.

One village shop.

None.

▲ **Figure 2** A hierarchy of settlements based upon services.

Functions of a settlement

Title	Function
Capital city	Political (centre of government); big capital cities like London and Paris are also financial, commercial, industrial, cultural (drama, music, sport) and tourist centres
Industrial town	Manufacturing industries such as car-making, engineering or assembling electrical and electronic goods are large employers
Port	Many specialize in one activity (e.g. container, ferry or fishing ports). Imports may lead to industries (e.g. oil in refineries and petro-chemicals)
Market town	Shops and services for people in surrounding rural areas
Mining/quarrying	Using a local natural resource, often the only reason for their existence
Tourist resort	Many are coastal, but some are inland in mountainous areas, or on lakesides, or are places with many historical remains

◀ **Figure 3** Examples of settlements classified by function.

The function of a settlement is its purpose – why it is there and the 'work' that it does. It can be assessed by looking at the occupational structure of the settlement. Towns can be classified according to their function, as shown in **Figure 3**. Many of the larger towns and cities are multi-functional, with a mixture of residential, industrial, commercial and educational functions.

Many settlements in the UK have changed their function over time. In some cases the original function no longer applies, such as the defensive function. Most settlements began with farming as their main function. Over time mechanization has reduced the need for farmworkers and therefore many villages, especially those close to large urban areas, now house workers in all types of employment who travel elsewhere to work. These settlements are called **commuter** or **dormitory villages**. They have often expanded with the addition of a new housing estate, and the residents commute to work in the cities nearby. In County Durham and South Wales many villages began as farming settlements. During the Industrial Revolution many new villages were created and some existing villages found themselves close to coal seams that could be mined. These villages expanded, with pit head workings, rows of colliery terraces, railway lines and new services, e.g. churches, public houses and schools. By the mid-1990s many village mines had closed. The earlier closures led to a decline in population and shops and services closed as people moved to find jobs in mines elsewhere. Today many of these villages have become dormitory settlements with the addition of a modern housing estate. **Figure 4** charts the history of Shincliffe, a former farming and mining village in County Durham.

Activities

1 Using a map of your local area and **Figures 1** and **2**, complete a copy of the table below.

Settlement hierarchy	Population size	Example from my local area	Functions
Conurbation			
City			
Town			
Village			
Hamlet			
Isolated farms			

2 **(a)** Define the term 'sphere of influence'.

  (b) Giving reasons, state where you and your family would go to buy:

  (i) a magazine and some sweets

  (ii) the weekly food shopping

  (iii) a new three-piece suite and clothes.

3 Study **Figure 4**.

  (a) How has the function of the village changed over time?

  (b) Describe and explain how the population of Shincliffe has changed.

Farming settlement — Middle Ages — Mine opened — Population growth — Mine closed — Population decline — Commuter settlement — 2000 — Population growth

▲ **Figure 4** A changing village: Shincliffe in County Durham.

Urban morphology

Urban land uses are the shops, industries, offices, housing, parks and open space found in larger settlements (towns and cities). Urban areas tend not to have a jumble of different land uses. Each type of land use usually clusters together to give distinctive **urban zones** such as the **Central Business District (CBD)**, where shops and offices are concentrated. Other zones are formed by industrial areas and the vast suburban housing areas. **Morphology** is the term used to describe the internal structure of a city. Various urban models have been developed to show the arrangement of land-use zones within cities.

An urban model for cities in MEDCs

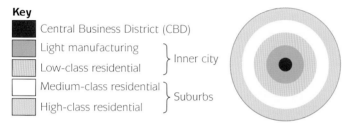

Key
- Central Business District (CBD)
- Light manufacturing } Inner city
- Low-class residential }
- Medium-class residential } Suburbs
- High-class residential }

▲ **Figure 1** The Burgess model.

The **Burgess model** (**Figure 1**), developed in the 1920s, has five concentric rings representing five land-use zones. The zones are arranged in a circular pattern around a CBD.

The Burgess model is based upon two main ideas:

1 Cities grow outwards from the original site and hence property becomes younger as the outskirts of the town are reached. The original site is generally where the CBD is located today.

2 Land costs are highest in the CBD where land is in short supply and, traditionally, accessibility is greatest. Only high-profit-making businesses can afford to locate in the CBD. Away from the city centre, costs decrease and more land is available, allowing industries and housing to locate here.

British cities are very varied and not all of them conform to the circular pattern of land-use zones as in the model. For example, old industrial towns often have lines of industry along river banks and canals, while coastal resorts are typically long and thin with hotels, guest houses and shops concentrated in a long strip

behind the beach. You need to remember that this is only a model. Although the Burgess model was based on Chicago in the USA, it still makes a good starting point for studying urban land-use zones in British towns and cities, especially if you are doing a transect from the centre to the edge of the built-up area.

An urban model for cities in LEDCs

The model in **Figure 2** is based upon cities in Brazil. As in all of the models, the CBD is located centrally. All CBDs tend to look very similar, with skyscrapers and Western-style shops. The model also has an industrial zone located along a road or railway. However, there are some striking differences between the models. The city in the developing world is much less regular, with more zones. The land uses are less well segregated. There is only a very small high-class sector, mostly located near the CBD or on prime sites, such as near a beach, as in Rio de Janeiro, or on high ground with a good view. There is no evidence of an extensive zone of middle-class housing similar to that found in the suburbs of the cities in the developed world. The largest zone is that of the shanty towns which may begin close to the CBD and stretch vast distances to the outskirts of the city. The shanties may occupy wasteland or swamps within the city, or some of the steeper slopes. The developing world city also has no traditional industrial areas. Industries are mostly modern factories in zones along lines of communication. Most of the industries have 'followed the people' to the city, unlike in the MEDCs where the people moved to the cities because work was available.

Key
- Central Business District (CBD)
- High-class expensive housing
- Medium-price old housing
- Low-class cheap housing
- Modern factories along main road
- Shanty towns/ squatter settlements

High-quality commercial spine develops

▲ **Figure 2** Land-use zones in a developing city.

Identifying urban zones on OS maps

The CBD can be recognized by the presence of historic buildings, the meeting of several main roads, a high density of buildings and bus and railway stations, often at the edge of the CBD. The **inner city** is an area of old terraced housing around the CBD with a regular, often grid-like, street layout. Industrial buildings in this zone stand out as large blocks, frequently located next to railways and waterways. In the **suburbs**, houses are usually built on estates, which have an irregular pattern of streets including circular avenues and cul-de-sacs. The amount of open space between the houses increases towards the edges of the city. In the **rural–urban fringe**, rural land uses such as farmland and woodland begin to appear among those connected to the city; these may include golf courses, airports, out-of-town shopping centres and modern industrial estates and business parks – all of which need more land than is available in the city.

Activities

Study **Figure 3**.

1 **(a)** Name the urban zone in each of the six grid squares outlined in black.

   (b) Describe the land uses in each of the six squares.

2 Describe and explain the pattern of land uses in Luton.

3 How well does the pattern of urban zones in Luton fit the Burgess model? Answer by describing where it fits well and not so well; then make a summary comment.

▼ **Figure 3** OS map of Luton, Beds. 1:50 000 (2cm = 1km).

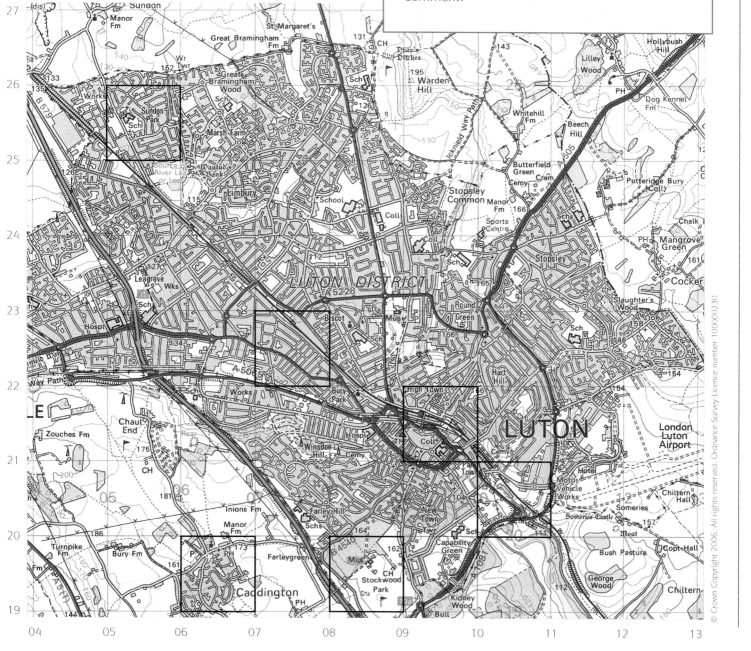

Transect through a typical British city

Central Business District   Inner city   Suburbs   Rural–urban fringe

▲ **Figure 1** The four urban zones from the centre to the edges in most British towns and cities.

A Central Business District (CBD)

Characteristics

This is located in the centre of the urban area, around the historical core – in some cities this is a cathedral or a castle. From a distance it is often easy to pick out the CBD by its concentration of skyscrapers and other tall buildings.

Compared with other urban urban zones, the CBD:

- has the largest offices and shops, including department stores

- has the widest variety of goods on sale

- has high land values, rents and rates (which helps to explain the many tall buildings built close together)

- is the main place of work by day, which leads to traffic congestion (especially in rush hours)

- is the most accessible location where the main roads meet and has the main railway station(s).

The main difference between the CBD and other urban zones is that few people live here.

B Inner city

Chararacteristics

Located next to the historical core, this is an area of old housing and industry suffering from urban decay. It is a zone of mixed land uses:

- old high-density terraced houses

- some are three or four storeys high and include basements and attics; these are often let out as flats and badly maintained

- old and sometimes abandoned factories and warehouses

- areas of derelict land around railway sidings, unused docks and canals

- high-rise flats (many built in the 1960s)

- pockets of smart new developments, in many cities around old docks such as the London Docklands, Albert Dock in Liverpool and Salford Quays in Greater Manchester.

The main difference between the inner city and other urban zones is its generally run-down appearance.

◀ ▼ **Figure 3** Two views of inner-city Birmingham.

▲ **Figure 2** CBD in Manchester.

C Suburbs

Characteristics

In most cities these cover the largest area. A suburb is part of the urban area which has grown outwards from the old centre across what was once countryside. This zone is predominantly residential:

*   along the sides of the main roads are inter-war semi-detached houses and small shopping parades

*   behind the main roads are more modern housing estates, mainly of semi-detached and detached houses

*   some are private estates, others were local-authority built, although many have now been bought by residents

*   the houses usually have gardens and garages with areas of open space between them

*   the more recent and expensive housing is in the outer suburbs, where the density of housing tends to be lower.

There is less change in this zone than in the other three; the houses are good for many more years and virtually all the land suitable for building has already been used.

D Rural–urban fringe

Characteristics

This is on and around the edge of the built-up area and is partly urban and partly countryside. It is another zone with a real mixture of land uses:

*   some are traditional rural land uses such as farmland and woodland

*   others are rural businesses targeted at people living in nearby urban areas, such as garden centres and market gardens with farm shops

*   recreation such as golf courses and stables for horse-riding, and public utilities such as water storage and sewerage farms

*   new urban developments include out-of-town supermarkets and shopping centres and business parks

*   new housing in villages leads to old settlements growing and becoming part of the urban built-up area.

This is a zone of many conflicts between developers who want to use the greenfield sites for building homes and commercial premises, and planners and conservationists who want to preserve as much countryside as possible.

◀ ▼ **Figure 4** Two different styles and types of housing in the suburbs.

▲ **Figure 5** Out-of town shopping centre.

Activities

1   Describe the characteristics typical of the CBD shown in the photograph in **Figure 2**.

2   Describe how the two photographs in **Figure 3** show that the inner city is a zone of mixed land uses.

3   **(a)**   Name the two types of house shown in the photographs in **Figure 4**.

   (b)   Which one is likely to be closest to the edge of the city? Justify your choice.

4   Explain why shopping centres of the type shown in **Figure 5** are more likely to be built in the rural–urban fringe than in other urban zones.

5   Investigation: Describe the main changes between the city centre and the edge of a built-up area for a town or city in your home area.

The Central Business District

The CBD of a city is not static; it is a dynamic area going through phases of growth and decline. Pass through any CBD of a large city to see areas in decay, with closed shops and a run-down appearance, and others that appear lively, smart and successful. The CBD also experiences problems of traffic congestion, parking and pollution as well as those caused by lack of space and shortage of land. Local planners have implemented a variety of different schemes to attempt to solve the problems of the CBD.

Main functions of the CBD

Shops

The CBD is usually at the top of the shopping hierarchy in a city. It has the widest range of shops and the largest department stores. Shops mainly sell comparison or high-order goods and they draw their customers from a wide sphere of influence. The highest land costs are in the centre of the CBD. Here, in the **core** of the CBD (**Figure 2**), are found large department stores and branches of many national chains of shops.

▲ **Figure 1** Leeds CBD.

Smaller, often privately owned, shops are located on the edges of the CBD in the fringe area called the **frame**. Some shops, such as clothing, shoe and jewellery shops, tend to cluster together to take advantage of competition, while others are more dispersed, such as newsagents and chemists.

Offices

Banks, building societies, solicitors, company headquarters, insurance companies and government offices occupy high-rise office blocks or the upper floors above shops in the CBD.

Culture and entertainment

Parts of the CBD 'come alive' at night as the theatres, cinemas, clubs, bars and restaurants attract customers. Certain parts of cities have become famous for their nightlife, such as London's West End and Newcastle Quayside.

Problems and attempted solutions

Traffic congestion is the major cause of air and noise pollution in city centres. Not only does it contribute to the stresses of working and living in urban areas, but it also aggravates diseases such as asthma and bronchitis. Banning heavy lorries from city centres has helped, as also does the use of clean fuel technology by public transport.

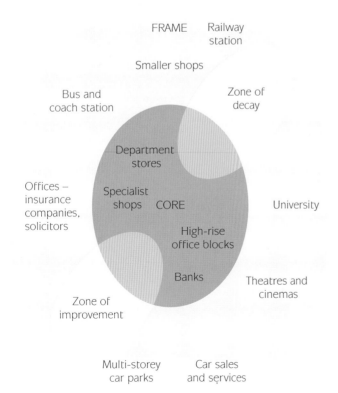

▲ **Figure 2** The core and frame in the CBD.

Traffic congestion

Towns grew and the street patterns were established before the motor car was invented. Now high car ownership and the concentration of shops, services and employment in the CBD create massive problems of congestion and parking in city centres. Roads are often narrow, with little pavement space, and in some cities the rush-hour traffic is in danger of causing 'grid-lock'. Attempted solutions include:

- ring roads and by-passes to divert traffic not going into the city centre
- urban motorways and flyovers
- public transport schemes such as 'park and ride', the Newcastle Metro (**Figure 3**), trams in Manchester
- multi-storey car parks
- pedestrianization of high streets.

Now, the emphasis is shifting from accommodating the motor car to schemes to ban or charge cars for going into the city centre. Environmentally friendly forms of transport are being encouraged, such as buses which run on gas, and electric cars. Integrated public transport schemes such as metro systems and trams are a high priority for many local authorities to reduce the number of cars in city centres, especially cars containing just one person. A congestion charge has been introduced on weekdays for traffic going into Central London.

▲ **Figure 3** The Newcastle Metro.

Lack of space and the high cost of land

Competition for land has led to high prices, and growing firms find it difficult to find space. In some CBDs the smaller retailers have been forced away from the city centre because of the high costs. Attempted solutions include:

- high-rise buildings to increase the floor area available
- new retailing areas in out-of-town shopping centres (**Figure 4**) in the suburbs or rural–urban fringe, in a process called **decentralization**.

Figure 4 The Gateshead MetroCentre.

Urban decline

Parts of some CBDs have declined. Shops and offices have closed down and the empty buildings are vandalized (**Figure 5**). City centres compete with out-of-town shopping centres to cater for the growing demands of shoppers. Attempted solutions include:

▲ **Figure 5** Decline in the West End of Newcastle upon Tyne.

- redevelopment of zones of decline in the CBD such as King's Cross and Covent Garden in London and Eldon Square in Newcastle
- expansion of the CBD into areas of the inner city – old factories and substandard terraced housing have been cleared, rehousing the occupants in the suburbs or New Towns and filling the space with new shopping and office developments.

Exam focus

1 a (i) Describe two characteristic features of a CBD shown in **Figure 1**. (2 marks)

    (ii) Does **Figure 1** show the core or the frame of the CBD in Leeds? Explain your answer. (2 marks)

b Study **Figure 2**.

    (i) State two differences between shops in the core and frame of CBDs. (2 marks)

    (ii) Explain why the land uses in the core of the CBD can afford higher rents and rates than those in the frame. (4 marks)

c Describe the appearance of zones of decay and decline within CBDs in the UK. (4 marks)

d (i) State three traffic problems in the CBDs of large cities in the UK. (3 marks)

    (ii) Describe how planners and city authorities are trying to solve one of these problems. (4 marks)

    (iii) Explain why traffic problems in CBDs are difficult to solve. (4 marks)

The inner city

Wealthy workers long ago moved out of most inner cities into the suburbs, small towns and country villages and became commuters. They were pushed out by the kinds of problems listed in **Figure 1**. The people who replaced them were poorer, with the result that many of the problems have got worse rather than better in recent years. Poverty and dilapidation are seed beds for crime, vandalism and drug-trafficking. From time to time since 1980, racial tension and general discontent have flared up into riots – for example, in Brixton in 1981, and more recently, in parts of Leeds, Bradford and Oldham in 2001.

Environmental problems

- Housing is either old decaying terraces or cheaply built tower blocks
- Many derelict buildings – factories, warehouses, churches, houses and flats – often vandalized and covered in graffiti
- Shortage of open space; most of what exists is wasteland

Social problems

- Above-average numbers of pensioners, singe-parent families, ethnic minorities and students
- Poorer than average levels of health, but higher than average levels of drug abuse and crime
- Difficult police–community relations

Economic problems

- Local employment declined as industries and docks closed
- Higher than average rates of unemployment, especially for the young and ethnic minorities
- High cost of land compared with the suburbs
- Low income and widespread poverty

▲ **Figure 1** The environmental, social and economic problems of British inner cities.

Inner-city problems were not caused only by the flight of the wealthy. Some of the actions of planners and local authorities between 1945 and the mid-1980s did not help. At first, the policy was slum clearance, which led to forced removal of families and whole communities, leaving behind large swathes of wasteland. From about 1960 the policy of comprehensive redevelopment was fashionable, which led to further clearances of terraced houses and their replacement by high-rise flats

(**Figure 2**). Many of these were badly built: residents complained about the cold and damp, broken lifts, open staircases taken over by gangs, criminals and drug dealers; pensioners and mothers with children felt like prisoners in their uncomfortable homes. Now it is acknowledged that this policy was a disaster. Many inner-city landscapes were blighted further by efforts to relieve traffic congestion in city centres, which was tackled by using land in the inner cities for building inner ring roads, flyovers and urban motorways.

▲ **Figure 2** High-rise flats – the 1960s answer to the problems of inner cities.

Improving inner cities

Attitudes of planners are changing. Inner cities contain larger areas occupied by **brownfield** sites (areas of previously built-up land available to be built on again) than any other urban zone. This reservoir of land is important at a time when new building land is at a premium and objections to the use of **greenfield** sites are increasing. Also it is now more common for local communities to be involved in the planning of changes, instead of having changes forced upon them as in the past. The insides of some tower blocks have been refurbished to make them more comfortable places to live, while their outsides have been painted and improved to make their appearance less formidable (**Figure 3**). After an unsuccessful redevelopment in Hulme in inner-city Manchester in the late 1960s, which produced highly unpopular, but award-winning, concrete crescents of high-rise flats, the Hulme City Challenge

▲ ▶ **Figure 3** Changes in apearance to flats in Glasgow.

A

▲ ▼ **Figure 4 A:** Salford Quays – land left abandoned by the closure of Manchester docks is now used for luxury homes, office, leisure and recreation. **B:** Castlefields – what was an area of abandoned canals, unsightly railway viaducts, decaying warehouses and factories has been transformed into a zone of commercial, residential and leisure developments that attracts over 2 million visitors a year.

B

launched in 1992 concentrated on providing 3000 low-rise homes accompanied by local services including community facilities.

Another trend is towards inner-city **gentrification** – the movement of wealthy people back into areas of former urban decay. When large old inner-city houses are bought up by wealthy people who have the money to improve them, formerly run-down areas such as Notting Hill and Islington in London once again become fashionable places to live. There has been a rise in the number of commercial redevelopment schemes in inner cities. An important part of many dock-, river- and canal-side redevelopment schemes is the conversion of old warehouses into good-looking modern apartments. At the same time the area around them is cleaned up and landscaped; Salford Quays and Castlefields in inner-city Manchester are examples (**Figure 4**).

How successful have these 'improvements' been? The series of government and city initiatives has released increased funds for inner-city use. Improvements have at least been noticed in some areas, even if overall inner cities remain the most disadvantaged and least attractive urban zone. The underlying problem is that too many inner-city residents are caught up in the wider poverty trap associated with lack of skills, low pay and no regular work. However, commercially sponsored redevelopment schemes such as Salford Quays and London Docklands have been successful in bringing economic activity and life back to formerly derelict areas, accompanied by tremendous improvements in urban landscapes.

Activities

1 **(a)** Write about the disadvantages for families living in high-rise flats.

   (b) Describe the improvements shown in **Figure 3**.

2 Choose one environmental, one social and one economic problem from **Figure 1**.

   For each one, explain:

   (a) why it exists

   (b) why it is difficult to overcome.

3 **(a)** What is meant by 'inner-city gentrification'?

   (b) Describe some of the improvements in inner-city Manchester since 1990.

The rural–urban fringe

The growth of cities has caused **urban sprawl** outwards into the countryside, engulfing small villages, farms and woodland. Land is in demand for housing, industrial estates, business parks, out-of-town shopping centres, bypasses, airports, recreational amenities and public utilities such as waterworks and sewage farms. Much of the growth in the past was haphazard and cities have gradually increased in size. Now planning authorities are actively involved in trying to control the growth of urban areas. There are many conflicts and issues surrounding the rural–urban fringe.

Motorways and bypasses

In the 1980s car ownership and road building were encouraged. The result has been enormous pressure on the existing road network, congested city centres and villages damaged by the weight and noise of traffic. The solution in the 1980s and early 1990s was to build more roads, but many of the schemes attracted considerable opposition from local people and conservation groups (**Figure 1**). By the late 1990s public opinion and government policy had changed. Increasingly schemes plan to reduce the amount of traffic by improving public transport, charging tolls on motorways, putting up the cost of fuel and car parking in town centres and encouraging more environmentally friendly forms of transport. Many new road schemes planned for the late 1990s and the early 2000s have been delayed or scrapped.

Carriageway duel to save the Blackdown Hills *14 June 1995*

Protesters Halt By-Pass Work Again *January 1996*

Newbury protesters told how to destroy *18 February 1996*

Bypass sparks new battle of Hastings *14 February 2001*

▲ **Figure 1** No more roads, please!

▲ **Figure 2** The Solar Building at Doxford International Technology Park on the outskirts of Sunderland.

Commercial and industrial developments

Developers like out-of-town locations on greenfield sites because they are cheaper and easier to build on than brownfield sites within the built-up area. These sites are equally attractive to companies locating in shopping centres and business parks because they offer:

- cheaper land and lower rates than in urban locations

- plenty of space for large one-storey buildings and car parks

- open space – easy to landscape and surrounded by pleasant countryside

- a cleaner, less congested environment

- easier access to motorways for deliveries and for worker access.

Conservation and green belts

Green belts are areas of land intended to be left open and protected from development. They have several aims:

- to check the sprawl of large urban areas and prevent neighbouring towns from merging

- to safeguard surrounding countryside from encroachment and destruction

- to assist in urban regeneration by encouraging reuse of derelict and other urban land (i.e. brownfield sites).

Figure 3 The main green belts in the UK, 2005.

From 1946 onwards, green belts were established around all the major conurbations and big cities (**Figure 3**) as well as several smaller towns, often historic ones with plenty to protect, such as Chester and Durham. Development within them is only permitted in exceptional circumstances.

In general the policy has worked well. For example, only 3 per cent of the new dwellings built between 1997 and 2000 were in green belts. However, pressures are increasing all the time. It would not have been possible to build the M25 around London without consuming

some of the green belt. Every time a green belt is under threat, local people and members of conservation groups protest, demanding that controls on development are strictly enforced. Councils are encouraged to redevelop brownfield sites to meet the rising demands for housing from population growth and a persistent increase in demand for one-adult homes (**Figure 4**).

UK total population	UK households	1961	2005	UK life expectancy
2005: 60 million	One person only	11%	30%	1951: 69 years
2025: 65 million	Single parent	6%	10%	2005: 79 years

Figure 4 Population data for the UK.

Pressures and issues

The pressure from developers and builders on planners and local authorities to allow more growth in the rural–urban fringe can only increase. The rural–urban fringe is a very attractive location, both for new housing and a range of commercial activities. Many of the benefits of living and working in the countryside can be enjoyed at the same time as those derived from being close to urban services and communications.

Plenty of farmers are willing to sell land. The price of land for building can be ten times higher than for farming; some farmers next to urban areas suffer problems from vandalism, sheep-worrying by dogs, theft and crop trampling. Many types of farming are less profitable than they used to be. The drive towards more intensive farming has been reversed; overproduction of farm produce led to the EU policy of 'set-aside', where farmers are paid not to farm their land (see page 167).

Activities

1 (a) Draw graphs of different types to show the data in **Figure 4**.

(b) Describe how **Figure 4** shows that more houses will be needed.

2 Study **Figure 5**.

◄ **Figure 5** Land in the rural–urban fringe next to a housing estate.

(a) Describe the land uses.

(b) Why is this an example of a greenfield site?

(c) NIMBYs (people saying 'Not in my backyard') on the housing estate are objecting to the farmer selling this land for housing.

(i) Suggest reasons for their objections.

(ii) Why are they called NIMBYs?

(d) Interested parties for the proposed housing development:

- Builder
- Council
- Environmentalist
- Farmer
- Planner
- Shopkeeper

(i) Make two columns. List those likely to be For the new housing and those Against it.

(ii) Justify your choice and explain why people have different views about building new houses.

Urbanization

Urbanization is urban growth which leads to an increase in the percentage of people living in urban areas. **Figure 1** shows past and projected urban population data. Present and expected future urban growth is overwhelmingly concentrated in LEDCs. What significant change occurred between 1970 and 1980?

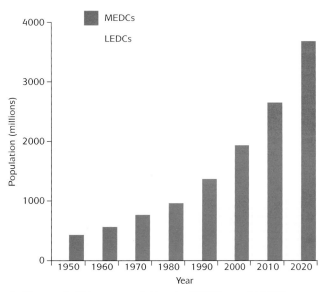

▲ **Figure 1** Urban population in MEDCs and LEDCs.

Urban growth in MEDCs

Towns and cities in MEDCs grew rapidly during the Industrial Revolution. In Britain this was mainly in the nineteenth century. At the time there was an Agricultural Revolution. New farm machinery meant less labour was needed on farms, so people moved to towns where there were plenty of new jobs available in new factories,

mines and shipyards. Urbanization was happening. Although growth was quite rapid, about 10 per cent per annum, there were enough jobs for people, and mine and factory owners built houses for their workers.

Towns and cities continued to grow into the twentieth century as a result of the push and pull factors referred to on page 130, causing rural depopulation, particularly from remote areas. As a result, almost 90 per cent of the UK population is now classified as urban; it is almost as urbanized as it can be. If anything, in the UK and in some other MEDCs, notably the USA, the movement is the other way. Big cities in particular are witnessing a loss of population as wealthy commuters and retired people are moving out into the country to live in villages and small market towns (page 131).

Urban growth in LEDCs

Phenomenal rates of urban growth continue to be recorded in LEDCs, for which three main causes can be identified:

* rural to urban migration, a result of both push and pull factors (page 126)
* high rates of natural increase among the youthful population of these cities
* the concentration of industry and all other modern economic activities, making the cities a natural magnet for young people looking for work.

Persistent urban growth in LEDCs has led to a change in the distribution of the world's largest cities. How does **Figure 2** suggest higher rates of urbanization in

▼ **Figure 2** The world's top ten cities in 1950 and 2000.

1950				2000			
Rank	City	Continent	Population (millions)	Rank	City	Continent	Population (millions)
1	New York	N. America	12.3	1	Tokyo	Asia	26.4
2	London	Europe	8.7	2	Mexico City	Latin America	18.1
3	Tokyo	Asia	6.9	3	São Paulo	Latin America	18.0
4	Paris	Europe	5.5	4	New York	N. America	16.7
5	Moscow	Europe	5.4	5	Mumbai (Bombay)	Asia	16.1
6	Shanghai	Asia	5.3	6	Los Angeles	N. America	13.2
7	Rhine-Ruhr	Europe	5.2	7	Kolkata (Calcutta)	Asia	13.1
8	Buenos Aires	Latin America	5.0	8	Shanghai	Asia	12.9
9	Chicago	N. America	4.9	9	Dhaka	Asia	12.5
10	Calcutta	Asia	4.4	10	Delhi	Asia	12.4

continents in the developing world? By 2000 there were 18 megacities (cities of more than 10 million inhabitants), 12 of which were in Asia. Between 1950 and 2000 the mean latitude of millionaire cities (cities of more than 1 million inhabitants) changed from about 40° to 30°; an average move of 10° towards the Equator confirms the faster urban growth rates in LEDCs, which are concentrated in the tropics on the southern side of the North–South line that separates the developed and developing worlds (page 226).

Advantages from urbanization

Although the disadvantages and problems from urban growth in LEDCs are more serious and receive most attention, there are benefits. These are summarized below.

- For the economy:

  - Big cities attract investment from overseas companies, encouraging modernization.

  - More value is added by processing and manufacturing than by exporting raw materials.

- For people's incomes:

  - The variety of employment opportunities increases and people have more chance of regular paid work.

  - Even self-help work in the informal sector often brings in more money than farming.

  - There are more commercial opportunities for farmers to sell produce in city markets.

- For people's quality of life:

  - Essential services such as safe water supply, sanitation and electricity are more likely to be available than in rural areas (page 235).

  - Often secondary education is only available in cities, opening up more chances of acquiring skills.

- Opportunities for improvement:

  - Improvements in shanty towns can be seen as a step on the ladder for future generations.

  - Even jobs in the informal sector may enable skills to be acquired, leading to better pay in the future.

  - Possibilities exist that are not present in the countryside, even if many will not be able to benefit from them.

Big cities are dynamic places in LEDCs, whereas the pace of change is slow in the countryside. Walk around some areas, especially in capital cities, and it is impossible to believe that you are in a poor country (**Figure 3**).

▲ **Figure 3** A residential area in Lima, the capital of Peru, a country with an average income of about US$ 200 per year. Does it look like it?

Activities

1 (a) Describe what **Figure 1** shows about urban growth (past and projected) in **(i)** MEDCs and **(ii)** LEDCs.

(b) Using values from **Figure 1** for 1950 and 2000, draw two pie graphs to show percentages of the total world urban population living in MEDCs and LEDCs.

(c) Give reasons why:

(i) a high percentage (90 per cent) of the population is urban in the UK, but the percentage is not likely to increase much in the future

(ii) a low percentage (under 40 per cent) is urban in an LEDC like China and the percentage is likely to keep on increasing.

2 Study **Figure 2**.

(a) (i) Add up the number of cities named for different continents in 1950 and 2000.

(ii) Explain how the results suggest greater urbanization in LEDCs.

(b) (i) On an outline map of the world, using different colours for 1950 and 2000, plot the locations of the cities named.

(ii) Describe how the pattern has changed between the two dates.

Problems of urban growth in LEDCs

Relentless population growth is leading to a multitude of urban problems without any hope of a breathing space to give governments and city authorities time to catch up. They are struggling to provide an adequate supply of essential services such as piped water, sewerage, electricity, health and education for existing city inhabitants without providing more for the constant flow of new arrivals. More would-be migrants are queuing up in the countryside, ready to move in, undeterred by the great economic and social problems faced by those who have already moved there.

Environmental problems

Some problems, such as traffic congestion, are faced by all big cities irrespective of whether they are located in MEDCs or LEDCs (**Figure 1**). However if anything, traffic congestion is more acute in LEDCs despite lower rates of car ownership. In Asian cities motorized transport must compete for space with pedestrians, rickshaws, scooters, donkeys and other animals. Congested streets are clogged all day with taxis, buses, lorries and cars, many of them old with inefficient exhaust systems, and this leads to high levels of air pollution and danger to health, especially in the form of asthma and bronchitis. Some cities are notorious. Mexico City is one: it is sited in a mountain basin where pollutants are easily trapped. Beijing is another: so furious has been the growth of cars (running on low-grade petrol) and industries (burning low-grade coal) that smog reduces visibility to a few hundred metres at times and gives a putrid smell to the air. Doctors are blaming dangerous levels of smog for rises in the cases of bronchitis, tuberculosis and lung cancer.

Rivers and seas are used as dustbins for human wastes, destroying all traces of plant and fish life. The sprawl of shanty towns and industries on the city edges destroys agricultural land, woodland and natural habitats. Underground water supplies are being used up at alarming rates; Mexico City has sunk over 7 metres during the last century as the aquifer below it has been emptied of water. The human problems are so great that there is little time or money to devote to environmental problems.

Economic problems

Unemployment and poverty are big problems. For the newly arrived migrant, the main economic problem is finding work: with few jobs available in the formal sector for non-skilled and illiterate rural people, most are forced to look for work in the informal sector, in petty roles such as street sellers, shoe shiners, human carriers, waste collectors and domestic servants. Survival in the city is tough for all newcomers. Most shanty-town dwellers are underemployed, their work taking up few hours in the week and earning very little. A few may find jobs in the industrial zones or in the city centre, but travelling there can be expensive.

Social problems caused by the growth of shanty towns

Social problems are mainly to do with housing and the effects on health and family life for those living

▲ **Figure 1** Traffic congestion, Cairo style.

▲ **Figure 2** Shanty town on a hillside in Rio de Janeiro, Brazil.

in squatter and shanty settlements. Most big cities in Africa, Asia and Latin America are surrounded by spontaneous and makeshift shanty towns (**Figure 2**), and it is common for 50 per cent or more of the city's population to live in them. They are known by different names in different parts of the world – *favelas* in Brazil, *bidonvilles* in North Africa and *bustees* in Kolkata (Calcutta).

The shanties are sited on any spare land the migrants can find. This includes steep slopes (as in Rio de Janeiro and Lima), swamps and rubbish tips. The areas used are often avoided by others because they are prone to landslides, flooding or industrial pollution. The shanty towns are illegal settlements and the people are squatters. The shacks and shelters are homemade, built from anything the people can find, including bits of wood, sheets of corrugated iron, cardboard, polythene and five-gallon oil drums. They are a real fire hazard. Typically, there are usually only one or two rooms where the family eats, sleeps and lives. Most shacks lack basic amenities such as electricity, gas, drainage, running water and toilets. In the *bustees* of Calcutta one water tap and one toilet may be shared among 30 people.

Sewage often runs down streets and pollutes the water supply, leading to water-borne diseases such as diarrhoea, typhoid and cholera. Diseases spread quickly because of the high density of housing (**Figure 3**). Also there is often no refuse collection, and any spare space becomes filled with rubbish, another breeding ground for disease. Infant mortality rates are high because babies are the most vulnerable to disease. Health care is often too expensive and too far away. Many families, and especially the children, suffer from malnutrition. Local shops and stalls, often fly-ridden and dirty, sell a limited range of poor-quality foods that lack the proteins, vitamins and calories needed for a healthy diet. The stress of living in shanty towns leads to frequent breakdown of marriages and increases in crime, mainly theft and robbery. In some cities there are large numbers of 'street children', who have either run away or been abandoned by family break-up.

The underlying cause of all the social factors is economic – lack of income and poverty. One of the first things that people who manage to find regular work and earn a reasonable wage do is build a better home. Money is needed to send children to secondary school to acquire skills and qualifications that are essential for them to gain higher-paid work to lift them out of the poverty trap. In the meantime, poor shanty dwellers continue to have more and more children, partly out of ignorance but partly in the hope that at least one of the family will get a job, either in the home city or overseas, that will allow the whole family to move out of the slum into a residential neighbourhood.

▼ **Figure 3** Overcrowded slum housing in Mumbai, the type of housing in which 8 million of its inhabitants live. Some of it is a lot worse than the housing shown here.

Activities

1 **(a)** Name and describe two environmental problems in LEDC cities.

   (b) Explain why they are worse than in MEDCs.

2 **(a)** From **Figure 3**, draw a labelled sketch of a shanty town.

   (b) Describe how some shanty towns are worse than this one.

3 **(a)** List the health problems of shanty-town dwellers.

   (b) Explain **(i)** why there are so many health problems and **(ii)** why infants and children are most at risk.

4 One person's view: 'The simple solution to urban problems such as shanty towns is birth control.'

   (a) Explain this person's view.

   (b) Why is it not the simple solution this person suggests?

Attempts to solve urban problems in an LEDC city

Cairo is an example of a big city with all the urban problems referred to on the previous pages (especially traffic congestion, sewage and housing). It is the largest city in Africa, with an official population of about 10 million but an estimated total of at least 16 million. Growth took off in the 1960s and shows no sign of slowing down (**Figure 1**). Despite the usual shortage of funds, the Egyptian authorities have implemented a number of projects to tackle its urban problems.

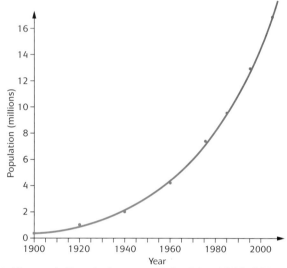

▲ **Figure 1** Population growth in Cairo 1900–2005

Traffic congestion

The scene in **Figure 1** on page 152 was taken after a massive new ring road was built around the city – imagine what it was like before! At least not everyone is forced to face the nightmare of travelling on Cairo's congested and fume-filled roads. The construction of the Cairo metro, the first in Africa, has been a great success (**Figure 2**). Two lines are already open and

a third is under consideration. The train services are well organized and quick, and offer air-conditioned comfort; the stations are clean and welcoming with televisions on the platforms to watch while waiting for a train. The metro is used by about 2 million commuters a day.

Sewage

The Greater Cairo Sewage project provides aid to repair the city's crumbling sewers, some of which date back to 1910, when the population was only half a million. The main work is to extend the sewers into areas currently without any; these are mainly the poor neighbourhoods where sewage on the streets is a common sight. At least Cairo has an efficient system of street collection. The Zabbaleen people, and their donkey carts, are the traditional collectors of rubbish from wealthy neighbourhoods (**Figure 3**). Their licence to operate has been officially extended as collectors and recyclers for some of the slum housing areas as well.

▲ **Figure 3** The Zabbaleen people _ official refuse collectors in Cairo's slums.

Housing

As elsewhere in the developing world, this is the biggest problem faced by urban authorities. The nature of the housing problems might be slightly different in Cairo from most other LEDC cities – there are no shanty towns as such, but brick-built houses are built illegally on state-owned irrigated green land next to the River Nile, which the government had reserved for food supply. These 'informal' houses now cover 80 per cent of the land area of Greater Cairo. But otherwise the problems

◀ **Figure 2** The Cairo metro runs underground in the centre but on the surface in the suburbs.

of overcrowding and numbers are the same, with an estimated 2–3 million people having set up homes in the 'Cities of the Dead' among the tombs in the cemeteries of Old Cairo and in homemade huts on the rooftops (**Figure 4**). Every Cairo citizen has an average living space of less than 2 square metres; the overall population density is 33000 persons per square kilometre. Urban sprawl, and accompanying air pollution, have already extended the borders of the city into the desert up to the Great Pyramids and Sphinx at Giza, which are being badly eaten away.

The long-term plan is for over 40 new settlements, many of them new towns, capable of housing 15 million people. They are to be located in the desert, away from the fertile irrigated land in the Nile Valley and Delta, which is needed for farming (**Figure 5**). One example is Sixth of October City, west of Cairo. To look at, it is not a very attractive place, dominated as it is by multi-storey apartments, many of them identical. However, included in the plans are open spaces and gardens, mosques, junior schools, shopping centres and industrial zones. It was considered essential to provide modern services and work opportunities if people are to be persuaded to leave Cairo. The Mercedes car factory has been established here for several years, and employs about 400 people, which is helping to attract others. As further attractions to new residents, loans and mortgages are offered on favourable terms, while a new super-highway provides a link to Cairo, only 30 minutes away in good traffic conditions. Some of the critics say that the plan favours the middle classes, who can afford the rents. They maintain that the government should not be spending vast sums on new settlements when more than 3 million homes in Cairo are poorly connected to public services.

▲ **Figure 4** Old Cairo is overcrowded.

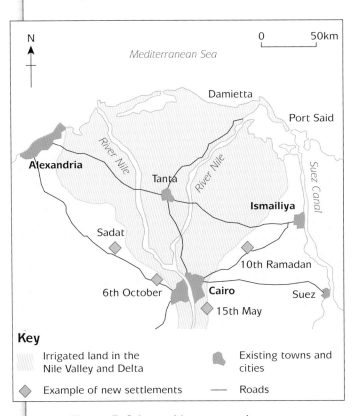

Key

▨ Irrigated land in the Nile Valley and Delta

◆ Example of new settlements

◆ Existing towns and cities

— Roads

▲ **Figure 5** Cairo and its new settlements in the desert.

Activities

1 **(a)** Describe what **Figure 1** shows.

(b)

	Average personal income per year (Egyptian £)	Infant mortality rate (per 1000)
Cairo	3461	50
Rural Egypt	2353	68

(i) Calculate the size of the differences between Cairo and rural Egypt for the figures in the table above.

(ii) Explain why there is likely to be no breathing space in growth to allow the authorities to deal with the already existing problems of Cairo.

2 **(a)** Describe **(i)** the housing problems of Cairo and **(ii)** the policies to solve them.

(b) Read the following comments from residents of 6th October city. What do they suggest about the success of the housing policy for Cairo so far?

I am pleased to have moved – I couldn't stand the pollution, traffic, crowds, and noise of Cairo.

All our good friends still live in Cairo

We were promised factories and work. There are none.

Apartment rents are too high. I am taking my family back to Cairo. I know we can stay more cheaply on a friend's rooftop.

Shanty towns – are self-help schemes a solution?

The idea of management of the housing problem is alien to some city authorities in LEDCs. From time to time, squatter and shanty settlements are demolished and cleared, making the residents homeless once again. Now it is more widely agreed that bulldozing these settlements is no solution; at the same time the authorities do not have the money to rehouse all the people. One policy is Aided Self-Help (ASH). Having noted that most shanty towns improve with time, it was realized that a few changes could help and speed up the process of improvement. ASH involves:

- giving shanty-town dwellers legal titles to the land

- connecting them to essential services such as water, electricity and roads

- providing building materials, technical help and loans.

Community participation is vital because families and neighbours do the work; they are responsible for building houses on allocated plots. Successful schemes encourage and reinforce the community spirit that often exists among people living in difficult circumstances. One example of this type of scheme is in São Paulo in Brazil, where simple single-storey homes with a water tank and indoor bathroom are built with cheap building materials such as breeze-blocks (**Figure 1**). The status of the housing areas improves and eventually they can become lower-class / low-income urban residential districts (**Figure 2**). The effects of some social problems decrease as people's health improves. For their success, such schemes depend upon cooperation and finance from city authorities, which some flatly refuse to contemplate.

POOR MIGRANTS FROM RURAL AREAS

↓

SQUATTER SETTLEMENT
Poor shacks

Gain work: have some income to improve houses

↓

SHANTY TOWN
Self-built houses

Regular work
Community action
ASH
Legal titles
Public services provided

↓

LOW INCOME RESIDENTIAL DISTRICT

▶ **Figure 2** How squatter settlements and shanty towns can change over time.

Electricity wires
Roofing tiles
Breeze blocks
Water tank
Bathroom with toilet
Sink
Streets improved
Living and sleeping quarters with concrete floor
Underground sewer
Improved roads

▲ **Figure 1** Self-help scheme in São Paulo, Brazil.

Exam focus

1 a State two push and two pull factors that are causing people to migrate from rural areas in large numbers in LEDCs.   (4 marks)

 b (i) Describe the main differences between a squatter settlement and a shanty town.   (2 marks)

 (ii) State two ways in which they are similar.   (2 marks)

 c (i) With reference to a case study, describe how self-help schemes for improving shanty housing work in LEDCs. (5 marks)

 (ii) State one advantage and one disadvantage of self-help schemes compared with the building of new towns (as in Cairo).   (2 marks)

 d (i) Give two reasons why many shanty towns are located on the edges of cities in LEDCs.   (2 marks)

 (ii) In MEDCs, why are the houses of wealthy people usually found on city edges?   (3 marks)

 e Explain why urban sprawl is a problem in all big cities, irrespective of whether they are located in the developed or developing worlds.   (5 marks)

Chapter 10

Agriculture

Wheat growing in the British countryside. Why did the farmer choose to grow crops on the land?

Key Ideas

The farm as a system:
• farming is a system with inputs, processes and outputs.

Agricultual activity varies from place to place:
• physical factors (relief, soils, climate) and human factors (market, capital, labour, politics) result in different farm systems
• farming systems may be commercial or subsistence, intensive or extensive
• there is a wide range of farm types in the UK.

Agricultural change can have advantages and disadvantages:
• food output has increased with the use of irrigation and high-yielding varieties of seeds
• intensification of farming has environmental consequences such as soil erosion and salinization
• for sustainable development in farming, greater use of techniques of soil conservation everywhere and appropriate technology in LEDCs are needed.

Farming systems, types and factors

The farm as a system

Farming is a business and it may be studied as a system with a series of inputs, processes and outputs:

INPUTS → PROCESSES → OUTPUTS

Inputs are what goes into the system. They may be physical inputs, such as the climate, relief and soils, or human inputs, such as labour, capital, machinery and seeds. **Processes** are the activities on the farm which turn the inputs into outputs. **Outputs** are the products of the system, such as crops, wool, meat and money. The outputs should be greater in value than the inputs so that a profit is made. **Figure 1** shows a range of possible inputs, processes and outputs in farm systems.

Inputs	Processes	Outputs
Physical	Planting	Cereals, e.g. wheat, corn
Relief	Ploughing	Vegetables, e.g. potatoes,
Soils	Spraying	cabbage
Climate	Harvesting	Oilseed rape, sugar beet
	Crop storage	Fruit, e.g. apples,
		strawberries
Human	Shearing	Eggs
Labour	Dipping	Milk
Machinery	Milking	Cattle, pigs, sheep
Fertilizers	Feeding stock	Wool
Animal feeds		Poultry
Money		Hay and straw
		Animal feedstuffs
		Wastes, including manure

▲ **Figure 1** The farm as a system.

Types of farming

From the names of the farming types it is usually possible to detect whether the farming is arable (growing crops), pastoral (keeping animals) or mixed (farming crops and animals). In the UK we are used to commercial farming – the farm outputs are sold either in a local market or to merchants and food companies. In many LEDCs, however, subsistence farming is more common; the farmer is working to feed himself and his family. If there is a surplus of produce for sale in a local market, that is a bonus (**Figure 2**).

Intensive and **extensive** are specialist farming terms. When farmers are attempting to obtain the maximum possible output from each hectare of land, the farming is described as intensive. This is done by a great use of

inputs, either by heavy financial investment or much use of labour, or both. Usually the quality of the farmland is too good to be wasted. In many countries very good land is in short supply; each farmer has only a small area to farm and has to make the most of it. In contrast, extensive farming is usually practised where large areas of low- or poor-quality land are found. A large area of land is required for the farmer to make a living.

▼ **Figure 2** Local market in a small town in Ecuador.

▼ **Figure 3** Summary of farming types.

Farming types	A	P	C	S	I	E	Key
Arable farming in East Anglia	✔		✔		✔		A = arable
Hill sheep farming in the Lake District		✔	✔			✔	P = pastoral
Market gardening/bulb growing in the Netherlands	✔		✔		✔		C = commercial
Shifting cultivation in the Amazon Basin	✔			✔		✔	S = subsistence
Rice growing in South East Asia	✔			✔	✔		I = Intensive
							E = extensive

Factors affecting farming

Deciding whether to grow crops or rear animals is affected by the physical conditions and human factors in an area (**Figure 4**). The physical conditions are often the most important. Sometimes the farmer may have several choices, especially in an area with fertile soils and a temperate climate. In other areas, the choice may be limited by infertile soils, lack of water or a short growing season.

PHYSICAL FACTORS

Relief

Mountainous areas have steep slopes and frequent rock outcrops. Farm machinery is difficult to use on steep slopes; the lowlands with gentle relief are more easily farmed.

Soil

Crops grow best on deep, fertile, free-draining soils, while less fertile soils prone to waterlogging are best used for pastoral farming. In the upland areas of the UK soils are thin, acidic and often rocky, and are unable to support crops, whereas in the lowlands the richer, fertile brown earths are more suited to agriculture.

Climate

Rainfall

Grass grows well in the west of the UK where rainfall is higher (above 750 millimetres); dairy farming is an important type of farming. Wheat and other cereals do not like too much rainfall; they grow best in eastern England.

Temperature

Since grass and most plants stop growing when temperatures fall below 6°C, the number of months with average temperatures above 6°C determines the length of the growing season. Grass grows longest in south-west England, favouring dairy farming. Warmest summer temperatures are in south-west England, favouring crop growing.

HUMAN FACTORS

Market

UK farms are commercial and proximity to market is an advantage. In the past, farms close to towns and cities in the UK often concentrated on dairying and market gardening in order to supply fresh foods quickly to the large urban population. Today, many farmers have contracts with supermarkets, e.g. Sainsbury's, and other food companies such as Bird's Eye which guarantee a market for their produce.

▲ **Figure 4** Factors affecting farming, with particular reference to the UK.

Finance

The profit a farmer makes affects the amount that can be invested in machinery, fertilizer, seeds and animals, and how much the farmer can afford to pay in wages for the family and any farmworkers.

Labour

This has become less important in the UK as machines for milking, harvesting etc. have become bigger and better. Certain types of farming still need large numbers of seasonal workers, e.g. to pick fruit and vegetables. In contrast, farming remains much more labour intensive in LEDCs.

Politics

The EU has a strong influence on farming according to how it distributes subsidies to farmers. At various times in the past it has supported the growth of different crops such as sugar beet and oilseed rape. Now the focus is being switched from farming support to farmers' stewardship of the countryside

Choice

Tradition is important in rural communities; many farmers practise the same type of farming as their parents did. Some, like hill sheep farmers, have little choice because of difficult physical conditions (**Figure 5**). Other farmers change according to market prices – what is most profiitable. Change can be expensive in terms of new buildings and machines.

▲ **Figure 5** Hill sheep farming dominates in uplands such as the Lake District where the poor physical conditions make crop growing difficult.

Activities

1 Draw a labelled sketch of a farming system for either arable or pastoral farming in the UK.

2 Explain why physical factors are usually more important than human factors in affecting types of farming in the UK.

3 **A** Some UK farms are businesses owned by companies.

   B On some UK farms farming is more a way of life than a business.

   (a) Which is likely to be the most important human factor for farms in **(i)** case **A** and **(ii)** case **B**?

   (b) Explain your answers.

Farming in the UK

In the UK many farmers combine a number of different activities on their farms. However, as **Figure 1** shows, in certain areas particular crops or types of livestock are dominant.

Arable and mixed

Important arable areas are overwhelmingly concentrated in the east and south of the UK (**Figure 1**). Physical conditions encourage cereal and root-crop cultivation. The **relief** (**Figure 2**) of the land is mostly less than 200 metres above sea level and flat or gently undulating. Large machinery such as combine harvesters can easily be used. **Soils** are mostly fertile boulder clay or loams, which are free-draining and easy to plough. The **climate** is a major influence on where cereals are grown. In the south and east of England summer temperatures are high, averaging over 16°C (**Figure 3**) and there is ample summer sunshine to ripen the crops. Where physical conditions are not quite so ideal for crop-growing further west in the Midlands and north in south-east Scotland, animal numbers (mainly beef cattle and sheep) increase, and mixed farming takes over.

Hill sheep

The main sheep-rearing areas in **Figure 1** show a close relationship with land above 200 and 400 metres in **Figure 2** in England, Wales and southern Scotland. Sheep are sturdy and sure-footed animals, able to cope with steep slopes and uneven terrain. Soils in the uplands are often thin, acidic and infertile, unsuitable for crop growing. In many areas the land is covered with

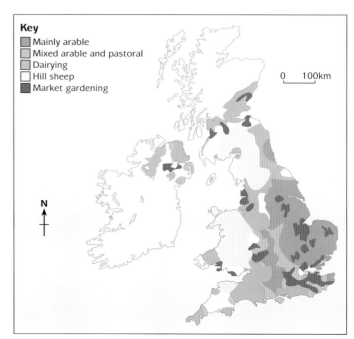

▲ **Figure 1** Five different types of farming in the UK.

moorland vegetation, including heather, bracken and rough grassland. The sheep graze on the open moors and hills where other livestock such as beef and dairy cattle could not survive because of the poor quality of the grassland. In the uplands, winters are cold, often with snow, and summers are also cooler than in the lowlands. Rainfall is high, often over 1000 millimetres a year, and there is much more cloud cover.

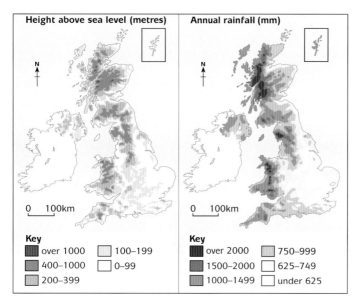

▲ **Figure 2** Relief and rainfall in the UK.

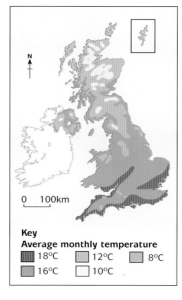

▲ **Figure 3** July temperatures in the UK.

Dairying

This is the farming type that dominates in certain lowland areas in the west (**Figure 1**). Some of the better-known English cheeses are named after counties located in the west like Cheshire and Gloucestershire. Here high rainfall (**Figure 2**) and a long growing season favour good grass growth. Cows can graze outdoors longer than in the east (see **Figure 2** on page 88), helping to keep milk yields high and reducing the amount of winter feed (usually silage and hay). The Friesian breed of cow shown in **Figure 4** is the specialist milk producer. Notice how the whole landscape is dominated by fields of grass, the main land use on farms in the western lowlands of Britain. Fresh milk is a perishable commodity; however, in these days of faster road transport and quicker distribution by supermarkets from dairies to stores, a nearby urban market is not as important a factor as it used to be.

▲ **Figure 4** Dairy cattle.

Market gardening

This type of farming is very intensive. Most market gardens are small in area but have very high inputs of labour and capital. There is a large labour input in planting, irrigating, fertilizing, spraying and harvesting the crops. Expenditure is also high on fertilizers, seeds, sprays, and heating and lighting for greenhouses. **Figure 1** shows that market gardening has the most varied and widespread distribution of the five farming types. This is partly because of the multitude of different vegetables, fruits and flowers that are grown, all with their own physical conditions for growth. The temperature needs of Scottish raspberries in late summer, for example, are very different from those of Cornish daffodils in early spring. Another reason is the highly perishable nature of the outputs (much greater than for fresh milk). Many need to reach shops and customers within 24 hours. Closeness to urban markets helps to explain their presence on the fringes of Greater London and Greater Manchester.

Overall, there are more market-gardening areas in the south and east of England than elsewhere, to take advantage of the combination of gentler physical conditions and the wealthiest and the most concentrated

market in the country. Flat low-lying land, a variety of deep fertile soils and warm sunny summers are favourable factors. Soils on the silty Fens around the Wash are light, fertile and stone-free; they warm up quickly in summer, encouraging seeds to germinate. Food-freezing and canning companies have factories dotted around the Fens and throughout East Anglia. All roads lead to the almost 8 million-strong London market.

▲ **Figure 5** Market gardening.

Case Study of farming in the UK

Arable farming in East Anglia

East Anglia (**Figure 1**) includes the counties of Norfolk, Cambridgeshire, Suffolk and the northern part of Essex. It is the most important arable farming region in the UK mainly because of its physical advantages of climate, relief and soils (**Figure 2**). The farming is both intensive and commercial. Most farms are large, over 200 hectares, and highly mechanized using huge combine harvesters and specialist machinery for sowing and harvesting oilseed rape and sugar beet. The crops are **cash crops**, sold for profit to the many local mills and factories. Wheat is bought by flour mills and used for cakes, flour and animal feed. Barley is used to make malt for the brewing and distilling industries as well as being sold as animal fodder. Carrots, peas and potatoes are sold to local freezing and canning factories. Sugar refineries buy up large quantities of sugar beet.

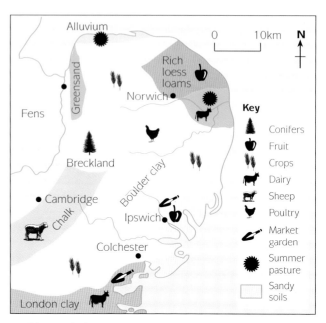

▲ **Figure 1** Soils and land use in East Anglia.
Source: Reproduced by permission of Edward Arnold

▼ **Figure 2** Physical conditions for farming in East Anglia.

Physical conditions	Impact on land use
Relief	
Flat or gently sloping	Easy to use machinery
Lowland, most less than 100m above sea level	Roads and railways have been easy to construct
Soils	
Mostly fertile boulder clay soils left behind by the ice sheets at the end of the Ice Age	Fertile soils suitable for growing cereals, sugar beet and potatoes
Loam soils	Very fertile, good for vegetables, fruit and cereals; retain plant foods and moisture, easily penetrated by roots
Waterlogged soils and marshland of the Norfolk Broads	A little dairying
Infertile sandy soils, e.g. Breckland	Some areas planted with pine trees. Others are heavily fertilized and reclaimed for cultivation
Climate	
Rainfall 500–700mm per year	Generally sufficient for crop growth
Warm summers, 17°C	Crops grow well and ripen
Long hours of sunshine in summer	Rapid growth; ripens cereal crops

East Anglia also benefits from some favourable human factors. It is situated in the east of England to the north of London. That densely populated region provides a large and wealthy market for the produce. In the south of the region more farmers specialize in market gardening and dairying. The flat relief has enabled a dense network of communication links to develop, such as the main East Coast railway, the M11 link to London and the A1, which allow the rapid transport of the produce. Since joining the European Union many of the farmers in East Anglia have benefited from the Common Agricultural Policy (CAP). They have received subsidies for growing certain crops, e.g. wheat, oilseed rape and linseed.

Recent changes and problems
The changes shown on the farm in **Figure 3** illustrate how greatly farming in East Anglia has changed since 1950. Increased farm unit size and specialization in crop-growing are associated with the growth of **agri-business**. Farming in East Anglia is now big business. Many farms are owned by large companies and run by a farm manager with an office packed with computers and other hi-tech equipment. Food-processing firms, supermarkets and insurance companies are among the farm owners. All businesses aim for maximum returns from their

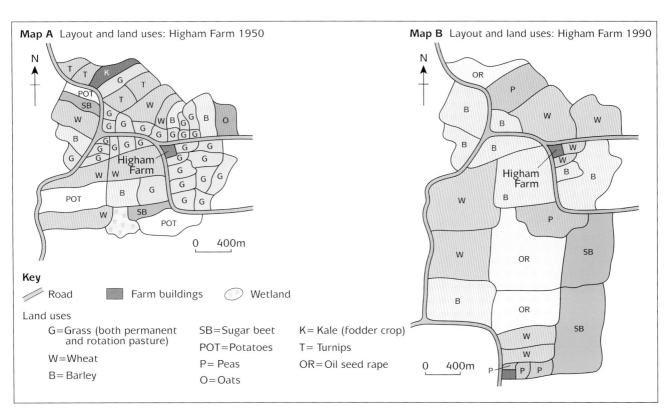

Map A Layout and land uses: Higham Farm 1950

Map B Layout and land uses: Higham Farm 1990

Key

Road   Farm buildings   Wetland

Land uses

G=Grass (both permanent and rotation pasture)

W=Wheat

B=Barley

SB=Sugar beet

POT=Potatoes

P=Peas

O=Oats

K=Kale (fodder crop)

T=Turnips

OR=Oil seed rape

0   400m

▲ **Figure 3** Changes on an East Anglian farm between 1950 and 1990.

investments. Therefore farming is intensive with high inputs of machinery and chemicals (fertilizers, pesticides and herbicides). Hedgerows were removed to allow access and easy use of ever-larger tractors, combines and other farm machinery. Farm units needed to be made larger to benefit from economies of scale. On some farms there are few workers – all the main tasks (ploughing, sowing and planting, spraying and harvesting) are undertaken by outside contractors, who can maximize the heavy investment in machinery by greater use.

There are few economic problems in farming in East Anglia: CAP subsidies have tended to favour farmers with large outputs (page 166). The same cannot be said about environmental problems (pages 166–7 and 174–5). The rapid drive towards increased agricultural productivity and output was made with little thought for the environmental consequences, such as loss of wildlife habitats from hedgerow clearances and contamination of ground and river water by chemicals used in pesticides and inorganic fertilizers.

Activities

1 (a) Draw a small labelled sketch map to show the main areas used for arable farming in East Anglia.

   (b) Describe the main features of relief, soils and climate in East Anglia.

   (c) Explain why physical (environmental) factors are better for growing cereal crops in East Anglia than in any other part of the UK.

2 Study **Figure 3**.

   (a) Describe the main changes in (i) farm size, (ii) field layout and size and (iii) land uses on this farm from 1950 to 1990.

   (b) State the evidence which shows that this farm was (i) a mixed farm in 1950 and (ii) an arable farm in 1990.

   (c) Explain (i) the economic advantages and (ii) the environmental disadvantages of the changes.

   (d) Suggest what further changes might have been made on this farm since 1990.

Case Study of farming in the UK

Hill sheep farming in the Lake District

Physical		Human	
Relief	– high mountains, steep slopes and little flat land.	Market	– a small local market and difficult access along narrow twisting roads with many steep gradients
Soils	– thin, rocky, acid and leached (podsols).	Capital	– there is often little profit to reinvest so farming methods stay the same
Climate	– temperatures fall by 1°C for every 160m above sea level. Temperatures are much lower than in the lowlands (14–15°C in summer). The growing season is short and there is high rainfall, over 2000mm on the Fells. There is more cloud cover, snowfall and hill fog.	Labour	– farmers can rarely afford more than one paid worker; many farms are family-run.
		Political	– farmers rely upon subsidies for a minimum standard of living, but these have been reduced for hill farmers in recent years
		Choice	– farming is very hard; it is a way of life and most are born into it. It retains the 'One man and his dog' image.

▲ **Figure 1** Factors affecting farming in the Lake District.

In the upland areas of the UK, a variety of *physical and human factors* (**Figure 1**) combine to make the grazing of sheep the main farming activity. Breeds of hill sheep such as the Swaledale (**Figure 2**) can survive the extremes of weather and the poor-quality pasture. The hill farmers make a living by selling the wool and the fattened sheep and lambs for meat. The hill farms cover a large area of land and use little machinery and labour. This makes hill farming one of the most extensive forms of pastoral farming.

Hill farms in the Lake District have three zones of land use. These are shown in **Figure 1** on page 62. During the summer the sheep graze on the **open** fell which begins about 300 metres above sea level. The farmer rarely owns this land but shares the grazing rights with other farmers in the parish. The land is open moorland with no fences or dry-stone walls. The coarse grasses mixed with heather, bracken and marsh give poor-quality grazing land.

The lower slope of the fell, called the **intake**, is often owned by the farmer. The land is enclosed by dry-stone walls and some of the pasture may have been improved by adding drainage and fertilizers. The slopes are still exposed and moderately steep, with acid soils leached by the heavy rain. The best land, the **inbye**, is on the valley floor but it is only a small part of the total land area. Here the soils are more fertile and the land more sheltered. The intake and the inbye are used to graze the sheep during bad winter weather when they are also fed supplementary feeds of hay, silage and turnips. The sheep are also here during lambing and shearing when the animals need to be closer to the farm. Parts of the inbye may be used to grow crops of hay and turnips which provide the winter fodder. Some hill farmers also keep small herds of dairy or beef cattle which graze on the valley floor or lower slopes. A milk herd is a useful addition to the farm, providing a weekly income to the farmer. On many hill farms the organization of the farm revolves around maintaining the sheep and cattle which together provide over 90 per cent of the farm income.

▲ **Figure 2** The hardy breeds of sheep are sure-footed on the rugged terrain in the Lake District.

Recent problems and changes

Hill farming is a marginal economic activity – it is not always profitable. A severe winter can deplete the stock of breeding ewes; changes to EU subsidies can lead to bankruptcy. In recent years farms have become less profitable as costs for inputs have increased (fuel, machinery and fodder) while for the main outputs lamb prices have stayed low and wool prices have collapsed. In 2004 the average farm in Cumbria made a profit of only £5000. As older farmers retire, fewer young people are willing to carry on the hill farming tradition. More attractive work and higher wages can be found in the towns and cities or in tourism in the Lake District National Park. The tourists also cause conflict with the farmers. The vast majority of visitors to the Lake District are sensible and follow the Country Code, but a small minority leave gates open, cause sheep worrying by allowing their dogs off the lead, break down the dry-stone walls, camp indiscriminately and leave litter. In addition, many farmers are at odds with the National Park Authority because they feel that they are not allowed to make progress and to invest in their farms because of the strict planning regulations.

Many farmers have seen their profits fall and feel threatened by sheep quotas and reduced grants. As a result of this, they have been forced to diversify.

In a National Park like the Lake District, farms have started to provide facilities for visitors, such as campsites, holiday homes in converted barns, bed and breakfast, farm visits, pony-trekking and farm shops (**Figure 3**). Other types of diversification also exist, such as fish farms, small workshops and farm woodlands.

Recent CAP reforms have seen further cuts to hill-farm subsidies. Instead, payments have been switched towards stewardship of the countryside, for example for farmers to restore dry-stone walls. They are also paid for keeping natural meadowland organic (free from chemical fertilizers and herbicides) and for delaying hay-making to allow time for flowering and seeding, as well as for birds to complete their nesting. This is to encourage greater **biodiversity** of plants and animals.

Some farms, especially smaller ones, have proved uneconomic and could not survive. Some were amalgamated with farms nearby while others were sold in separate lots. The buildings were sold for high prices as second homes while the land was bought by other farms or by the forestry companies for plantations.

▲ **Figure 3** Diversification in the Lake District?

Grow crops?
Oilseed rape, sugar beet, but in this climate?

Organic fruit and vegetables?
A growing market

Recreation?
Horse riding, pony trekking, bike scrambling

Plant conifers?
Good demand for paper and pulp, but no income until after 20 years

Stewardship?
Repair dry stone walls, cut meadows for hay not silage

Tourism?
Camping and caravan site, B&B

New animals?
Goats for milk, deer for venison

Farm shop/café?
Make own ice cream, meat pies

Activities

1 Describe the physical problems for farming in the area shown in **Figure 2**.

2 (a) From **Figure 1** on page 62, draw a cross-section to show the three parts of a typical Lake District farm.

   (b) Explain how and why land uses are different between the three parts.

3 Why do Lake District farmers struggle to make a good living from hill sheep farming?

4 (a) What is meant by 'diversification'?

   (b) From **Figure 3**, choose three types of diversification that are useful to Lake District farmers. For each one, explain how a farmer might be able to make money out of it.

   (c) Name one type of diversification from **Figure 3** that would be of little use to a Lake District hill sheep farmer. Explain your choice.

Farming issues in the UK

Agriculture is more important to the UK than its contribution to employment (2 per cent of workforce) and to national wealth (0.7 per cent of the total economy) would suggest. Food is an essential commodity and it is not surprising that the government considers agriculture important enough to have its own department DEFRA (Department for Environment, Food and Rural Affairs). Farming enjoys high levels of government financial support; DEFRA administers support policies that provide around £3 billion a year to UK agriculture from the EU budget under the CAP.

Intensification of farming and environmental damage

Until about 2000 CAP policy was focused upon increasing food output in order to make the EU as self-sufficient in food as possible. Farmers were given a guaranteed price for their crops; they were encouraged to produce more and more in the knowledge that the EU would buy up surpluses that could not be sold on the open market. This led to serious problems of overproduction, made famous by the cereal, butter and beef mountains and the wine and olive-oil lakes, which were at their highest and deepest in the early 1990s.

Cereal farmers in East Anglia had most to gain from CAP policy and they responded by farming even more intensively and treating farming as a business. Cambridgeshire had lost more than half its hedgerows between 1945 and 1960, due to the increased concentration on crop-growing and increased use of machinery (**Figure 1**). Many of the others disappeared over the next 20 years. What had once been a common sight in East Anglia has since become something of a rarity. Instead there is now an open landscape of large fields, a process known as 'Prairieization' because of its resemblance to the big open spaces on the Prairies of North America (**Figure 3**).

Hedgerows take up valuable agricultural land and limit the size of machinery that can be used (**Figure 2**). Apart from their beneficial effects upon landscape appearance, hedgerows are important wildlife habitats. Intensification of farming had other adverse environmental effects. Not only did the increased use of chemical sprays destroy insect and animal life, but the increased use of chemicals for fertilizers and pesticides is blamed for the pollution of the land and water supplies. Nitrates and chemicals, when used in large enough quantities, can seep into streams, speeding up the natural process of **eutrophication**, which eventually kills stream life, damaging the whole ecosystem (see also pages 174–5). Groundwater stores, upon which people in eastern England depend most for water supplies, are being increasingly contaminated with biological and chemical wastes.

▼ **Figure 2** A surviving hedgerow in East Anglia; there are not many of them.

▼ **Figure 3** The monotonous East Anglian landscape has been likened to that of the Prairies.

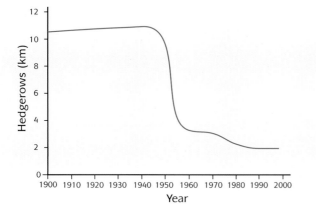

▲ **Figure 1** The decline in hedgerows in Cambridgeshire, East Anglia, 1900–2000.

Countryside stewardship – the new policy

When the new government department of DEFRA was created, significantly its role was widened to include environmental protection for wildlife and the countryside in general. Also the CAP has been partly reformed: many of the subsidies supporting the price of cereals and other crops have been taken away or much reduced. For many years the policy of set-aside has been in operation. Under this scheme arable farmers are paid for not growing crops on some of their land, typically between 5 and 20 per cent of the farm. Fields are left fallow (empty) and restrictions are placed upon when and how often weeds can be cut back (**Figure 4**). Although this policy was introduced for economic reasons, to reduce the large food mountains and lakes, it has been of enormous benefit to wildlife (birds, small animals and insects). Increasingly since 2000, CAP funds are being used directly to support environmental initiatives.

Under the Countryside Stewardship scheme, farmers sign agreements to manage their land in an environmentally beneficial way for up to 10 years, in return for annual payments. Additional grants are available towards associated capital works, such as hedge-laying and planting and repairing dry-stone walls. Those farms in the process of becoming organic (farming without chemicals, using only organic fertilizers such as animal manure), are given financial help in the conversion process. On organic farms insects increase; as these are near the bottom of the food chain, they allow an expansion in the number and variety of birds higher up the chain.

Arable farmers who still have hedgerows are paid to leave a wider margin between them and the crops, because hedgerows offer food and nesting sites to birds and homes to small mammals. Under the stewardship scheme, the cutting of hedgerows is restricted to once every three years; cutting must be done at the end of winter, after all food sources have been eaten by the birds. Environmentally friendly farm practices are rewarded as well. Leaving the stubble in cereal fields provides seed-eating birds with a food supply through their toughest time in mid-winter. Upland farmers are rewarded for cutting the meadows on the valley floors for hay instead of silage: hay-making is a later activity and gives birds time to finish nesting and flowering plants time to produce flowers and seeds. Flowering meadows have all but disappeared in most parts of England under the chemical onslaught, but for some farmers in upper Teesdale in the Pennines, who never adopted intensive modern methods, 80 per cent of farm income is now from grants for looking after wildlife. This leaves only 20 per cent from traditional sheep-rearing.

▲ **Figure 4** A field in set-aside; this farmer is being compensated for not growing crops in this year, thereby not adding to the food surpluses.

In the past, farmers were rewarded for being more efficient and growing more food. This could only be achieved by planting a new crop in a field as soon as harvesting was completed, and by using the whole area right up to the edges, which meant keeping any remaining hedgerows well cut back. Now all farmers are being given the chance to farm in a way that will help wildlife. Some farmers are cutting off squares from grazing within pastures. These are known as 'skylark blocks'. One field left at rest from cultivation for a year, as in set-aside, has been described as a 'giant bird table' because it provides bird feed all year round.

Activities

1 **(a)** Describe the landscape changes in East Anglia since 1945.

   (b) State one advantage and one disadvantage of removing hedgerows.

2 **(a)** What is meant by set-aside?

   (b) Describe how **Figure 4** shows it.

   (c) Does the policy of set-aside make sense? Explain your view.

3 **(a)** What is meant by countryside stewardship?

   (b) Describe some of the policies associated with the scheme.

   (c) Explain why the UK government and the EU support the scheme.

Market gardening and bulb growing in the Netherlands

The Netherlands is one of the most *intensively* farmed countries in the world. In part this is due to the skill of the Dutch farmers, but it also reflects the high land values. The Netherlands is a densely populated country where land is scarce. Farming must be profitable to compete with other land uses.

Market gardening and bulb growing are concentrated in a strip of land behind the coastal sand dunes in the west of the Netherlands (**Figure 1**). The relief is low-lying and flat, and soils are very light and sandy. The soils warm up and dry out quickly, which helps seeds to germinate; free drainage is important for bulbs such as tulips and hyacinths. The most important area for market gardening is Westland (**Figure 1**). Seen from the air it is a 'sea of glass' broken by slender chimney stacks. The area is famous for the production of tomatoes, cucumbers and lettuce, although in recent years over 60 per cent of the farmers have diversified into cut flowers and pot plants, which are more profitable (**Figure 2**). The change was necessary to cover the rising costs of heating the greenhouses, and severe competition in the EU. A considerable proportion of the flowers, bulbs and plants are *exported* to other European countries and to Canada and the USA. Although market gardening in the Netherlands occupies only

▲ **Figure 2** Bulb growing in the Netherlands – the flowers attract many tourists in the spring but they are cut off quite early so that the bulbs are not exhausted.

5 per cent of the farmland area it accounts for 25 per cent of the value of Dutch agricultural exports.

Market gardening is both *labour-intensive* and *capital-intensive* and many holdings are less than one hectare in size. The farming is successful due to the intensive methods, the high value of the crops and the use of technology. It is a sophisticated, **high-tech** type of farming. Growers are supported by scientific research and advisory services. There are many cooperatives which provide credit for the growers to buy fertilizers, irrigation equipment and other products. Cooperative auctions exist to sell the growers' output. The cooperative auction at Aalsmeer is the world's largest flower auction.

▼ **Figure 1** Farming in the Netherlands.

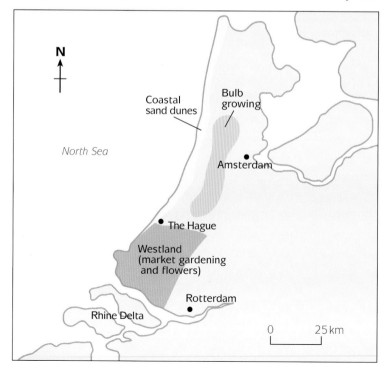

- N
- Coastal sand dunes
- Bulb growing
- North Sea
- Amsterdam
- The Hague
- Westland (market gardening and flowers)
- Rotterdam
- Rhine Delta
- 0   25 km

Activities

1 Using **Figure 1** and the text, complete a table showing factors affecting market gardening in the Netherlands. Use the headings Physical factors and Human factors.

2 What is intensive farming?

3 Why is farming so intensive in the Netherlands?

Shifting cultivation in the Amazon Basin

Shifting cultivation is a traditional form of subsistence agriculture once found in many areas of tropical rainforest, such as the Amazon Basin. Today this extensive form of agriculture is only found in the inaccessible and least 'exploited' areas of rainforest.

Characteristics of the farming

In the Amazon Basin, the Amerindians use **machetes** to clear about one hectare of forest. After time enough to allow the vegetation to dry, the trees and undergrowth are burned, the ash being used to fertilize the soil. This is also known as 'slash and burn'. In the clearings, called *chagras*, the women plant and grow crops such as manioc, yams, beans and pumpkins (**Figure 3**). The farming is all done by hand with only simple tools. The men supplement the diet by hunting for tapirs and monkeys, fishing and collecting fruits from the forest.

The soil very quickly loses its fertility. Once the tree canopy has been removed, the source of humus for the soil has gone and the heavy rains can strike the bare ground. This causes soil erosion and the leaching of minerals down through the soil (see **Figure 5** on page 107). Harvesting the crops removes more nutrients from the soil and after 4–5 years yields decline and the plots have to be abandoned. The tribe will then 'shift' to another part of the forest to begin the cycle all over again.

Shifting cultivation needs a high labour input and large areas of land to provide enough food for a few people. The abandoned clearings need to be left for over 20 years or more to allow the forest to regenerate and the soils to recover. Traditionally, shifting agriculture was in harmony with the environment. It is a wasteful method of farming but less harmful to the environment than permanent agriculture.

Changes

Recently, new developments in the rainforest have forced the tribes into smaller and smaller areas of land or into reservations. The Indians are being forced to clear for a second time patches of land which have not yet fully recovered. The soils are being damaged beyond repair, making it impossible for the rainforest to re-establish. Large numbers of Indians have also died from 'Western' diseases or been killed by developers. Many have decided to leave the rainforest and join other migrants in the shanty towns of the large urban areas. Therefore what was for thousands of years a sustainable type of agriculture is becoming unsustainable, due to commercial pressures on Brazil's rainforests from outsiders (pages 108–9).

▼ **Figure 3** Clearing the rainforest; it is difficult to see that crops such as manioc and bananas have been planted.

Activities

1 (a) Explain how shifting cultivation is
    (i) extensive and **(ii)** subsistence farming.

   (b) Show this type of farming as a system in a table or as a diagram, as on page 158.

2 (a) Explain why farmers practising shifting cultivation have low incomes.

   (b) Why is it difficult for them and their families to break out of the poverty trap?

3 (a) Why is shifting cultivation an example of a sustainable economic activity?

   (b) Describe the social, economic and environmental problems caused by changes to shifting cultivation.

   (c) Put together a five-point plan which would ensure that shifting cultivation remains a sustainable form of farming in the future.

Case Study of farming in LEDCs

Intensive subsistence rice growing in South East Asia

South-east Asia is one of the most densely populated areas in the world and rice is the **staple** or main food crop. Wet (padi) rice cultivation is uniquely linked to the monsoon climate and the high densities of population. Most farmers are *subsistence* farmers who grow the rice to feed themselves and their families. Rice grows best in areas with heavy **monsoon** rains or where ample **irrigation** water can be provided. The flood plain and delta of the River Ganges in India and Bangladesh (**Figure 1**) have many advantages for rice growing.

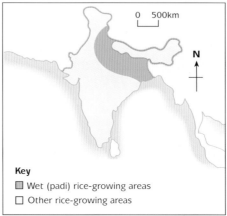

Key

⬜ Wet (padi) rice-growing areas
⬜ Other rice-growing areas

▲ **Figure 1** Rice-growing areas in India and Bangladesh.

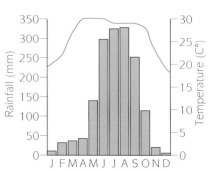

▲ **Figure 2** The monsoon climate.

The growing of rice is hard work and usually involves the whole family (**Figure 3**). In some years the farmer grows a second crop on the same land, either beans, lentils or peas. The farmers may also keep a few chickens for eggs and meat.

Physical factors

• A monsoon climate (**Figure 2**) with heavy rains and high temperatures provides ideal conditions for rapid growth of rice.

• Heavy alluvial soils provide an impervious muddy layer.

• Flat flood plains make the flooding of fields easier; terraced hill slopes can be used for 'dry rice' in areas such as Assam.

• There is a water supply from the River Ganges and from wells.

▲ **Figure 3** Planting rice, a labour-intensive activity.

Problems of rice growing

The River Ganges frequently floods and under normal conditions the floodwaters are useful, providing the water and fertile silt for the padi fields. Sometimes the floods are catastrophic, and they destroy the rice crop. In some years the monsoon 'fails'; rainfall is lower than expected and the rice crop is ruined.

Large areas are owned by big landowners and there are many landless peasants. In addition, traditionally when a man dies his land is divided among his sons.

Human factors

• Rice gives high yields per hectare, which helps to feed the large population.

• Water buffaloes are used for work and as a source of manure for the fields.

• Rice seeds are stored from one year to provide the next year's crop.

• Rice growing is labour-intensive.

• Many people can work in the padi fields ploughing, planting, harvesting and threshing.

The units become too small to support a family. India has a high rate of population growth and new methods of farming are needed to feed the population.

Changes to increase output

Both India and Bangladesh need to keep increasing rice output to feed their growing populations. One long-established method is using irrigation water to overcome the problem of irregular and sometimes inadequate monsoon rains. Several different techniques, both traditional and modern, are used to irrigate the land.

The monsoon rains in India are irregular and sometimes inadequate for rice to grow. The new HYVs and multiple cropping also require far more water. This has increased the importance of irrigation to water the land artificially. In the Ganges valley several different techniques, both traditional and modern, are used to irrigate the land.

1 Wells

Wells are a traditional method of irrigation in the Ganges valley. Holes are dug to reach the water table and each well can irrigate one or two hectares of land. It is a very cheap method of irrigation for many of the poor farmers. The Green Revolution is replacing traditional methods using the *shaduf* or *sakia* with modern electric or diesel pumps. Deep, modern **tube wells** are being dug. Electric pumps are used to bring water to the surface, and up to 400 hectares of land can be irrigated from one tube well.

2 Inundation canals

Inundation canals are used in the Ganges valley. These are canals dug on the sides of a river to lead water into the fields when the river level rises and floods. The canals are cheap to build and maintain, and they are able to bring valuable nutrients from the floodwater to the fields, reducing the need for expensive fertilizers. However, irrigation canals also cause problems of waterlogging and salinization through the upward movement of salts in the soil. The canals attract disease-carrying insects, and evaporation rates are high.

More recent, since the 1960s, is the use of high-yielding varieties (HYVs) of rice such as IR8. Their introduction was part of what is known as the Green Revolution (see page 173), the large increase in food production from the use of new seeds with yields three to six times greater than the old seeds. Some varieties of the new seeds grow faster. By ripening in four months instead of five, a second crop can be grown each year. This is called double or multiple-cropping. Many HYVs have shorter and stiffer stems, which make them more resistant to wind and rain, allowing more to be planted per unit area.

The Green Revolution has also been associated with modernization of farming such as:

- a greater use of fertilizers to ensure a continuation of high yields

- the replacement of the slow water buffalo by tractors and mechanized ploughs.

Grant and loans were made available to poor farmers to purchase new seeds and equipment. The up-side has been significant increases in rice output. The down-side is a widening gap between rich and poor farmers as successful farmers, often with larger than average areas of land, grow richer, while small farmers remain poor and saddled with debts they can never pay off.

Activities

1 State the evidence from these two pages to show that wet-rice growing is **(a)** subsistence and **(b)** intensive farming.

2 **(a)** Draw a labelled sketch of the farming scene shown in **Figure 3**.

   (b) State the differences between the methods used for growing rice in India and cereal in East Anglia.

3 **(a)** Describe the main features of the monsoon climate shown in **Figure 2**. (Use what is on page 85 to guide you about what to look for.)

   (b) Explain why the monsoon climate is a good climate for growing wet rice.

   (c) Describe the ideal relief and soils for wet-rice cultivation.

   (d) State where the best physical conditions for growing wet rice are found.

4 Describe **(a)** the economic advantages and **(b)** the environmental disadvantages of using irrigation water and high-yielding varieties of seeds on rice farms in India.

Agricultural change to feed the world

Despite many gloomy predictions, enough food is grown to feed the world's 6 billion plus people. The fact that in some parts of the world many people are suffering from malnutrition and some are dying from starvation is more to do with lack of food availability in the places where it is needed. Food surpluses exist in the developed world in Europe, North America and Australasia. Disruption of local food supplies is often only temporary for either economic or physical reasons, or a combination of the two (see pages 232–3). **Figure 1** shows how food production has been able to keep ahead of population growth, rapid though this may have been. What agricultural changes explain the great growth in global food output?

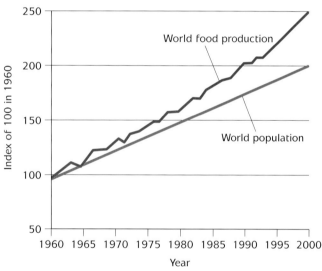

▲ **Figure 1** Trends in world food production.
Source: World Resources Institute

Agricultural changes to increase output
1 Irrigation – artificially adding water to farm land to increase output and extend the area of cultivation.

In the UK this is sometimes used by arable farmers in East Anglia to increase the yield of root crops such as potatoes; however, in a hot desert country like Egypt farming would be impossible without irrigation. The fact that Egypt can feed its 70 million people (and growing) is entirely due to control of the Nile at the Aswan High Dam in the south, which allows 8 per cent of the country to be farmed intensively. In India irrigation water is used to overcome the unreliability of the summer monsoon; it also allows a second or even third crop to be grown during winter and spring (page 171).

▼ **Figure 2A** The Aswan High Dam.

▼ **Figure 2B** Advantages of the Aswan High Dam.

* assured water supply throughout Egypt

* desert reclaimed for farmland

* cultivated area doubled from 4% to 8%

* 2 or 3 crops per year instead of 1

* no longer any risk of summer floods

* electricity supply for the whole country

▼ **Figure 2C** An oasis of cultivation in the Egyptian desert, impossible without the Nile water.

2 Chemicals – these are available for every farming need.

- Inorganic fertilizers replace natural shortages of one or more essential soil nutrients.

- Herbicides destroy weeds that compete with crops for water, light and nutrients.

- Pesticides kill crop-eating insects and animals.

Now it is often possible to specialize in growing just one crop on the land, whereas previously change was need to allow nutrients to recover naturally. Plantation farming is widespread in the tropics (**Figure 3**).

3 High-yielding varieties of seeds (HYVs) – mainly

cereals such as rice, wheat and maize, staple foods for billions of people. These were first developed by researchers and scientists in laboratories in the developed world from the 1950s onwards. Their widespread use in LEDCs since the 1960s, especially in Latin America and Asia, led to spectacular increases in food output which were called the Green Revolution.

Figure 3 Banana plantation in ▶ Costa Rica (Central America) with a much greater output per hectare than on small Caribbean farms (page 240).

4 Capital investment – money for agricultural improvements and machinery.

With modern machinery, farm tasks such as ploughing, planting, spraying and harvesting using tractors and combines can be completed more quickly and in a more controlled manner than by hand labour using primitive tools. Farms can take maximum advantage of periods of good weather. However, nothing comes free. This is why output has increased most on farms in MEDCs, aided by subsidies from governments notably in the USA and EU. Here they led to overproduction and food surpluses. In LEDCs increased output has mainly been achieved on large farms; not only are farm owners more wealthy, but they operate on a large-enough scale to make investment worthwhile. The gap between rich and poor farmers is widening (**Figure 4**).

However, the impressive increases in world production shown in **Figure 1** have not been achieved without heavy costs to the environment. Some of these are listed in **Figure 4**. The world's natural ecosystems are being disrupted and destroyed as more forests are cleared to make space for farmland. In the opinion of many people, one of the worst environmental impacts has been the clearance of tropical rainforests to be replaced by cattle ranches and plantations (pages 108–9).

Successes	Failures
For farmers	**For farmers**
• Farmers who could afford the HYVs and fertilizer increased yields by three times. • Faster-growing plants allowed multiple cropping each year. • Increased output created a surplus to sell in the cities, raising the farmers' income and standard of living and reducing the costs of imports.	• Many poor peasants could not afford to buy HYVs and fertilizer so their yields did not change. • Poorer farmers who borrowed money could not pay it back and their debts increased, forcing some of them to move to the city.
For the country	**For the country**
• Food production is 'revolutionized' and in some areas surpluses for export are created. • Less dependence on imported food and reduced risk of food shortages. • Reduced rates of rural–urban migration due to higher living standards in rural areas.	• Environmental problems increase from use of fertilizers and pesticides. • Salinization is a problem in areas of widespread irrigation.

▲ **Figure 4** Successes and failures of the Green Revolution.

Activities

1 (a) Draw a graph to show world population growth (billions).

     Actual: 1900 1.6; 1930 2.0; 1960 3.0; 1990 5.2. **Expected**: 2020 7.8; 2050 9.0.

(b) Describe what your graph shows.

2 Give the reasons for increased food production in **(a)** Egypt and **(b)** India.

3 Explain why the increases in food production have been greater **(a)** in MEDCs than in LEDCs and **(b)** on big farms than on small farms.

Human and environmental issues from increased food production

In the drive to intensify farming and produce more food, impacts on the environment are rarely considered. This happened in East Anglia (page 166). However, subsistence farmers, like the shifting cultivators in the Amazon referred to on page 169, must be environmentally aware in order to survive. Instead of contaminating the land by using inorganic fertilizers and compressing the soil with heavy machinery, subsistence farmers maintain soil fertility and reduce the risk of soil erosion by using animal manure and practising crop rotation. They expect some losses from pests (insects, birds and animals). For them, the problem is low food output, insufficient to support rapidly increasing numbers of people.

▲ **Figure 1** Subsistence farming in Peru – what are its advantages and disadvantages?

Inappropriate technology

Rising populations and changes such as the Green Revolution brought with them the need for more irrigation water and cheap sources of power. Some LEDCs followed the example of the developed world and built large dams for water supply and electricity. Although many benefits followed, as in Egypt (see **Figure 2** on page 172), a lot of these schemes had major disadvantages for local farming communities. Some of these are summarized in **Figure 2**; examples of those associated with the recently built Three Gorges Dam in China are given on page 210. The benefits from irrigation water and electricity supplied from large schemes are rarely felt by people living close by. Subsistence farmers cannot afford to buy or operate the electric or diesel pumps needed to gain access to the new water supply without going into debt; any increased yields are unlikely to be big enough to pay off loans, due to high interest rates.

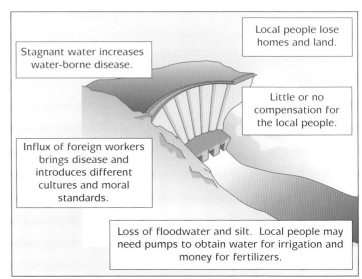

Stagnant water increases water-borne disease.

Local people lose homes and land.

Little or no compensation for the local people.

Influx of foreign workers brings disease and introduces different cultures and moral standards.

Loss of floodwater and silt. Local people may need pumps to obtain water for irrigation and money for fertilizers.

▲ **Figure 2** Disadvantages of large-scale dam projects.

Environmental impacts

Some of the negative effects of agricultural changes for increased output are summarized in **Figure 3**. What they all have in common is that they are caused, or made worse, by human abuse of the land and soil. Soil erosion is a good example to use to illustrate this.

Loss of topsoil by wind and water is a natural process – it is happening all the time, especially on steep slopes and in dry climates. In most wet climates it happens only slowly, due to the protective covering of vegetation. Generally new soil formation from weathering of surface rocks can keep pace and replace what is being lost. Therefore soil erosion is really 'accelerated soil erosion' speeded up by damaging human activities. The human activities that cause soil erosion include:

- clearances of forests for farmland so that there are no tree roots to hold the soil in place and no leaves and branches to break the force of falling rain

- bad farming practices such as ploughing up and down the slope because it is easier, thereby creating channels for easy rainwater flow down the slope

- overexploitation of land by monoculture (growing only one crop), or by grazing too many animals for the capacity of the land; all the grass is eaten, leaving patches of bare ground.

Misuse of soil leads to weakening of the soil structure, which makes its removal by wind and rain more likely.

A
Overuse of chemicals

Excess concentrations of nitrates in ground and river water used for drinking water supplies.

Enriched nutrient flows in rivers and streams, causing faster algal growth, which consumes oxygen supplies in the water, leading to the death of fish and other aquatic life.

B
Overcultivation and overgrazing of farmland

Soil erosion and degradation of land surface, making the land unfit for farming.

C
Overuse of irrigation water

Salinization is caused by salts drawn up to surface as moisture evaporates; after further evaporation, salt is deposited around plant roots and on the surface; most crops cannot tolerate high salt levels and die.

▲ **Figure 3** Adverse environmental effects of modern agricultural practices.

The situation goes from bad to worse once all plant remains are removed from a soil (see **Figure 4** on page 111).

One example of an area suffering from **salinization** (an increasing proportion of salt in a soil) is the Indus delta in Pakistan. A series of very big dam schemes in the upper parts of the Indus valley, intended to boost crop production, take too much water out of the river. By the time it reaches the delta the availability of irrigation water is too low to offset surface accumulation of salt, in a region with regular summer temperatures above 45°C. Rice farming is no longer possible in some delta villages, due to the hard crust of salt on the surface of once-rich farmland.

Activities

1 **(a)** Define the term 'subsistence farming'.

   (b) Describe the advantages and disadvantages of the farming in **Figure 1**.

   (c) Is it an example of sustainable farming? Explain your answer.

2 Explain why the risk of soil erosion in East Anglia is greater today that it used to be.

3 **(a)** Explain how **(i)** soil erosion and **(ii)** salinization can lead to cropland having to be abandoned.

   (b) Why do both occur widely in LEDCs with hot desert climates?

4 'Soil erosion and salinization are examples of natural processes speeded up by human actions.' Explain this statement.

◀ **Figure 4** The results of salinization in the Indus valley in Pakistan.

Sustainable options to feed the world

Behind the concept of sustainability for the future is the idea that, while tackling today's food needs, we should not be piling up environmental debts that will burden and limit the options for future generations. The future should not be jeopardized for short-term gains. Many believe that **appropriate technology** (a level of technology suited to use by local people in LEDCs) is the key, and that projects should be:

- small scale

- in line with the levels of wealth, skills and need of local people

- well suited to the local climate and environment

- giving long-term (i.e. sustainable) benefits.

For agriculture, using small dams and individual wells for water supply in the cultivation plots in the fields and in the vegetable gardens next to the houses is one example (**Figure 1**). Outside help is often needed to set up the scheme. If it is in harmony with the local environment and relies only on local resources and the skills of local people rather than on modern technology, it increases the chances that long-term sustainability will be achieved. Projects to harness renewable energy sources from wind, solar power or biogas (**Figure 2**), instead of expensive electricity and oil, support its aims.

One example of a successful local scheme, from a village in north-west India, is given in **Figure 3**. This project involved planting trees on bare slopes, harvesting (collecting) rainwater and trapping water in a surface river behind a small earth dam – measures which provided surface water and replenished groundwater. Local farmers and village populations in LEDCs need only relatively small amounts of water, which is why a 'mini-scale' like this one is an appropriate solution for their needs. Other examples of appropriate technology in agriculture include:

A *Trickle-drip irrigation* – water is directed at plant roots, which means that more is used by the plant and less is evaporated (**Figure 4**).

B *Organic fertilizers* – either animal manure or green manuring from compost in rice- and other cereal-growing areas.

C *Natural predators* – encourage an increase in the number of birds that feed on crop-damaging insects and small animals.

D *New seeds and plant varieties* – there are opportunities for research into a wider range of new seeds for crops other than cereals which are widely grown in Asia, Latin America and Africa, such as millet, legumes (beans and cowpeas) and roots (yams and cassava). Some argue that researchers in biotechnology have yet to make enough use of gene banks of local seed varieties for the benefit of people living in the developing world.

◀ **Figure 1** Appropriate technology in action.

▲ **Figure 2** A biogas plant.

Thunthi Kankasiya village	Before 1991	By 2000
Perennial drinking-water wells	0	23
River dams	0	1
Months of water availability	4	12
Land under cultivation (hectares)	85	135
Number of crops per year	0–1	2–3
Agricultural production (quintal/ hectare)	900	4000
Migration rate (15–40 year olds)	78%	5%
Average period of migration (months)	10	2
Income per household (rupees per year)	8600	35 600

▲ **Figure 3** Micro-watershed development in a village in Gujarat, India.
Source: *New Internationalist*

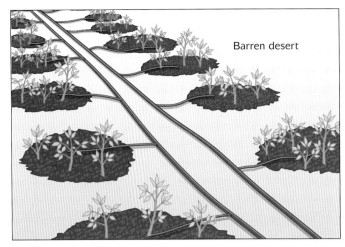

Barren desert

▲ **Figure 4** Trickle-drip irrigation – how does it reduce the risk of salinization?

Soil conservation

The aim is retention of fertile topsoil. **Figure 5** illustrates one example from each of the three categories below.

1 Mechanical methods

The most widely used method is terracing on slopes. Soil is held in place on the flatter land on the tops of the terraces, which are also easier to farm. On a smaller scale, embankments can be placed along the bottom of steep slopes to hold back soil and water.

2 Changes in farming practices

Contour ploughing (working around the slope) is one example: the ridges created by ploughing impede the free movement of water (and therefore soil) down the slope. Mixed cropping helps, especially if bush or tree crops are planted between ground crops – they continue to give the land in between some shelter from the wind at times when no crop is growing above the surface.

▶ **Figure 5** Methods of soil conservation – name the three ways shown.

3 Planting trees

Tree planting on slopes next to farming areas may need to be on too large a scale to be undertaken by individual farmers; a local or national government may have to take responsibility. In flat, low-lying areas lines of trees are often planted as windbreaks; the direction should be at 90° to the prevailing wind for maximum effect.

Exam focus

1 a State the differences between:

  (i) pastoral and arable farming   (2 marks)

  (ii) intensive and extensive farming.   (3 marks)

b Name one example of each of the following types of farming:

  (i) intensive arable, **(ii)** intensive pastoral, **(iii)** extensive arable and **(iv)** extensive pastoral.   (4 marks)

c Choose one example from the farming types named in **(b)**.

  (i) Describe the main characteristics of the type of farming.   (3 marks)

  (ii) Explain why it is intensive or extensive.   (3 marks)

d Study **Figure 1**, which shows a farming area in an LEDC.

▲ **Figure 1** Colca valley in the Andes of Peru.

  (i) State two reasons why this is an area with a high natural risk of soil erosion.   (2 marks)

  (ii) Explain how farmers in this area might increase the risk.   (3 marks)

  (iii) Describe what farmers have done to reduce the risk.   (2 marks)

  (iv) Is farming here intensive or extensive? Explain your answer.   (3 marks)

2 a Describe fully what is meant by the Green Revolution.   (4 marks)

b Three summary comments about the Green Revolution are given below.

  A Overall the Green Revolution has brought great benefits.

  B As with all changes, some people have benefited more than others.

  C As farming becomes more intensive, its environmental impact becomes stronger.

  Choose and explain two of these comments.   (6 marks)

c (i) Explain how irrigation water can be used to increase farm output.   (3 marks)

  (ii) Describe one environmental problem that results from overuse of irrigation water.   (2 marks)

  (iii) Explain how this problem can be reduced.   (2 marks)

d Study **Figure 2**, which shows farming in an area previously covered by tropical rainforest.

▲ **Figure 2** Farming in Costa Rica, Central America.

  (i) What suggests that this area was once covered by forest?   (1 mark)

  (ii) Describe the farming and the land uses shown.   (4 marks)

  (iii) How great are the environmental risks from farming in the area shown? Explain your views on this.   (3 marks)

Chapter 11
Industry

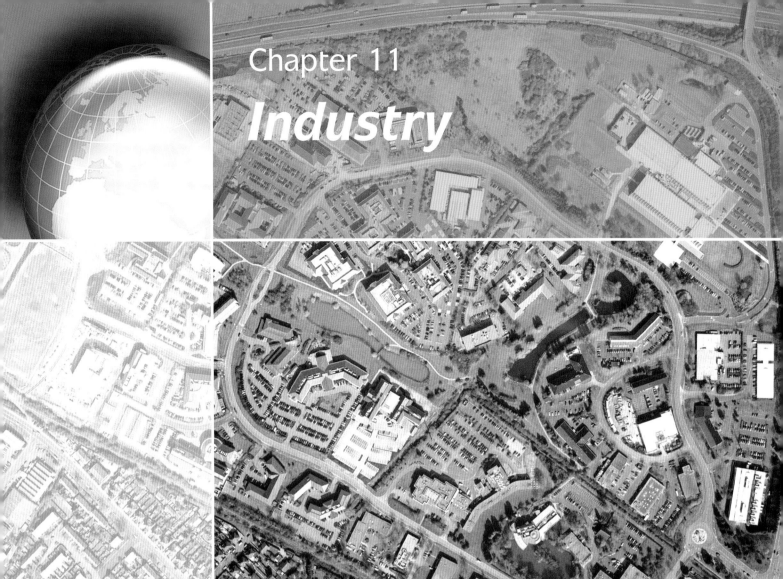

Cambridge Science Park, an example of a modern development on a greenfield site.

Key Ideas

Industrial activity can be classified:
• primary, secondary, tertiary and quaternary are the four types of industry.

Industry is a system:
• as with all systems there are inputs, processes and outputs.

Industrial location is influenced by many factors:
• locational factors include raw materials, energy, capital, labour, transport and government
• the relative importance of these varies according to the type of industry.

Industrial changes have both advantages and disadvantages:
• globalization has led to the growth of newly industrializing countries (NICs) and transnational corporations
• change in the UK is encouraging growth in the South East and in out-of-town locations
• global and national changes have economic, social and environmental effects.

Types of industry

Four types are usually recognized – primary, secondary, tertiary and quaternary. These mean simply first, second, third and fourth, and refer to the order in which they developed.

1 Primary industry

Primary industries extract **raw materials**, which are natural products untreated by people, from the land or sea. Although many are truly natural products, some are farm products from plants and animals. Mining, quarrying, forestry, farming and fishing are all examples of **primary industries**. People who work in these industries have occupations in the primary sector.

▲ **Figure 1** Herding in Egypt – starting young in primary industry.

 Information

Examples of raw materials

Minerals, e.g. crude oil
Metals, e.g. iron ore and bauxite
Vegetation, e.g. wood
Plants, e.g. cotton
Plant products, e.g. cocoa and rubber
Animal products, e.g. wool and hides

---

2 Secondary industry

When raw materials are manufactured, they are *changed* by one or more manufacturing processes – by manual (hand) labour, by machines, by being heated, by having water or chemicals added – until they are made into something different. These manufactured products are secondary products. People making goods in factories work in the secondary sector, because manufacturing is classified as a **secondary industry**. Secondary industries also include building and construction.

▲ **Figure 2** A factory making dyes in Leicester – an example of secondary industry.

3 Tertiary industry

Many industries neither produce a raw material nor make a product. Instead they provide services to people and to other industries. The long list of services in the Information Box on page 181 still misses out some important services, such as education (schools and teachers) and broadcasting (TV and radio). The list includes essential services used every day, such as water and electricity, and others equally essential but used less frequently, such as doctors. Others are optional services, such as tourism, leisure, sport and entertainment, that make life more enjoyable and interesting. These services are examples of **tertiary industries**. These tertiary industries increase and multiply as countries and people become more wealthy.

Information

Examples of manufacturing industries

Raw material	Manufacturing process(es)	Manufactured goods
Crude oil	Refining using heat	Petrol, plastics, polythene
Wood	Hand carving	Wooden ornaments
Cotton	Spinning and weaving	Clothes, sheets, towels made by machines
Oranges	Squeezing with machines and adding water	Fruit juice

ⓘ Information

Types of services

Health, e.g. doctors and dentists

Local Council, e.g. refuse collection and care services

Local services, e.g. water, gas and electricity

Communications, e.g. phone and cable companies

Transport, e.g. buses and trains

Retail, e.g. supermarkets and stores

Financial, e.g. banks and building societies

Sport and leisure, e.g. sports centres and tennis clubs

Tourism, e.g. hotel staff and tourist guides

▲ **Figure 3** The downtown Central Business District (CBD) of New York, where the offices of many well-known companies and big banks are located.

4 Quaternary industry

Quaternary means fourth. This industry is based upon high-tech. It is number 4 because it is newer than the other three types of industry. High-tech now plays a big part in the lives of people and companies in MEDCs. Computers are being used for more and more tasks.

High-tech industries which invest heavily in research and development of new products are quaternary industries. They include companies researching into new computer systems, telecommunications (including mobile phones), aerospace and biotechnology (including new drugs and healthcare). Their workers are highly skilled and include many graduates. In the UK many jobs in quaternary industries are found in:

- the 'high-tech corridor' west of London near the M4

- over 60 'science parks' attached to universities (for example the Cambridge science park shown on page 179).

In the USA, they are concentrated in 'Silicon Valley' in California.

Activities

1 State the three primary sector occupations for obtaining the raw materials listed in the first Information Box on page 180.

2 Name two different manufactured products which can be made from each of the following raw materials:
- wool • iron ore • trees.

3 Name the main raw material used for manufacturing each of the following:

- bread • cheese • beer • wine
- soft-drinks cans • electrical wire.

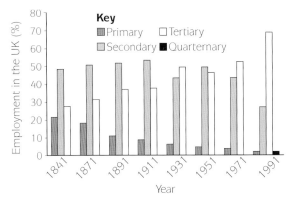

▲ **Figure 4** Changes in employment in different types of industry in the UK between 1841 and 1991.

4 (a) From **Figure 4** state the year:

  (i) when primary employment fell below 10 per cent for the first time

  (ii) when secondary employment was at its highest

  (iii) after which tertiary employment has always been the highest.

(b) State the evidence from **Figure 4** for the following features of employment, which happen as a country becomes more economically developed.

  (i) The primary sector decreases in importance.

  (ii) The tertiary sector increases in importance and becomes the most important.

  (iii) A small quaternary sector develops.

(c) (i) Employment data for 2001 (percentages):

   Primary: 2   Secondary: 18
   Tertiary: 78   Quaternary: 2

   Draw a graph to show these percentages.

  (ii) Describe the changes since 1991.

Industry as a system

All systems have inputs, processes and outputs. A simple version of the factory system is given in **Figure 1**. Raw materials (an input) go into the factory at one end. They are worked upon and changed (processes) in the factory. Manufactured goods (output) come out of the other end of the factory and are transported away to market. A factory is also a business. For it to survive, the factory needs to make a profit. The combined costs of obtaining the inputs and undertaking the processes must be lower than the value of the outputs that are sold (**Figure 2**).

Of course, the system is more complicated than this. Many factories use more than one kind of raw material. Some factories, such as in the car industry, use many component parts made in other factories. There are several inputs; a fuller list is given in **Figure 3**. The company owning the factory pays for the inputs, although in some cases the cost may be reduced by government grants. Although the real interest is in goods for sale, most factories produce waste as a by-product of processing raw materials. Some wastes are produced as gases and burnt off into the atmosphere, but there are also solid and liquid wastes which need to be treated and disposed of, and that costs money. When factories make profits, they go to the owners or shareholders. However, most companies re-invest some of the profits back into the factory for greater profits in the future.

◀ **Figure 1**
A simple factory system.

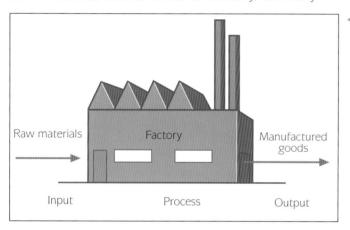

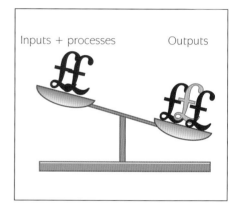

▲ **Figure 2** A profitable factory.

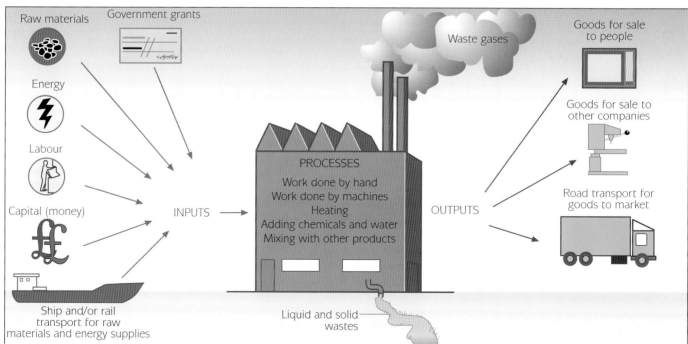

▲ **Figure 3** Factory system for a manufacturing industry.

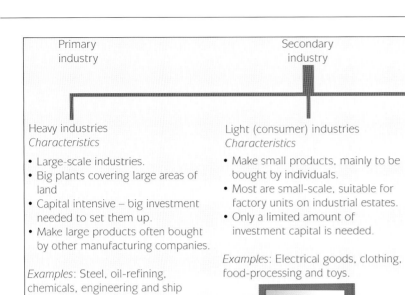

Primary industry	Secondary industry	Tertiary industry	Quarternary industry

Heavy industries
Characteristics

- Large-scale industries.
- Big plants covering large areas of land
- Capital intensive – big investment needed to set them up.
- Make large products often bought by other manufacturing companies.

Examples: Steel, oil-refining, chemicals, engineering and ship building.

Light (consumer) industries
Characteristics

- Make small products, mainly to be bought by individuals.
- Most are small-scale, suitable for factory units on industrial estates.
- Only a limited amount of investment capital is needed.

Examples: Electrical goods, clothing, food-processing and toys.

Hi-tech industries
Characteristics

- Make high-value products using modern technology.
- Heavy capital investment in research and development.
- Many are sufficiently small to be housed in units on business parks.

Examples: Computers, business systems, microprocessors and communications equipment.

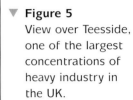

◄ **Figure 4** Types of manufacturing industry.

▼ **Figure 5** View over Teesside, one of the largest concentrations of heavy industry in the UK.

Types of manufacturing industry

Manufacturing industries make goods from raw materials and/or assemble parts made by other companies. A three-fold division of manufacturing industry is often used – heavy, light and high-tech (**Figure 4**).

Some heavy industries refine minerals, such as crude oil, into many different products, which can then be used by other industries to make a multitude of smaller products. Others smelt metals, such as iron ore, into steel, and these are produced in many different shapes and forms depending upon what they are going to be used for. The works may cover large areas of land; pollution from their chimneys may be obvious many kilometres away. They do not add to the scenic beauty of the landscape, but they are necessary industries because they provide the materials many light industries use for making consumer goods (**Figure 5**).

Light and high-tech industries can be combined under one heading – **footloose industries**. The term 'footloose' refers to their greater freedom to choose a location compared with heavy industries. This freedom comes from having low raw material demands and the need for only small factories, which means that suitable sites are widely available. The final products, being high in value and low in weight, can be easily distributed anywhere by road. New factories typically have pleasant locations on the edges of towns surrounded by green areas.

Exam focus

1 a (i) Define 'manufacturing industry'. (2 marks)

(ii) State one similarity and one difference between light (consumer) industries and high-tech industries. (2 marks)

(iii) Describe two differences between heavy and light industries. (4 marks)

b Explain why:

(i) waste is an output from the factory system (3 marks)

(ii) ship and rail transport are used more for inputs than outputs. (3 marks)

c Study **Figure 5**.

(i) Draw a labelled sketch to show the location and features of the heavy industries. (5 marks)

(ii) Describe the disadvantages for people living in the housing areas in Teesside. (4 marks)

(iii) A housing estate is located within 1 kilometre of these works. Suggest two reasons why people continue to live so close to this area of heavy industry. (2 marks)

Factors affecting industrial location

If a manufacturing company is going to be profitable, the directors need to consider the location for their factory carefully. Most consider the costs of inputs for locations in several different areas. After deciding upon the area with the greatest number of favourable factors, they look for a suitable site. Sometimes there are two or three equally attractive locations where costs are similar. In these cases the availability of government financial aid for one of the locations may be the deciding factor. Sometimes when advantages are similar the choice of final location can be based on quite irrational human decisions such as nearness to a good golf course or a successful football team.

The type of manufacturing industry strongly influences decisions about location. Another simple example can be used (**Figure 2**). If the three locational factors – nearness to raw materials, markets and labour force – are considered to be of equal importance, point A is the logical position for the factory. However, for Factory 1 nearness to raw materials is the most significant factor with the greatest pulling power. The position of Factory 2 suggests that market is the most important factor in choosing a location. Factory 2 may make goods that are perishable, or of low value and heavy to transport. What goods regularly bought in the shops could be made in Factory 2?

Raw materials

Materials from which goods are made in the factory.

For examples see Information Box on page 180.

Heavy industries most likely to have raw material locations.

Energy

Needed to change raw materials into manufactured goods.

Used for operating machines.

Important for heavy industries, which use the most.

Electricity is available everywhere, allowing light industries to be footloose.

Capital

Money invested in buildings and machinery to set up and run the factory.

Government (or EU) may help in areas of high unemployment.

Large amounts needed by heavy industries and high-tech companies.

Labour

Quality: workers with necessary skills.

Cost: lower wages in LEDCs than MEDCs.

High-tech industries need educated, skilled workers most.

Cheap labour matters most for labour-intensive industries e.g. textiles.

Container ship

Government

Positive role by offering grants, cheap loans and industrial sites.

Negative role by refusing planning permission and levying taxes.

Can attract industries to areas of high unemployment.

Heavy and large industries with high set-up costs are most attracted.

Market

Size: largest in urban areas.

Wealth: greatest in MEDCs.

Important for all industries, especially consumer industries.

KEY

Information about the factor

Effect on location

▲ **Figure 1** Factors affecting the location of manufacturing industries.

LABOUR

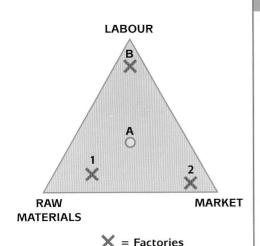

X = Factories

▲ **Figure 2** Factory locations.

Site needs

Availability and cost of land are important.

Flat land preferred.

Heavy industries need large areas of flat, cheap land.

Light industries need less land; good appearance may be important.

Transport

Ship, road and rail.

Containers are interchangeable between all three.

Heavy industries use bulky and heavy raw materials; they need access to cheap transport.

Light industries need speedy delivery and easy access to markets, usually by road.

Other communications

Dramatic growth in local and global communications.

By fax, phone, cable, on-line computer systems, e-mail and video.

Allowed global expansion of companies.

Possibility of rural locations and staff working from home.

Activities

1 (a) What shows that Factory 1 in **Figure 2** is located in a position associated with heavy industries?

(b) Name two examples of consumer goods which are **(i)** perishable and **(ii)** low in value but heavy.

(c) (i) What type of industry would you expect at point **B** in **Figure 2**?

(ii) Justify your choice.

2 Figure 3 shows the relative costs of transporting iron ore by sea.

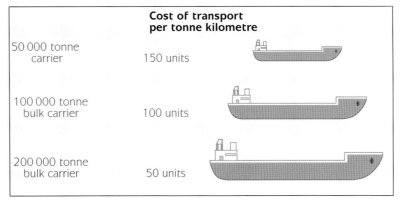

▲ **Figure 3**

(a) Describe what **Figure 3** shows.

(b) Why are many modern steelworks next to deepwater harbours?

(c) State the advantages of using container ships for transporting manufactured goods.

3 Decision-making. The directors of an oil company have decided to locate a petrochemical works in the area shown in **Figure 4**. The letters A, B, C and D indicate the positions of four sites that are being considered. The supply of crude oil will come by super-tanker.

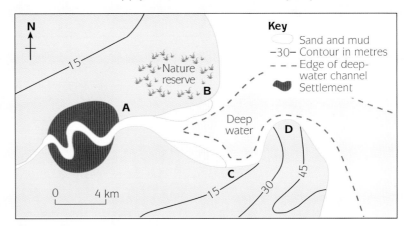

▲ **Figure 4**

(a) From the evidence available on the map, which site would be most suitable for the location of a petrochemical works? Give reasons to support your choice.

(b) Explain why you considered the other three sites less suitable.

(c) What other information about this area, not given on the map, could have influenced the decision about choosing a site?

Heavy industry

Heavy industries have been located around the estuary of the River Tees for more than a century and they dominate the landscape, as Figure 5 on page 183 shows.

Steel-making

Iron and steel industries began to develop along the south bank of the Tees between Middlesbrough and Redcar in the nineteenth century because all the **raw materials** and **fuel supplies** were locally available (**Figure 1**). A location near to these made economic sense since they weighed a lot more than the iron and steel products, although large markets in the engineering works and shipyards of Tyneside were not far away.

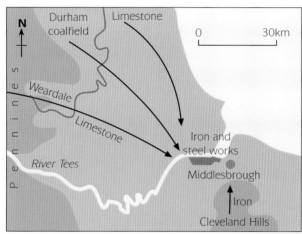

▲ **Figure 1** Sources of supply for the early iron and steel works.

Although most of the early advantages disappeared as coalmines and shipyards closed, British Steel still chose Teesside as the location for a modern steelworks in the 1980s – the Redcar works (**Figure 2** and square 5625 on **Figure 3**).

▲ **Figure 2** Redcar steelworks.

Teesside still has many locational advantages for a steelworks. **Transport** is the key factor. The Tees estuary is wide, sheltered and deep, which makes it navigable for large bulk ore carriers. Coal can be delivered to the steelwork's own terminal (squares 5425/5525) from anywhere in the world where the price is right. The River Tees is equally useful for exporting steel by sea to EU and other overseas **markets**, necessary to offset the decline in the UK's big steel-using industries like shipbuilding. A railway runs next to the works for distribution within the UK.

The long history of steelmaking ensures that a **labour** supply with the necessary skills and experience exists. **Site** factors are ideal. Flat land is in plentiful supply; wetland on the sides of a river estuary has few other uses. Land is cheap, an important consideration when a large site is needed.

Heavy industries on Teesside – the future

The future of steel-making on Teesside is not secure. By 2005 the workforce was well below 4000, and falling all the time (compared with more than 25000 in the 1970s). Corus, the present owner, has to compete in global markets with Asian producers, who benefit from significantly lower labour costs (pages 192–3). The long history of steel-making and continuing locational advantages no longer guarantee survival in today's global business world.

The story for the chemical industry on Teesside is similar. The early works, set up by ICI, were based on local raw materials such as rock salt, gypsum and coal, but now these operate on a much-reduced scale. More important today are oil-based industries. The OS map (**Figure 3**) shows the banks of the Tees lined with jetties, oil storage tanks, refineries and chemical works. A pipeline brings North Sea oil ashore at Seal Sands and super-tankers can dock in the Tees. ICI's Wilton Works, the largest industrial complex, produces petroleum-based products, such as plastics and synthetic fibres, much in demand from light industrial firms to make consumer goods. Its future looks brighter.

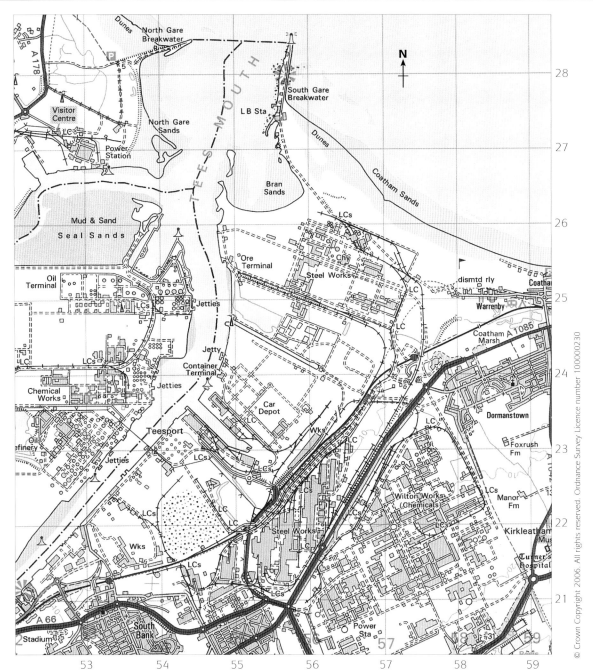

▲ **Figure 3** OS map of the Tees estuary at a scale of 1:50 000 (2cm = 1km).

Activities

Use **Figure 3** for answering questions 1–3.

1 On a sketch map:

  (a) show where the steelworks are located

  (b) label their site features.

2 (a) Describe the physical features of the River Tees which allow it to be used by large ships.

  (b) (i) Name three human aids to help ships navigate the river.

    (ii) Describe how each one is useful.

3 (a) Describe the pattern of roads and railways.

  (b) Give one physical and one human reason for the high density of roads in the area.

4 'Teesside – a case study of a region of heavy industry.' Using information from these pages and page 183, make notes using the headings:

- Examples of industries
- Where they are located
- Factors influencing their location
- Likely environmental issues and problems
- Future economic prospects.

Footloose industries

New industrial regions, in which the industrial structure is dominated by footloose industries, are most likely to be located next to motorways. The growth industries they contain are often referred to as **sunrise industries**. There are many of them along motorway corridors, such as the M11, M23, M3 and M4 (**Figure 1**). The greatest concentration of all is along the M4 corridor in the section between London and Reading. **Light industries** abound, including electrical goods, car parts and many food companies. Many of the **high-tech companies** are engaged in research and processing involving micro-electronics for computers and telecommunications equipment. High-tech companies here include Digital, Oracle and NEC.

ports for access to the increasingly important EU market. They also benefit from nearness to London's three main airports for international business links.

Specific advantages for high-tech industries

Labour is a key locational factor. The availability of highly skilled research scientists and engineers is very important. The presence of several universities has helped to provide a pool of graduates, and universities offer research facilities as well. The long-established presence of aerospace research in the Bristol area, undertaken by companies such as Rolls-Royce and British Aerospace, has been a further attraction for some companies. Another important consideration is where these

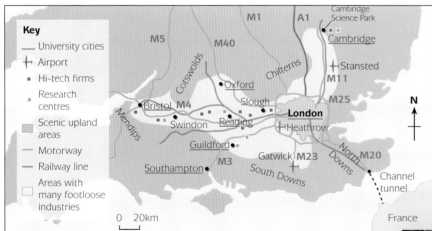

◀ **Figure 1** Industrial corridors along motorways out of London.

General advantages for the growth of light industries

Industries next to motorways would suggest that transport is the most important factor in deciding their location. Since they use or assemble parts of no great weight or bulk made by other industries, transport to market is more significant than raw material assembly. The wealthiest and most concentrated market in the country is in Greater London and the surrounding counties. Being close to this market is an enormous locational advantage. At the same time, there are motorway connections to the rest of Britain, and a high-speed rail link follows the M4 corridor between London and South Wales. Some parts of these corridors benefit from closeness to the routes to the Channel Tunnel and

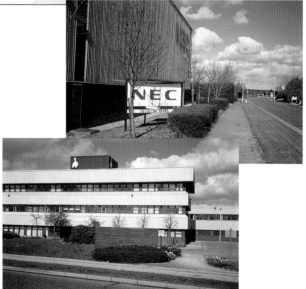

▲ **Figure 2** Both food and high-tech companies are located in the industrial estate in square 7169 on the OS map in **Figure 3**.

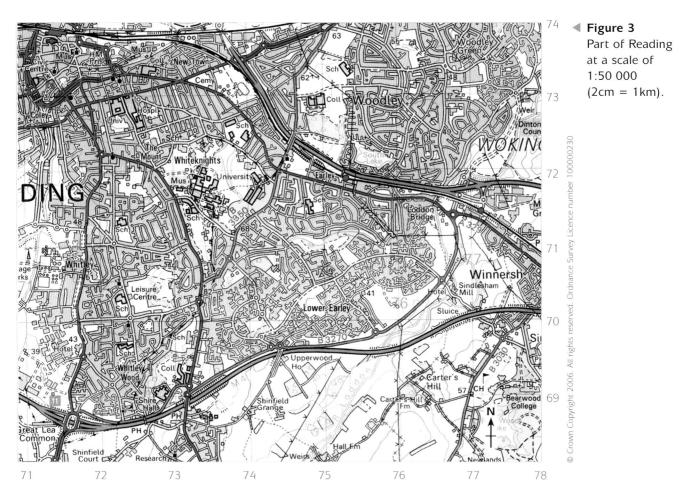

Figure 3
Part of Reading
at a scale of
1:50 000
(2cm = 1km).

specialist workers prefer to live. Areas of pleasant countryside are close enough to be accessible at weekends, they are near airports for holidays abroad, and everywhere is within easy reach of London for big sporting events, exhibitions and West End shopping and shows.

Activities

1 (a) 'The M4 corridor – case study of an area important for footloose industries': make notes using the headings: Examples of industries; Factors which favoured their growth.

Use information from the text and add specific information to it (such as numbers for motorways, names for universities and airports, etc.) from **Figure 3**.

(b) Draw a labelled sketch map to show the transport advantages of the area.

2 Find square 7169 on the OS map, in which the industrial estate at Worton Grange is located. Two companies located there are shown in **Figure 2**. The large building next to the M4 is part of a brewery.

(a) Describe **(i)** its site and **(ii)** its situation in relation to the rest of Reading.

(b) Explain the advantages of its location for light industries such as food and drink.

3 Reading is attracting people as well as industries. Look at **Figure 4** which was taken in square 7570 in 1995.

▲ **Figure 4**

(a) Describe the photograph and map evidence that Lower Earley is an area of recent growth.

(b) Using **Figure 3**, suggest why planners do not want Reading to grow any further to the east and south.

The Rhine–Ruhr region of Germany

North Rhine–Westphalia is the richest and, with 18 million people, the most populous of the German states. The River Rhine flows through it towards its western edge. The largest town is Cologne (Köln) with a population of about 1 million (**Figure 1**). Within this state is the single greatest concentration of industry anywhere in the EU. The Rhine–Ruhr region covers the Ruhr coalfield and nearby areas in the Rhine valley.

Types of industries

For over a hundred years the region's economy was dominated by coalmining (a primary industry) and by the secondary industries which depended upon coal – textiles, steel-making and smelting metals, heavy engineering and chemicals. Today heavy industries remain important and oil-refining has been added to the list; however, consumer industries have also grown, with food-processing and making cars the most important of these.

Factors for the growth of industry

The Ruhr coalfield is Europe's largest **energy** supply, with massive reserves. It has a variety of types of coal, including high-quality coking coal. The amount and the quality of the coal attracted heavy industries. Heavy industries are large consumers of energy; by locating in the Ruhr they made considerable savings on transport costs. For the chemical industry coal was also a **raw material** because various products can be manufactured from it.

Such was the concentration of steel and metal-smelting industries in the area between Duisburg and Dortmund that they created their own market. Industries using the metals, such as heavy engineering, and industries needing machines for their factories, such as textiles, set up in the Rhine–Ruhr region as well. The smaller companies and lighter industries benefit from a location in the state with the richest and largest **market** in Germany. The region also occupies a central position within the EU, and benefits from a wider market of over 400 million people.

The River Rhine is used for water **transport** (**Figure 3**). It is a busy highway for barge traffic and gives a link to the North Sea and the world shipping lanes. Convoys of push-barges can move up to 9000 tonnes at a time. Moving bulky and

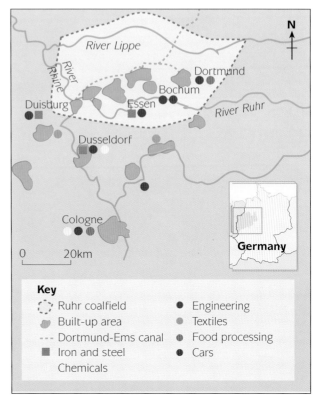

▲ **Figure 1** Manufacturing industries in the Rhine–Ruhr region.

heavy raw materials by barge is cheaper than using rail or road. The Rhine waterway therefore reduces the disadvantages of the region's inland location for heavy industries. Canals, such as the Dortmund–Ems canal, lead from the Rhine to serve steel towns such as Dortmund.

Industrial change since 1950

Industrial decline in the Rhine–Ruhr region has not been as rapid as in many other coalmining and heavy industrial regions in Europe because of:

- the amount and quality of Ruhr coal
- the productivity and prosperity of German industry
- the central position of the region within the EU
- the water links along the River Rhine.

However, there have been changes.

1   There has been a decline in the number of workers in the heavy industries and coalmining. For example, the number of miners dropped from over half a million in the 1950s to well under 100 000 by 2000. Numbers have only been kept up by government subsidies and guaranteed markets for coal in electricity and steel.

2 Some non-traditional manufacturing industries are now well established. Motor vehicle production is now one of the top five industries, and making telecommunications equipment is of growing importance.

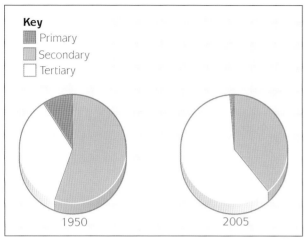

Key
- ▨ Primary
- ▨ Secondary
- ☐ Tertiary

1950   2005

▲ **Figure 2** Employment structure in the Rhine–Ruhr region, 1950 and 2005.

3 The state government has pursued a policy of diversification away from manufacturing industry into services. Cologne has become a major telecommunications and multimedia centre with many radio and TV stations as well as music recording and video-making companies. It is an attractive city on the banks of the Rhine with good transport links (**Figure 3**).

▲ **Figure 3** View from Cologne Cathedral over the Rhine busy with barge traffic.

The main problems

1 Socio-economic – unemployment is a major problem. It has not been possible to replace all the jobs lost from the mines and heavy industries. Both still need to lose more workers to remain competitive with companies in other parts of the world. In many towns there remains an over-reliance on just one or two big industries.

2 Environmental – the landscape is badly scarred from the effects of over 100 years of mining and heavy industry. Air and water pollution are major problems, as also are the large areas of old housing, unsightly heavy industry and the absence of green spaces. Up to half the land is often derelict in recently abandoned mining areas. The Ruhr Planning Authority is active in landscaping waste tips and pit heaps, cleaning up water courses, planting trees and preserving green wedges between built-up areas, but it faces a massive task. Industry itself is required to invest in environmental protection to the tune of over £250 million a year; much of the money goes into controlling sulphur and nitrogen emissions from coal-fired power stations.

Activities

1 (a) On a sketch map, show the location of five different manufacturing industries in the Rhine–Ruhr region.

(b) Explain why the largest concentration of heavy industries in Europe developed here.

(c) Describe some of the environmental problems which are caused by mining and heavy industry.

(d) State and explain two reasons why the decline in coalmining and heavy industry has not been as rapid as in many other EU countries.

2 (a) Describe the changes in employment structure shown in **Figure 3**.

(b) Why did these changes occur?

(c) Suggest reasons why modern manufacturing and services industries choose to locate in Cologne instead of in towns in the Ruhr valley such as Essen.

Industries in less economically developed countries

The economies of many LEDCs are still dominated by primary activities, mainly farming. This is reflected in their employment structures; typically more than 50 per cent of the workforce are employed in agriculture (**Figure 1**).

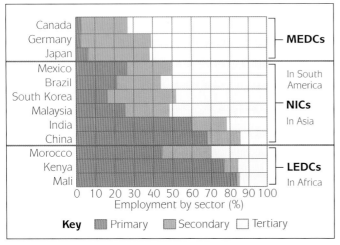

▲ **Figure 1** Employment structures for selected countries with different levels of economic development.

In countries in sub-Saharan Africa, farming is the way of life for almost everyone. In Mali, for example, trading and a tiny amount of craft industry and food-processing are confined to the capital city, Bamako, or to desert trading outposts such as Timbuktu. The employment pattern is the reverse of that in MEDCs.

On the scatter graph (**Figure 2**) the percentage employed in agriculture has been plotted against the country's wealth. This has been done for the ten main South American countries. Whilst the relationship is not perfect, you would have no problem drawing in a 'line of best fit'. A negative relationship is clearly shown – as the wealth of a country increases, the percentage employed in agriculture decreases.

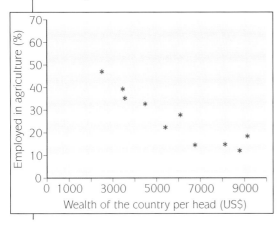

◀ **Figure 2** Relationship between wealth and percentage employment in agriculture in South America.

Newly industrializing countries

Figure 1 shows that certain LEDCs are classified as NICs (**newly industrializing countries**). In these countries there has been sufficient industrial growth for it to make a substantial contribution to the economy; this is accompanied by significant increases in percentages of workers in the secondary and tertiary sectors. Even though world manufacturing industry remains overwhelmingly concentrated in MEDCs, the proportion contributed by LEDCs is increasing (**Figure 3**). By 2005 it was approaching 30 per cent. However, the shares in manufacturing output are very unevenly distributed between LEDCs in different parts of the world (**Figure 4**). The significant overall increase in LEDCs from 1995 to 2005 was largely due to industrial growth in China. The comparative industrial backwardness of sub-Saharan Africa (that is, all of Africa except for the five countries with a border on the Mediterranean Sea) is apparent in **Figure 4** and illustrated by the employment structures for Kenya and Mali in **Figure 1**.

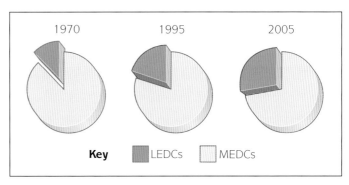

▲ **Figure 3** Percentage share of world manufacturing output.

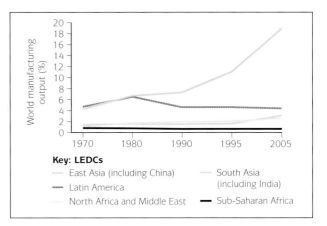

▲ **Figure 4** Shares of world manufacturing output of LEDCs in different parts of the world, 1970–2005.

The advantages of East Asia for industrial growth

1 Undoubtedly the most important is cheap **labour** supply. Wages are low by world standards (**Figure 5**). The textile industry is used as an example because it is labour-intensive. Asian workers are reliable and work hard for long hours, often in factories that would not meet all the health and safety standards of those in the West.

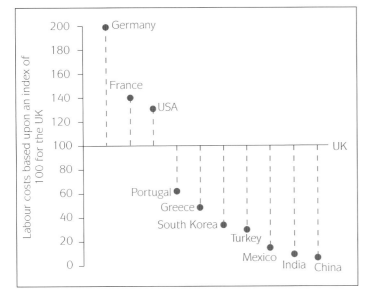

▲ **Figure 5** Examples of labour costs in different countries compared with the UK.

2 **Transport** is important. All countries in the region have access to the main shipping lanes. The use of containers (see **Figure 1** on pages 184–5) has reduced the cost of transporting manufactured goods by sea, as well as making it much easier and more secure. Sea has always been the cheapest way to transport goods over long distances.

3 The great push towards industrialization in NICs would not have been possible without **government backing** and a commitment to economic development. Governments in the region encouraged the import of capital and technology by overseas companies to establish factories and provide employment. They were also for responsible for providing the political stability necessary for successful industry and trade.

4 **Market** is an advantage of increasing importance. Although many factories were set up to export all their products, home markets are increasing as people become more prosperous. Asia is the most populous continent; the future potential of Asian markets is enormous as economies grow and personal wealth increases.

▲ **Figure 6** Singapore looks like an MEDC.

The countries that developed first and fastest were Hong Kong, Singapore, Taiwan and South Korea; they were referred to as the 'East Asian tigers'. They are far enough along the road of economic development to have many of the characteristics of MEDCs (**Figure 6**). Other NICs in the region are Thailand, Malaysia, Indonesia and the Philippines. They are known for a wide range of industries, from heavy industries (steel, ships and petrochemicals) to light consumer industries (clothes, trainers and electrical goods), as well as high-tech industries and the electronic and electrical goods that they generate (micro-chips, semi-conductors, computers and digital goods). More recently Asia's two population giants, China and India, have begun to stir (pages 195 and 197).

Activities

1 **(a)** On an outline map of East Asia, shade and name the newly industrializing countries.

(b) Add labels for goods in your home that were made in Asia.

2 **(a) (i)** Describe what **Figure 4** shows about East Asia's share of world manufacturing output.

(ii) State how it is different from that of LEDCs in other regions.

(b) Describe what **Figure 5** shows about differences in labour costs between Europe and Asia.

3 Explain fully why many clothes, sports goods and electrical appliances on sale in shops in the EU are manufactured in East Asia.

Globalization and the rise of transnational corporations (TNCs)

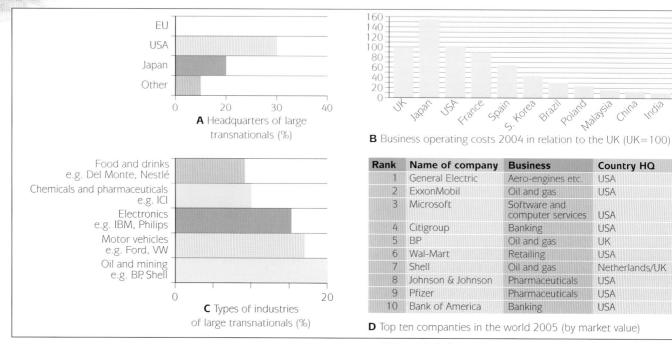

A Headquarters of large transnationals (%)

B Business operating costs 2004 in relation to the UK (UK=100)

C Types of industries of large transnationals (%)

Rank	Name of company	Business	Country HQ
1	General Electric	Aero-engines etc.	USA
2	ExxonMobil	Oil and gas	USA
3	Microsoft	Software and computer services	USA
4	Citigroup	Banking	USA
5	BP	Oil and gas	UK
6	Wal-Mart	Retailing	USA
7	Shell	Oil and gas	Netherlands/UK
8	Johnson & Johnson	Pharmaceuticals	USA
9	Pfizer	Pharmaceuticals	USA
10	Bank of America	Banking	USA

D Top ten companies in the world 2005 (by market value)

▲ **Figure 1** Information about transnationals.

Globalization refers to the *increasing* importance of international operations for people, companies and governments. People in one country or region of the world are being affected *more and more* by economic events and decisions made in other parts of the world, often thousands of kilometres from where they live. Both manufacturing and service companies are operating in ways that are increasingly international. The result is that few countries, people or industries are untouched by what happens in other parts of the world.

The growth of transnational corporations

TNCs are large multinational companies with business interests in many different countries. They include BP (based in the UK), Nestlé (Switzerland), Ford (USA) and Sony (Japan). They are truly global companies; some information about them is given in **Figure 1**. Most have their headquarters in MEDCs; this is where important decisions about global operations are made. They set up operations in MEDCs and LEDCs alike, wherever there is an opportunity for profit. They can take advantage of the fact that overall business operating costs are much lower in some countries than in others. Although more transnationals are based in Europe than in the USA, the majority of the ten biggest are based in the USA. The business interests of transnationals can be many and varied and include mining, plantation farming and manufacturing, and many of their names are well known to the general public.

Advantages and disadvantages of transnationals

Countries in which these companies operate have benefited from economic growth. Those countries with no attractions for transnationals, mainly in Africa and Asia, are among the world's poorest countries. However, the transnationals are motivated by profit rather than by a desire to achieve economic development for their host country; they are the world's

Advantages	Disadvantages
These are mainly **economic**	These are a mixture of **economic**, **social** and **environmental**.
• They bring **capital**, **modern technology** and **skills** which the counrty does not have.	• **Economic** disadvantages include low wages, tax avoidance and the fact that profits are taken out of the country and sent back to MEDCs. Also, TNCs can leave a country as quickly as they came.
• The country's **infrastructure** (e.g. transport and energy supply) is improved for them or by them.	
• They create **jobs** which increase exports; if these are manufactured goods, the dependence on low-value primary products is reduced.	• **Social** disadvantages include poor working conditions and lower safety standards than would be allowed in MEDCs.
• There may be other benefits such as a **multiplier effect** for service-sector jobs and economic development.	• **Environmmental** disadvantages include water and air pollution, because local pollution controls are either weak or ignored.

▼ **Figure 2** Summary of the advantages and disadvantages of transnationals.

most competitive companies and will leave a country as quickly as they came if there is no profit to be gained.

Countries have had such different experiences with transnationals that it is difficult to generalize. For example, those areas in which transnationals set up manufacturing industries (e.g. South East Asia) seem to have profited more. Those in which plantations cover large areas of the best farmland bring little reward to the workers and the country's economy (e.g. Central America). The information in **Figure 2** is a very general guide to the advantages and disadvantages of the presence of transnationals in LEDCs.

The rise of China

Industrial growth has been nothing short of spectacular. With manufacturing growing by 5–10 per cent per year, China is now the world's third-largest manufacturing country after the USA and Japan. It is already responsible for 7 per cent of global manufacturing industry, which some people think will rise to 25 per cent by 2025.

Undoubtedly the main reason for the rush of transnational companies into China is low production costs, as shown in **Figure 1B** on page 194 and in **Figure 5** on page 193. China can undercut almost every other Asian country on costs. What is more, there is no shortage of workers. With the world's largest population (1.3 billion) and rapid rural–urban migration, there is a great urban pool of industrial workers. Since the Chinese government signalled its willingness to allow inward investment from transnationals and encourage economic growth, the flood of overseas companies setting up businesses in China has been almost unstoppable. Another reason is the potential size of the home market. The demand for consumer goods is rising in urban areas. Can any major company afford not to have a presence in the country that has one fifth of the world's population?

Exam focus

1 a **(i)** What is meant by a 'newly industrializing country'? (2 marks)

(ii) Describe how its employment structure is different from that of other LEDCs. (3 marks)

b Read report **A**.

Bhopal, Central India, December 1984

A leak of gas from a pesticide plant, owned by Union Carbide, an American transnational corporation, killed over 2500 people. It affected the health of up to a quarter of a million people, most of whom were slum dwellers. They had been attracted to the city by its rapid industrial growth and were unaware of the risk of living next to a factory producing toxic chemicals.

The cause of the accident was put down to design faults in the plant's safety systems, made worse by a decline in maintenance standards. Significantly, in the USA plants producing these chemicals are required to be located 50km away from settlements.

A

(i) State three characteristics of transnational corporations like Union Carbide. (3 marks)

(ii) How could Union Carbide have reduced the risk of a gas leak? (2 marks)

(iii) Why was there a greater loss of life in India than there would have been from a similar incident in the USA? (3 marks)

c Read report **B**.

Twenty years on and Bhopal still bears the scars

Bhagwan Singh and his family live in the shadow of the most notorious factory in India. Their modest house lies only a stone's throw from the disused Bhopal chemical plant, where in 1984 a catastrophic toxic leak claimed the life of their baby and more than 3000 of their neighbours.

Today the factory site lies abandoned and closed to the public. But city investigators say that thousands of tonnes of toxic waste are still stored there. In the monsoon season, the rains wash the chemicals into the ground water, contaminating local wells. Campaigners claim that almost 20,000 have since died from the effects of the disaster and that 150,000 continue to suffer from the symptoms of chemical poisoning. They say these include cancer, anaemia, infertility and birth defects.

For families such as the Singhs, finding uncontaminated water to drink is a pressing priority. Residents say that water piped in or delivered by tankers is inadequate. People are still drawing stinking water from condemned wells. The cycle of death threatens to engulf another generation.

Source: *Sunday Telegraph* 2004

B

(i) Describe the long-term effects of the Bhopal disaster. (2 marks)

(ii) Explain why these effects are still being felt 20 years later. (2 marks)

(iii) Would this have been the case in a MEDC? Explain your answer. (2 marks)

d At the time the plant was being planned, many Indians were in favour of it being built. Suggest reasons why **(i)** the people of Bhopal and **(ii)** the Indian government were in favour. (6 marks)

Case Study – South Korea

Industry in an LEDC

Total population: 48 million

Capital city: Seoul (12 million)

Industrial output: US$175bn (9th largest in world)

Top four industrial corporations (called chaebols):
Samsung; LG (Lucky Goldstar); Hyundai; Daewoo

Main industries and world rankings:

Shipbuilding	1	Cars	6
Petrochemicals	5	Electronics	6
Textiles	5	Steel	6

◄ **Figure 1**
South Korea.

During the last 50 years South Korea has changed from a poor country with few resources to one of the world's top ten industrial countries. Why?

It shared the same general reasons for industrial growth – cheap labour, transport, government and market – that were given for East Asia on pages 192–3. In the beginning, wages were some of the lowest in the world; the workforce was efficient and willing to work long hours. Almost everywhere on the peninsula is within 100 kilometres of the sea, giving easy access to cheap water transport. The government put up many barriers against the import of manufactured goods that would compete with its own industries. The size of the home market continues to increase; the main markets for its exports are in the Pacific Rim (Japan, USA and Australia).

The special factor in South Korea is the way in which its industries are organized. Four large chaebols (the Korean name for large corporations controlled by Korean families) are responsible for 80 per cent of the industrial output. Each one is active in several areas of industry, as the example of Samsung shows (**Figure 2**). Samsung is the Korean giant; it is four times larger than LG, its nearest rival. All four have used their size to support aggressive export drives; they are examples of LEDC-based transnational corporations with investments in MEDCs in Europe and North America.

Samsung has increased the gap between itself and the other chaebols by responding more quickly to global changes. South Korea is no longer a very low-cost producer of consumer goods and competition from China is formidable and increasing. Therefore Samsung is switching to digital products; by 2004 the company had become the world's largest producer of two of the main components of digital devices: liquid crystal displays (LCDs) and memory chips.

Electrical and electronic consumer goods
• Microwaves
• TVs
• PCs

Heavy industries
• Chemicals
• Petrochemicals
• Shipbuilding
• Engineering

Samsung business interests

High-technology digital goods
• LCD flat-panel TVs
• Mobile phones internet enabled

Service industries
• Leisure orientated (e.g. hotels and theme parks)
• Finance and insurance

▲ **Figure 2** Samsung business interests.

ℹ Information

Problems

• *Environmental* – terrible air pollution from traffic and industrial fumes in Seoul, leading to health problems; water pollution from domestic and industrial wastes

• *Economic* – wages and business costs are higher than in other LEDCs (see **Figure 1B** on page 194)

• *Social* – South Koreans are demanding improved working conditions and a higher quality of life

IT and call-centre growth

One aspect of globalization is the movement of services and support industries from MEDCs to LEDCs. The main advantages for companies are the same as for manufacturing: cheaper labour and lower operating costs. One of the countries benefiting from this switch is India. This huge country of over 1 billion people produces 2 million graduates a year, 80 per cent of whom are fluent in English. Although call-centre jobs have received the greatest media attention, the range of work transferred is broader than this and includes computer programming and data processing. This is having a multiplier effect on service-sector employment, because each new phone and IT job is reckoned to support another job such as driving, catering or cleaning.

The Indian city which is benefiting the most is Bangalore (Karnataka), the 'IT capital' of India. Although it all began when Texas Instruments set up a successful design centre here in the mid-1980s, it only took off during the late 1990s after the Indian government adopted a more welcoming attitude to investment by overseas transnationals. This southern city, home to 6 million people, houses more than 250 high-tech companies, located in impressive new technology parks that have sprung up around the city. One of the largest and most popular with overseas companies is Electronics City (**Figure 3**). It was built for companies working in information technology, software development, telecommunications and financial services. Here are located a mixture of Indian companies (Infosys, Wipro) and international companies (Digital, Siemens, Motorola).

Nice work, if you happen to live in Bangalore

Fancy earning twice as much as a doctor and being driven to and from work every day? At the office there's a free canteen, a fully equipped gym and a pool table. Now here's the snag. You'll be earning £150 per month and may have to work until 2.30a.m. And you'll be living in India.

According to the Communication Workers Union, 33 large companies, including Barclays, British Airways, Lloyds TSB, Prudential and Reuters have together outsourced 52,000 jobs serving UK customers to India. Trade Union Amicus has predicted that 200,000 jobs could be lost overseas by 2010.

It is not hard to see why. It has been estimated that British companies usually save a minimum of £10m a year for every 1,000 jobs they move overseas. India produces more graduates per year than any other underdeveloped nation in the world. Call centre advisers are paid around £150 per month, while a junior doctor or teacher earns £60.

Norwich Union, Britain's biggest insurer, is one example. It has quickly built up a workforce of 3,700 people in India and plans to have 7,000 by 2007. According to the company, cheaper labour is not the only reason for turning to India. 'The work ethic is tremendous. It is partly cultural and reinforced by the education system. They have a high competitive attitude and status is important.'

One Indian customer manager from Bangalore said 'My parents (a civil servant and teacher) walk to work each day. My company send a car for me. I avoid the heavy traffic of old rickshaws, bicycles and dirty cars, plus the odd cow'.

Like all boom cities, Bangalore is not without its problems. Most of the problems, typical of big cities in LEDCs, are present and in pressing need of attention, including shanty housing, clogged and fume-ridden roads, frequent power cuts and shortages of clean water supplies. People continue to flock into the city.

▲ **Figure 4** Newspaper report from January 2005.
Source: *Daily Telegraph* 2005

Activities

1 In what ways is growth of manufacturing industry in South Korea **(a)** similar to and **(b)** different from that of other NICs in East Asia?

2 Explain fully why service-sector work is being transferred from the UK to India. Use values from this and previous pages to support your answer.

3 **(a)** Describe the economic and social benefits to India and its people of the growth of call centres and IT-related work.

   (b) Are there any disadvantages? Describe two possible disadvantages for Bangalore of rapid growth supported by investment from overseas.

4 **(a)** Why are service-sector jobs from the UK more likely to go to India, whereas manufacturing jobs go to China? Try to explain as fully as you can.

   (b) What is the 'China effect' upon other Asian countries – how and why is China's growth affecting neighbouring countries?

◄ ▲ **Figure 3** Electronics City houses over 100 technology companies.

Industrial change in the UK

Since 1950 the UK has experienced **de-industrialization** – decline in the relative importance of manufacturing industry, resulting in a sharp decline in its percentage share of the total workforce from around 50 per cent to less than 20 per cent (see **Figure 4** and Activity 4 on page 181). The greatest job losses have been in traditional industries such as iron and steel, engineering, shipbuilding and textiles. Many were located on or near coalfields during the Industrial Revolution, which explains why most were located in the Midlands and North of England and in South Wales, Scotland and Northern Ireland (**Figure 1**). These are referred to as **sunset industries**; they used to employ thousands of people, but from 1960 they were increasingly unable to compete on costs and delivery times with producers in newly industrializing countries in eastern Asia.

Closure of coal mines coincided with decline in heavy industries, which led to massive job losses in places where there was little chance of alternative employment. This led to severe socio-economic problems in industrial towns and among mining communities, including high rates of unemployment, poverty and deprivation, made worse by falling property prices as whole streets of houses were boarded up. Abandoned houses, factories, mines and railway sidings further blighted the appearance of these areas, already suffering from environmental problems such as polluted rivers and canals, waste tips and derelict land, caused by many years of heavy industry and mining. Despite many years of financial help from UK governments and the EU regional fund, the number of new jobs has not kept up with the number lost, so that migration from North to South continues.

Manufacturing industry is still important in the UK – after all, it still employs about 4 million people. However, as the focus has switched from heavy and traditional industries to footloose consumer and high-tech industries, a location around London and in the South East has become more and more attractive. For example, over 75 per cent of companies connected with the micro-electronics industry have chosen sites within 100 kilometres of London, citing many of the factors referred to on pages 188–9, notably a wealthy market, highly skilled labour and easy international links.

The growth of out-of-town locations

The traditional location for manufacturing industry is near to the centre of towns and cities. The canal-side location of the dye works shown in **Figure 2** on page 180 is an example. If clearance creates space for new building, these areas are called **brownfield sites** – ones that have already had urban land uses on them. These sites are not particularly attractive to modern footloose industries. Their preferred location is on **greenfield sites** – rural open land which has never been built on, out of town, usually in the rural–urban fringe.

Advantages of greenfield over brownfield sites include:

- cheaper land
- more space
- closer to motorways
- less traffic congestion
- easier to landscape to create a pleasant environment.

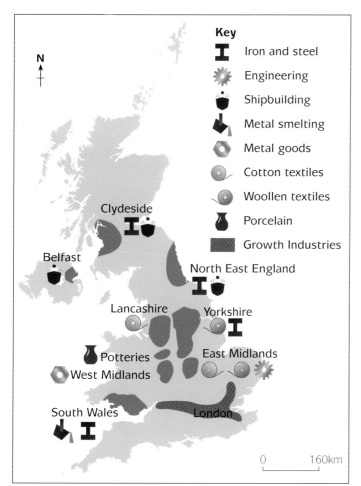

Key

- **I** Iron and steel
- Engineering
- Shipbuilding
- Metal smelting
- Metal goods
- Cotton textiles
- Woollen textiles
- Porcelain
- Growth Industries

N

Clydeside

Belfast

North East England

Lancashire

Yorkshire

Potteries

East Midlands

West Midlands

South Wales

London

0   160km

▲ **Figure 1** Traditional and new industrial areas in the UK.

Many urban land uses – factories, offices, shopping centres, houses and roads – compete for greenfield sites. Not everyone is happy with these trends, which raises further issues, some environmental and some socio-economic. What about the continuing loss of wildlife habitats? Isn't preservation of the green belt, stopping more of the countryside from being covered by concrete, important any longer? Should village communities be protected and preserved from urban sprawl?

A variety of names is used for the new industrial areas created in and around urban areas. **Industrial estate** is a term used where many of the tenants are manufacturing companies. Where there is a mixture of two or more of the following – light manufacturing industries, service industries, retail outlets, leisure complexes, distribution warehouses, administrative offices and research establishments – they are called **business parks**, particularly if offices and research industries dominate. **Figure 2** shows that Worton Grange is carefully designed with a low building density and set among landscaped surroundings (see also page 188). This is not easily achieved on brownfield sites, which is why pressures on the rural–urban fringe are so strong.

▲ **Figure 2** Worton Grange on the edge of Reading.

ⓘ Information

Cambridge science park

Opened in 1972

Established by Trinity College, Cambridge

Over 50 hectares landscaped with lakes and trees

About 65 tenants employing over 2000 people, many involved in scientific and research developments

---

The idea of **science parks** came from the USA. Unlike business parks, these always have a direct link with a university. Universities are places where research is undertaken. By providing a pleasant working environment, the intention is that research and development can be linked to successful business possibilities. There are about 60 science parks in the UK, of which one of the first, largest and most successful is the Cambridge Science Park. An aerial view of the park can be seen on page 179 and its location is shown on page 188 (**Figure 1**).

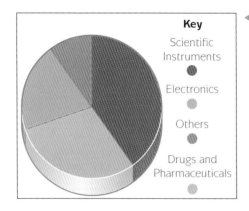

Key
Scientific Instruments
Electronics
Others
Drugs and Pharmaceuticals

◀ **Figure 3** Types of companies in Cambridge science park.

Activities

1 Choose one region of industrial decline in the UK and give the reasons for its decline.

2 (a) Make a list of the problems caused by industrial decline using the headings Economic, Social and Environmental.

   (b) Describe how the problems are different in areas of industrial growth such as the South East.

3 'Case study of a science park – Cambridge'; put together information using the following headings as guidance:
   • Definition of 'science park'
   • Location
   • Appearance
   • Types of companies
   • Reasons for growth.

Industry and the environment

Industry and the environment

Industries and people are concentrated in urban areas, which are centres of economic growth. Therefore urban areas are great producers of wastes onto the land and into rivers and the atmosphere. These include:

- toxic wastes from metal smelters, e.g. lead and nickel

- toxic metals leached from land dumps, e.g. arsenic and copper

- radioactive wastes from nuclear power stations

- smoke, gases and solid particles from chemical works and coal-fired power stations.

During the early stages of industrial development, environmental consequences are rarely considered. However, governments and city authorities in MEDCs are under increasing pressure from environmental groups to clean up.

◀ **Figure 1** A smog mask is an essential item of clothing for people working outdoors in Bangkok, which is one of the world's most notorious cities for air pollution.

Strategies for environmental improvement

One strategy is to use urban zoning to keep industrial and residential areas separate, especially in MEDCs where planning controls are strong. Another is to encourage the use of cleaner fuels such as natural gas instead of low-grade coal. Industries can be forced to increase the height of chimneys and install scrubbers to reduce toxic gas emissions. In the UK, the Clean Air Act of 1956 and the creation of smokeless zones have prevented a repeat of the London smog of 1952 which caused 12 000 deaths. **Figure 2** shows what can be done when regulations against pollution are passed and enforced and nature is allowed to repair the damage to the river that was described as the most polluted in Scotland. For industrial and economic development to be sustainable, strategies like these need to be adopted by governments in LEDCs, but there are major issues and question marks related to cost, government priorities and implementation of laws. Is rapid economic growth impossible without environmental damage?

1800	River reasonably clear
1810	
1820	Rapid growth in industry and population
1830	River becomes an open sewer for industrial and domestic waste
1840	
1850	Last recorded fish
1860	
1870	Described as foul and stinking
1880	The most polluted river in Scotland
1890	
1900	
1910	
1920	River poisoned from dye works, chemical works, metal works, distilleries and toilets
1930	
1940	
1950	
1960	Laws passed for clean-up. Clean-up begins
1970	Biological survey – no fish found First fish found in river
1980	18 species of fish counted
1990	42 species of fish counted
2000	Much river flora and fauna restored

▲ **Figure 2** Time line for pollution on the River Clyde in Glasgow.

Activities

1 Study **Figure 2**.

  (a) For how long was the Clyde without river life?

  (b) How many years did it take for life to return after the clean-up began?

  (c) Has this example of the River Clyde any relevance to rivers flowing through industrial areas in LEDCs today? Explain your answer.

2 **(a)** Describe:

    (i) the causes

    (ii) the effects

    (iii) the strategies to reduce the effects of air pollution in big cities in LEDCs.

  (b) State three reasons why air pollution is much lower in UK cities.

Chapter 12
Managing resources

Antarctica – the last great wilderness on Earth without permanent human inhabitants. Mineral resources, including oil, are known to exist and tourist numbers are increasing, which is why the existing international agreements are essential to conserve it as a wilderness.

Key Ideas

Management of resources is crucial to sustainable development:
- some natural resources are non-renewable and finite
- use of finite resources is increasing with adverse consequences for the environment
- to achieve sustainable development, conservation of non-renewable resources is needed
- a greater use of resources that are renewable is also needed
- global warming is an increasing threat unless strategies for its reduction are introduced more quickly.

Tourism is a growth industry in both MEDCs and LEDCs:
- varied environments offer different opportunities for tourism
- tourism can lead to economic development, but it may also bring social and environmental problems
- for tourism to be sustainable, stewardship and conservation are needed and green tourism is to be encouraged.

Resources

The Earth provides many **natural resources** without which human survival is impossible. These are summarized in **Figure 1**. They provide for all human needs. Fresh *water* and *heat* come from the atmosphere. Soil formed from the breakdown of rocks supports the growth of trees and plants; these are the main *food* providers, either directly through fruit and crops, or indirectly by eating the meat of animals that feed on plants. Wood and stone are used for *shelter*. *Clothes* for warmth are made from plants (such as cotton), animal wool and skins or from synthetic materials derived from minerals (especially oil).

Many of these essential natural resources are naturally replaced after use; in other words, these resources are **renewable** (**Figure 2**). Others, mainly minerals and metals, are described as **non-renewable** because they can only be used once. Common minerals in rocks in the Earth, such as iron ore, will last for hundreds of years, but many

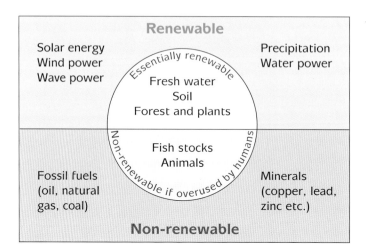

Renewable

Solar energy
Wind power
Wave power

Essentially renewable
Fresh water
Soil
Forest and plants

Precipitation
Water power

Non-renewable if overused by humans
Fish stocks
Animals

Fossil fuels
(oil, natural
gas, coal)

Minerals
(copper, lead,
zinc etc.)

Non-renewable

◀ **Figure 2** Some of the Earth's natural resources.

▶ **Figure 3** Life expectancies for certain minerals, calculated in 2000.

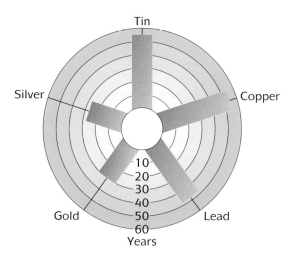

widely used others are expected to run out more quickly (**Figures 3** and **4**). The 'life expectancy' of a mineral is the number of years known reserves are expected to last at present rates of use. Both figures suggest that many minerals will be in short supply by 2050.

In the centre of **Figure 2** is a circle of uncertainty about resources – whether they are renewable or non-renewable. Left to nature without significant human interference, forests, soils, fresh water, animals and fish stocks are all examples of renewable resources. However, they can be destroyed by human over-exploitation to the point where they can no longer be replaced naturally. For example, destruction of forests can expose the surface to soil erosion and soil loss during heavy rains (as shown in **Figure 5** on page 107). Cod stocks in the North Sea are now so low from years of over-fishing that they may be below the biological limit necessary for recovery.

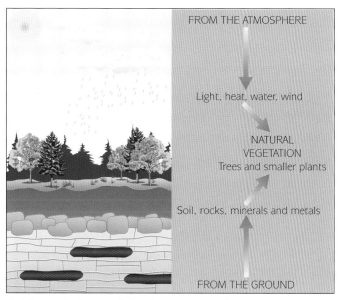

▲ **Figure 1** Some of the Earth's natural resources.

Energy resources

A Fossil fuels

Which are the main ones?
Coal, oil (petroleum) and natural gas.

What makes them *fossil* fuels?
1. They have been formed from the decomposition or remains of plants and animals.
2. Their formation began many millions of years ago – about 300 million years ago for the UK's coal deposits from the Carboniferous period.
3. Massive numbers of plants and animals were needed to form just small amounts of fuel.

Why are they *fuels*?
When they are burnt, they give off heat which can be used directly as a source of power. For example, heat from coal furnaces smelts metals, petrol from oil drives car engines and natural gas heats the water in central heating systems.

Why are they described as *non-renewable*?
It takes a very long time to form even small deposits of fuel. It will take millions of years to replace the fuels that have already been used. One estimate is that the world consumption of fossil fuels in 1 year is the equivalent of 1 million years of formation.

Why are they also described as *finite*?
Fossil fuels being used today will not be replaced for millions of years. We have the deposits already discovered, which are called **reserves**. Although other deposits of coal, oil and gas are still to be discovered, there is only a limited (or finite) amount of each fuel on or close to the Earth's surface. There is not an unlimited amount – every time a tonne of fuel is used, that is a tonne less for future people on Earth to use.

For how long will the fossil fuels last?
One estimate is shown in **Figure 4**. Estimates keep being revised based upon amount consumed and amount discovered, so they should be treated only as a guide.

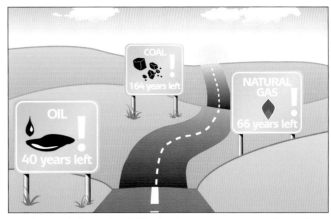

▲ **Figure 4** Length of life for known reserves of fossil fuels at present rates of consumption.

B Other fuels

Are there fuels which are not fossil fuels?
Wood and animal waste (dung) are two examples. They give off heat when burnt and are therefore fuels. They are growing or being produced now and can be used immediately. They are renewable for as long as there are trees and animals.

C Sources of energy

Energy is a broader term than fuel. It includes the fossil fuels, which give off energy when burnt. This is called **primary energy**. Also included is electricity, which must be generated from another source. This is called **secondary energy**.

D Uses of energy

- Cooking, lighting and heating.
- Power for operating machinery in factories.
- Fuel source for most modern forms of transport.

Activities

1. **(a)** Outline the differences between:

   (i) renewable and non-renewable resources

   (ii) fuels and energy.

   (b) **(i)** What is meant by a 'resource'?

   (ii) Make a list of the Earth's natural resources under three headings:
   - Fossil fuels
   - Other fuels
   - Renewable resources.

   (c) **(i)** Why can some natural resources be described as both renewable and non-renewable?

   (ii) Name one example and explain more fully why it can be described as both.

▲ **Figure 5** Bingham copper mine in the Rocky Mountains (USA).

2. **(a)** Describe what is shown in **Figure 5**.

   (b) What effects does resource extraction have on the environment? Answer in relation both to **Figure 5** and to mining in general.

Increased use of resources

There are two underlying reasons for increasing pressure upon the Earth's natural resources: world population growth and higher levels of economic development.

World population growth

World population has increased and the rate of increase has speeded up, particularly in the second half of the twentieth century and in the LEDCs (see Chapter 8). It took almost 100 years for world population to double from 1 to 2 billion, but then took less than 50 years for it to double from 2 to 4 billion (**Figure 1**).

1830	1 billion	1974	4 billion
1927	2 billion	1984	5 billion
1960	3 billion	1999	6 billion

▲ **Figure 1** Total world population.

Higher levels of economic development

Modern economic development dates back to about 1750 in the UK and the start of the **Industrial Revolution**. The technological breakthrough came with the invention of the steam engine, which was capable of driving machinery. It led to the growth of factories and towns. **Coal** was the natural resource upon which the Industrial Revolution was based; the Industrial Revolution soon spread to other countries that had their own coal. Germany and the USA industrialized from 1870 onwards. During the last 100 years the number, range and diversity of manufactured goods have increased, leading to today's **consumer-orientated society** in MEDCs. Economic growth since 1750 has been accompanied by a massive increase in consumption of natural resources of all types.

People everywhere are seeking an improved standard of living. Homes in MEDCs are full of electrical and electronic goods; some do jobs that were formerly done by hand, such as washing and washing-up, while others are for leisure and entertainment, such as TVs, stereo systems, computers and ipods. Private cars use more energy per person than public transport, especially when occupied by only one person. People in LEDCs are desperate to follow their example: once a home is hooked up to an electricity supply, the minimum requirements are a TV and fridge. Some LEDCs, particularly in South East Asia, have experienced rapid economic development since the 1970s (pages 192–3). Factories, offices, homes and transport are consuming more and more energy. China (almost 1.3 billion people) and India (over 1 billion), the world's population giants, are experiencing record rates of economic growth (pages 195 and 197).

All of these factors come together to explain the continued rise in world energy consumption to the record highs shown in **Figure 2**. Note that oil has more importance than coal today. Why do you think this is?

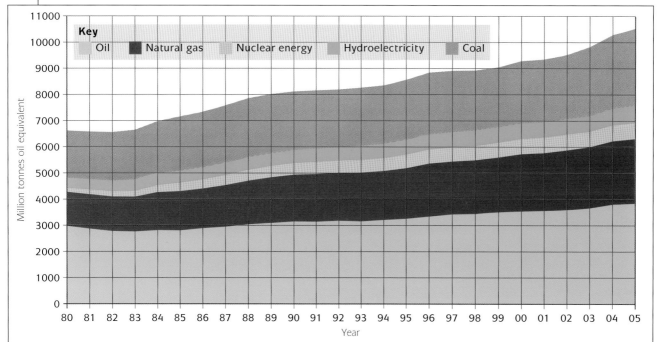

◀ **Figure 2** World commercial energy consumption, 1980–2005.

Source: *BP Statistical Review of World Energy* 2006

Global primary energy comsumption growth slowed in 2005 but still exceeded the 10-year average. Asia accounted for nearly three-quarters of global growth, with China alone accounting for more than half. In the past decade, natural gas and coal have increased their shares of the total at the expensed of oil, nuclear energy and hydroelectricity.

Country	Energy consumption per head (million tonnes oil equivalent)
Canada	9.94
USA	8.15
South Korea	4.62
Russian Federation	4.61
France	4.38
Japan	4.06
Germany	4.02
UK	3.78
China	1.10
India	0.37

◀ **Figure 3** Energy consumption per head, 2004 (million tonnes oil equivalent).

▶ **Figure 4** Differences in energy consumption, 2004 (million tonnes oil equivalent).

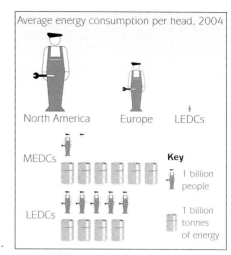

Average energy consumption per head, 2004

North America   Europe   LEDCs

MEDCs

Key

1 billion people

1 billion tonnes of energy

LEDCs

World variations in resource use

Some people, societies and countries make low demands upon the world's resources. There is, for example, no direct relationship between total energy consumption and total population in a country. When a value of energy consumption per head is calculated, wide variations emerge (**Figure 3**). **Figures 3** and **4** highlight the high consumption of energy resources in MEDCs, and in North America in particular. Average consumption per head in North America is double that in Europe and more than 800 times that of an LEDC.

Differences in levels of economic development explain some differences in energy consumption. Indeed, energy consumption per head is regarded as a reliable indicator of a country's level of economic development (Chapter 13). Much higher levels of economic activity are associated with well-developed manufacturing and service sectors, which create the need for high levels of transport of goods and many movements of people. People have higher incomes which allow them to travel by private car, take holidays and buy appliances. Each one has high energy demands. Compare the American lifestyle you have noticed in films and TV programmes with the tiny energy demand of an African farmer living in a village without electricity. Almost all tasks in the village are done by human energy; often fuelwood for cooking is the only fuel used.

Climate is another factor. Canada, Russia and many parts of the USA experience very cold winters. What would be called a 'big freeze' in the UK is normal winter weather in many places in North America. In contrast, summers in the centre and south are very hot; 'heatwaves' occur every summer which make air conditioning a necessity in a way that it is not in Europe.

However, waste of energy is also a relevant factor. The American love of big cars is waning but has not been lost. There is not the same incentive to turn to smaller, more energy-efficient cars when the price of petrol is less than one-third of the European average. North Americans tend to overheat their buildings in winter, while in summer the air conditioning is quite severe.

▲ **Figure 5A** Rural Peru.
▼ **Figure 5B** Urban India (Delhi).

Exam focus

1 a Using **Figure 2**:

   (i) Describe the changes in total world energy consumption between 1980 and 2005.
   (3 marks)

   (ii) Name the fossil fuels shown in **Figure 2**.
   (1 mark)

   (iii) Of the total world consumption of energy in 2005, how much was contributed by fossil fuels? Show your working. (3 marks)

b (i) From **Figures 3** and **4**, give information that supports each of the following statements:

   A More people live in LEDCs than in MEDCs. (1 mark)

   B More energy is consumed in MEDCs than in LEDCs.
   (3 marks)

   (ii) Explain why MEDCs use more energy.
   (5 marks)

c Study **Figures 5A** and **B**.

   (i) Describe the sources of energy that are being used. (2 marks)

   (ii) Why are the scenes more typical of LEDCs than MEDCs?
   (3 marks)

d Energy consumption in LEDCs is likely to increase during the next 50 years. Give two reasons for this.
   (4 marks)

Conserving resources and reducing pollution

We have already seen how fossil fuels dominate world energy consumption. Although they are likely to continue to do so for many years to come, this cannot go on for ever. A range of strategies is needed to ensure that future economic development is achieved in a more sustainable manner with less damage to the environment. Three strategies are:

- conservation of existing resources by **recycling** and greater efficiency in use

- establishment of pollution controls to reduce atmospheric pollution and global warming

- increased use of renewable sources of energy as alternatives to fossil fuels (pages 208–11).

Conserving resources

Recycling is the recovery and conversion of waste products into new materials. One product commonly recycled is the glass bottle (**Figure 1**). Next to the bottle banks near supermarkets, there are often bins for waste paper (pulped down and made into new paper goods), clothes and textiles (converted into upholstery and blankets), and aluminium cans (melted down and manufactured into new containers). However, recycling is only worth while if it saves the natural resources without consuming a large amount of energy for reprocessing. Recycling aluminium cans is particularly worthwhile because reprocessing only consumes 5 per cent of the energy needed to produce aluminium from its raw material (bauxite). **Resource substitution** may be useful as well, for example making goods out of aluminium instead of from metals that are less easy to recycle. We can all contribute to reducing energy consumption, by switching off lights when they are not needed, using low-energy light bulbs, and not using the car when it is possible to walk, cycle or use public transport. Small measures from many people can have a big impact.

For **energy efficiency**, however, the role of governments and companies is often more important. The UK Government has toughened up the energy efficiency standards for new buildings so that less heat is lost through walls, roofs and windows. You may have noticed energy-efficiency labels on electrical goods in the shops, such as fridges, which is a government initiative. Combined heat and power schemes are encouraged. Next to some power stations are glasshouses heated by the water which has passed through the cooling towers. Supermarkets use the heat given off by their freezers to heat other parts of the store.

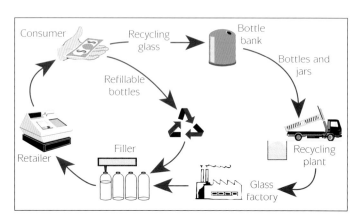

▲ **Figure 1** Recycling glass bottles.
Source: WWF, Data Support Sheet for Education 24

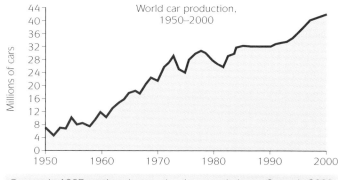

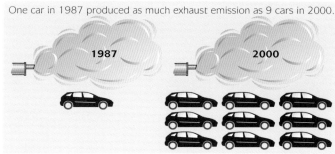

One car in 1987 produced as much exhaust emission as 9 cars in 2000.

▲ **Figure 2** More but cleaner cars.
Source: Adapted from *New Internationalist*

There is no evidence that people in MEDCs will stop using their cars, and people in LEDCs also have the same desire to own a car. At least cars can be made more energy-efficient and cleaner (**Figure 2**). Today's petrol car in Europe is 90 per cent cleaner than its ten-year-old counterpart. The car companies have been highly successful at reducing emissions of pollutants, such as oxides of nitrogen, carbon monoxide and hydrocarbons, mainly through the use of catalytic converters. Today the pressure is on them to achieve greater energy efficiency by reducing fuel consumption using improved engine technology.

Reducing pollution

Industrialization and modern economic development were fuelled by burning fossil fuels. Fossil fuels are convenient to use and remain relatively cheap, despite significant increases in the price of oil and gas during 2005. Unfortunately, fossil fuels are great environmental polluters.

Coal-fired power stations and vehicle exhausts are the main sources of sulphur dioxide (SO_2) and nitrogen oxides (NO_x) responsible for causing **acid rain**. The presence of these pollutants in the atmosphere increases the acidity of falling rain, which in turn increases acidity levels in soils, rivers and lakes, killing trees and causing the deaths of fish and aquatic plant life. Once in the atmosphere, pollutants are carried by prevailing winds to other places, and can be carried long distances, sometimes to other countries. Acid rain is a major problem in Scandinavian countries (Norway and Sweden) because prevailing south-westerly winds carry the pollutants from the UK's power stations, especially from those which are coal-fired (**Figure 3**).

▲ **Figure 3** Drax power station in Yorkshire, one of the UK's largest. The air pollution is coming from the chimney, not from the cooling towers, which are giving off steam. From this station, most of the sulphur is 'scrubbed' out from the flue gases by using a mixture of limestone and water, which converts the sulphur dioxide into calcium sulphate. This process is called flue gas desulphurization.

Pollution which moves across national boundaries requires international directives for its control and reduction. Both the UN and the EU are involved (**Figure 4**). They recognize that it is impracticable to achieve a reduction in emissions to desirable levels overnight; therefore they operate by setting targets for reductions in emissions for the UK, over a series of dates. The targets set for 2010 will not be easily met. Strategies for reduction cost money and increase the cost of electricity production, which is passed on to consumers (both domestic and industrial). A potentially greater worldwide pollution problem is global warming from emissions of carbon dioxide (CO_2), which is also released when fossil fuels are burnt. This is discussed on pages 212–13.

UNECE (United Nations Economic Commission for Europe)
- Convention set up in 1979; has issued a number of Protocols to reduce emissions of pollutants
- The latest was the UNECE Gothenburg Protocol 1999, which set targets for reduction in SO_2 and NO_x for 2010 (see table below).

NECD (EC National Emissions Ceilings Directive)
- Agreed in 2001 for countries within the EU
- The ceilings for the UK in 2010 were cut further (see table below).

UK emissions – recorded in 1990 and 2000 and targets for 2010 (thousand tonnes)

	1990	2000	UNECE target for 2010	NECD target for 2010
SO_2	3721	1165	625	585
NO_x	2763	1512	1181	1167

▲ **Figure 4** International Protocols for controlling acid rain pollution.

Strategies for reduction of emissions causing acid rain
- Fit giant scrubbers (FGDs) to coal-fired power stations.
- Switch power stations from coal to natural gas, which has a much lower sulphur content).
- Selective catalytic reduction in power stations by adding ammonia and passing it over a catalyst to produce nitrogen and water.
- Compulsory fitting of catalytic converters to vehicle exhausts.

Activities

1 **(a)** Draw a bar graph to show the values in the table within **Figure 4**.

   (b) Describe how they show a big reduction in emissions from 1990 to 2000.

   (c) State three reasons for this reduction.

   (d) Why is achieving the targets set for the UK in 2010 likely to be more difficult?

2 Make a large version of the table below and fill it in.

	Definition and examples	Advantages	Problems
Recycling			
Energy efficiency			
Reducing pollution			

Renewable sources of energy

A **renewable** resource is a natural resource that will never run out. Renewable resources can provide **sustainable** sources of energy that people will be able to use long after fossil fuels have been used up. Natural resources such as sun, wind and water will always be available. The main examples of renewable sources of energy are hydroelectric, wind, solar, geothermal (from the heat of the Earth, for example in areas of volcanic activity) and tidal power. At present they contribute only a tiny percentage to world energy supplies (**Figure 1**). They also come under the heading **alternative sources of energy**, which usually refers to an energy source that can be used instead of fossil fuels. Nuclear power is also included under this heading.

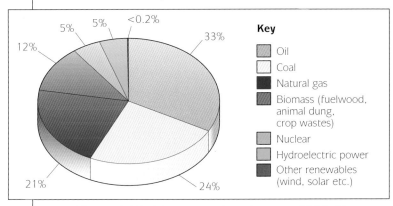

Key

	Oil
	Coal
	Natural gas
	Biomass (fuelwood, animal dung, crop wastes)
	Nuclear
	Hydroelectric power
	Other renewables (wind, solar etc.)

33%   24%   21%   12%   5%   5%   <0.2%

▲ **Figure 1** World consumption of energy, 2004. The graph shows the relative importance of all sources of energy, both commercial and non-commercial (such as biomass).

Some of the advantages of using renewable sources of energy are shown in **Figure 2**. They make impressive reading – no atmospheric pollution and complete sustainability are the two big advantages that renewables enjoy over fossil fuels. In LEDCs the potential for local availability is a very important factor because the cost of electricity transmission by wire over large distances is expensive and requires a substantial infrastructure. A small dam or a few wind turbines may be able to generate enough electricity for village needs and lead to a great improvement in quality of life.

Why do renewable sources of energy make a low contribution to world energy supplies?

Figure 1 shows that **hydro-electric power** (HEP) is the only renewable of significant importance among the sources of commercial energy supply. Even its further expansion is limited by the shortage of sites with the required physical conditions (page 210). Similarly,

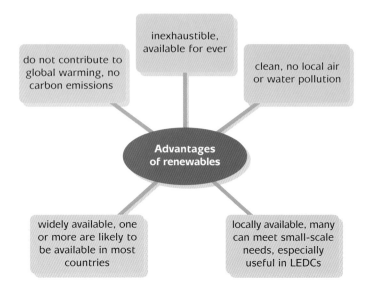

Advantages of renewables

inexhaustible, available for ever

do not contribute to global warming, no carbon emissions

clean, no local air or water pollution

widely available, one or more are likely to be available in most countries

locally available, many can meet small-scale needs, especially useful in LEDCs

▲ **Figure 2** Advantages of renewable sources of energy.

▲ **Figure 3** Examples of renewable sources of energy. **A**: Solar in the Caribbean – heat from the light of the Sun using solar panels and photovoltaic cells (PVs). **B**: Wind turbines along the coast in Cumbria. Coastal locations, both onshore and offshore, are popular locations because of higher average wind speeds.

opportunities for setting up **geothermal** power plants are restricted to a few places in the world, such as Iceland and New Zealand, where volcanic activity continuously produces large quantities of boiling water. Some renewables depend upon weather conditions that cannot be guaranteed; there are cloudy days when the Sun does not shine and calm days when the wind does not blow. Even if all the UK were to be covered by **wind turbines**, it would still not be possible to close any of the thermal power stations because they would be needed during times of high pressure and calm weather.

◀ **Figure 4**
El Tatio geysers in Chile. The physical requirements for geothermal power exist here, but at 4000 metres above sea level in an uninhabited part of the Andes, the location is too remote for commercial use.

Wind power has benefited from improvements in technology, which have increased the efficiency of modern turbines. Wind turbines, like those in **Figure 3B**, are becoming a common sight in certain countries in Europe, particularly Denmark, Germany, Spain and the UK, where average wind speeds are high. Large numbers cover hillsides in California where environmental groups have a stronger voice than in other parts of the USA. Although they do not pollute the atmosphere, renewable sources still have some negative environmental effects. Wind turbines are noisy and can interfere with TV reception; some people claim they are unsightly and ruin the natural beauty of scenic uplands. Photovoltaic panels (PVs) that convert sunlight into **solar energy** use toxic substances such as cadmium sulphide. Harnessing the **energy of the tides** involves building a barrage across an estuary, thereby interrupting natural water flows, possibly affecting marine life and changing rates of erosion and deposition.

However, the main factor that has held back their greater use is simply cost (**Figure 5**). What is obvious is that fossil fuels are cheap relative to other energy sources; they are still cheap, even after the big rises in oil prices in 2005. Years of low oil prices and continued new discoveries of oil and gas with improved technology have not encouraged companies to spend money on research into alternative sources. Research is a gamble. New technology is expensive to develop without any guarantee of a successful outcome: all the money invested can be lost.

Activities

1 (a) Draw a spider diagram like the one in **Figure 2** to show five disadvantages of renewable sources of energy.

(b) Number the disadvantages according to what you consider to be their order of importance, with 1 = most important disadvantage and 5 = least important. Justify your order.

2 One person's assessment of **nuclear energy** is given below.

Renewable	✗
No carbon dioxide emissions	✓
No air pollution	✓
No local environmental problems	✗
Safe	✗
Cheap	✗
Known technology	✓
Simple technology, suitable for use in LEDCs	✗
Always available; does not rely on weather	✓

(a) Make a table and give your own assessments like these for **(i)** fossil fuels and **(ii)** solar power.

(b) What are the significant differences in assessment between fossil fuels and solar power? Comment on these differences.

Technological breakthroughs are being made, however slowly; today's PVs and giant wind turbines are much more efficient than earlier versions. The potential energy from ocean waves around the British Isles is enormous: it is estimated that Atlantic wave energy could satisfy 20 per cent of the UK's energy needs by 2020. So far, discovering wave machines that can harness this energy has proved elusive. However, a new wave machine that 'snakes' across the sea surface, and floating buoys that are rocked by the waves to generate electricity, are moving from the experimental to the operational stage. Some will be in operation off the coast of Cornwall in 2007. Visit websites (see Hotlinks, page iv) to keep abreast of developments. The companies aim to bring down the cost of marine energy per kw/hr to a level closer to that of fossil fuels; they are being helped by high oil prices bringing up costs of thermal energy to meet them. Wave machines and buoys have the advantage over wind turbines that the general public will not be able to see where the electricity is coming from.

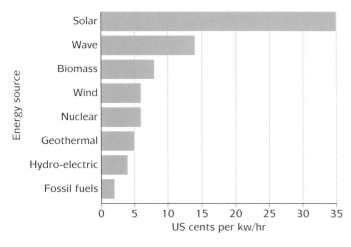

Note: these are average values; there are great variations according to country, site and scheme size. Prices in the UK tend to be slightly higher than these averages (see Activity 3 on page 211).

◀ **Figure 5** Average costs of producing electricity from different energy sources in 2003.

HEP and wind power – examples of renewable sources of energy

HEP (hydro-electric power)

Running water is used to drive turbines – the greater the volume and force of water, the greater the amount of electricity generated. Certain favourable physical conditions are essential, for example:

- fast-flowing water such as waterfalls
- high rainfall spread throughout the year
- a lake or natural water store, or a narrow and deep valley to create a reservoir as an artificial store of water.

Once the turbines are in operation, the natural flow of water allows continuous production without any pollution to the environment. From large schemes, costs per unit of electricity can be as low as, and sometimes lower than, those from thermal power stations. The water is not consumed and can be used for other purposes such as water supply for homes and industries and for irrigation.

There are some drawbacks and problems. Sites with the required physical conditions are more likely to exist in mountainous areas than in lowlands close to centres of population. Transmission lines are expensive to build and maintain, and the longer the distance, the more electricity is lost in transmission. The plant is a high-cost investment. Although large dams are environmentally friendly when producing power, building them can have adverse environmental effects such as flooding forests, releasing greenhouse gases (such as methane and carbon dioxide from the rotting vegetation) and destroying wildlife habitats. People may need to be forcibly moved from the land to be flooded, and relocated, often to areas of inferior land with poorer living conditions. The larger the dam, the greater the size of any problems, but also greater is the amount of fossil fuels saved by the electricity generated.

Wind power in the UK

Two decades of research has produced a modern, electronically-controlled wind turbine which stands over 30 metres high and has fibreglass blades 35 metres or more across (see **Figure 3B** on page 208). The technology is now well tested. Wind turbines are located in open, exposed places, mainly on hilltops but increasingly along the coast. The investment needed to set up a wind farm is high and has been financially possible only because of government support. Otherwise the turbines are cheap to run; they are pollution-free and the ground between them can still be used for farming. Local people are often less enthusiastic about them because they are noisy and can disrupt TV reception.

The Three Gorges Dam and HEP scheme in China

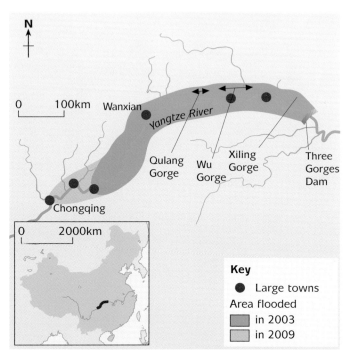

▲ **Figure 1** Location of the Three Gorges Dam.

Facts
- dam height 185m
- dam length 2000m
- length of reservoir 650km
- cost US$ 30bn
- start date 1992
- completion date 2009

Benefits
- massive power generation
- 18000 megawatts supply to national grid
- will support China's rapid economic development
- will reduce reliance on coal, a dirty fuel
- will bring water to the drier north of China
- flood prevention in lower Yangtse valley
- improved navigation above the dam
- symbol of China's ability to tame nature

Known drawbacks
- more than 320 villages and 140 towns flooded

- at least 1.2 million people forced to move from their homes
- 40000 hectares of good farmland under the water
- promises of new farmland and homes not kept
- destruction of river habitats of dolphins, sturgeon and alligators
- destruction of habitats of rare land species – tigers and giant pandas

Possible future drawbacks
- increased risk of landslides (and perhaps earthquakes) upstream
- parts of Yangtse become clogged with sediment due to less flow
- only a fraction of the promised power will be generated
- scheme may be too large for effective management

Many turbines are needed to produce the same output as one thermal power station, and on calm days no power at all is produced.

The UK has more potential for generating electricity than most of its European neighbours. Strong westerly winds are a feature of its maritime climate, and being an island country gives it a greater length of exposed coastline. Many of the upland areas are towards the north and west, where westerly gales are most frequent. The government is supportive; it has international and self-imposed targets for cutting carbon dioxide emissions, and is keen to encourage renewable initiatives that will cut down on fossil fuel use. Wind power is seen as a big hope because the technology is further developed than for tidal, wave and solar power.

Record year for wind-farm construction

Construction of wind farms hit a record in 2004, with 250 megawatts of new capacity, double that of 2003. Even so, the development of offshore schemes must accelerate if the government is to meet its climate-change targets. About 8000 megawatts, enough to power about 6 million homes, will need to be built if the country is to achieve its aim of generating 10 per cent of electricity from renewable sources by 2010.

Wind power is expected to represent about three-quarters of the target. This would require almost 3000 more turbines to be erected compared with the 1000 currently in place. Meeting the government's targets would depend upon offshore schemes going ahead as planned. Developers remain concerned about possible planning delays and opposition from the military authorities over radar interference. Scotland, which has some of the best wind resources in Europe, accounts for 70 per cent of the current onshore schemes being considered.

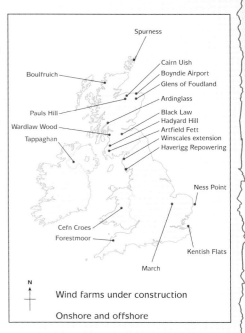

▲ **Figure 2** Newspaper report from November 2004.
Source: Text adapted from *Financial Times* 22 November 2004; map British Wind Energy Association

Activities

1 Give an assessment (as in Activity 2 on page 209) of HEP as source of energy.

2 Make notes for a case study of a large HEP scheme using these four headings: Location, Details of the scheme, Reasons for setting it up, Social and environmental problems from dam and reservoir construction.

3 Cost of electricity generation for different technologies in the UK in 2004 (pence per kilowatt hour): coal 3.2p; photovoltaic 45p; onshore wind 4.2p; offshore wind 5.1p; wave and tidal 14p; nuclear 4.2p; gas 2.5p.

  (a) Show these values on a graph.

  (b) Describe what they show about wind power in the UK as a source of energy compared with **(i)** fossil fuels and **(ii)** other alternative sources.

4 (a) Describe the distribution of wind farms under construction in 2004 from the map within **Figure 2**.

  (b) Suggest reasons for the distribution shown.

  (c) Are there any wind turbines in your home area? Explain why or why not.

5 (a) Arrange the comments about wind power in **Figure 3** under three headings: For, Against, and Neutral.

  (b) Write a paragraph supporting the use of wind power in the UK.

  (c) In your view, how strong are the arguments for wind power? Explain your point of view.

▲ **Figure 3** Comments about wind power and wind farms – from people and organizations within the UK.

Global warming and the future of the planet

There is no doubt that the Earth is currently warming up, that global warming is happening (**Figure 1**), although there are arguments about its causes. Similarly there is no doubt that many of the Earth's finite natural resources are being consumed by humans more quickly than they are being replaced. Oil is a good example (**Figure 2**). Equally, everyone knows that the Earth's atmosphere, water and land surfaces are now more polluted than ever: although nature has amazing powers of recovery, it is asking too much for it to quickly repair continuing pollution damage from over 6 billion humans.

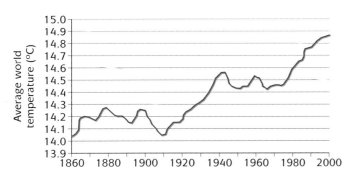

▲ **Figure 1** Change in world temperature, 1860–2000.

Total global oil

994 billion barrels extracted to date

764 billion barrels remaining in known fields

142 billion barrels yet to find

Discovery vs demand

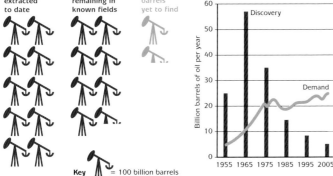

Key = 100 billion barrels

▲ **Figure 2** Oil reserves are running out.
Source: © Guardian Newspapers Limited 2005

Global warming and the greenhouse effect

The greenhouse effect is a natural process. Water vapour and carbon dioxide in the atmosphere absorb long-wave (heat) radiation from the land surface. This delays radiation loss from the Earth into space and keeps the

Earth warmer (by some 30°C) than it would otherwise be. This is what makes the Earth such a pleasant place to live.

General world temperatures have been rising since the end of the Ice Age 10000 years ago, clearly as a result of natural causes, because the human population was small and levels of technology were low. However, during the twentieth century the rate of increase changed, most notably during the last 25 years (**Figure 1**). Although some scientists and Americans remain unconvinced, the majority have come to the view that the current rate of warming is too rapid for it to be due to natural causes alone. Most agree that people are to blame, at least in part. Human activities from the Industrial Revolution onwards have sent levels of greenhouse gases in the atmosphere soaring, making it likely that they are creating what is known as the 'enhanced' or 'accelerated' greenhouse effect. Burning fossil fuels and deforestation are most responsible (**Figure 3**).

Greenhouse gas	Percentage contribution to the greenhouse effect	Number of years it stays in atmosphere
Carbon dioxide	64	up to 200
Methane	18	12
CFCs	14	1000 or more
Nitrogen oxides	4	120

▲ **Figure 3** Contribution to the greenhouse effect made by different gases.

Who are the main culprits? Although enormous compared with LEDCs, emissions per person in Europe and in other MEDCs such as Japan seem quite modest when compared with those of North America. The USA by itself, with less than 5 per cent of the world's people, is responsible for 25 per cent of world emissions. In output, that is equal to that from all LEDCs (which are home to 80 per cent of world population).

Consequences of global warming

Global warming is already having impacts that can be observed and measured. The Arctic ice cap is thinning by 10 centimetres a year; icebergs of record sizes are breaking off the Antarctic ice sheet during summer. Glaciers in the Alps and other high mountain areas are retreating to higher levels. Sea levels are already

18 centimetres higher than they were 100 years ago. The sequence of events shown in **Figure 4** is already happening and is expected to continue and become more widespread in its effects. Island countries like the Maldives and the Seychelles in the middle of the Indian Ocean are particularly alarmed at the prospects of further global warming, as also are delta countries such as the Netherlands and Bangladesh. One gloomy prediction is that a complete melting of the Greenland ice sheet would result in a sea level rise of 7 metres, enough to drown many of the world's big cities, including London. What is more, some scientists believe that global warming is already causing major changes in atmospheric circulation to the point where world patterns of weather are being affected. Extreme weather events, such as hurricanes, storms and droughts, appear to be happening more frequently and to be more intense with serious consequences for people in affected areas.

Strategies for the future

The Kyoto conference in 1997 focused on climate change; it proposed compulsory reductions of 5 per cent for carbon dioxide emissions for all MEDCs by 2010, compared with levels in 1990. The **benefits** have been limited. The US Senate refused to ratify the agreement and since that time US emissions have increased by 3 per cent per year. LEDCs were deliberately excluded so as not to hinder their economic development, although some of them, such as China and India, are industrializing rapidly and rely heavily on their own large coal deposits. Anyway, environmentalists argue that the agreed level of reduction was not large enough. To have any real impact, they suggested that a reduction of 60 per cent, applied worldwide, was the minimum reduction needed.

Many EU countries are struggling to meet even the modest Kyoto target. The widespread adoption of clean renewable sources of energy to replace dirty fossil fuels is seen by most people as the best strategy for a cleaner planet, but at the moment these are more expensive (page 209) and rely upon further technological breakthroughs. In this respect a time of rising oil prices is useful; look again at **Figure 2**, and there does not seem to be much chance of a big fall in oil prices in coming years. As the cost advantage of fossil fuels is reduced by high oil prices, research into alternatives will be encouraged because of improved chances of commercial success.

There are hopes for the future. Some technologies, such as wave power from 'wave machines' and 'power buoys' floating on the moving sea surface, are moving from the experimental to the operational stage. There is a greater public awareness of energy-saving measures. There are also good reasons for despair. People in MEDCs want to drive more, fly more and use more electrical and electronic equipment in their homes; 'gas-guzzling' 4x4s are fashionable in urban areas. People have hardly begun to change their behaviour. Not surprisingly, carbon dioxide emissions in the UK increased by 5.5 per cent between 1997 and 2005.

Activities

1 **(a)** Draw a line graph to show total cumulative carbon emissions from 1860 to 2000.
   Billions of tonnes of carbon
   1860: 40   1900: 60   1940: 130   1980: 270
   1880: 50   1920: 90   1960: 180   2000: 380

   (b) Describe the similarities and differences between your graph and **Figure 1**.

   (c) Explain how the two graphs might be related.

2 **(a)** Study **Figure 3**. Describe the evidence which shows that carbon dioxide and CFCs are more important greenhouse gases than methane and nitrogen oxides.

   (b) Explain how the following can reduce greenhouse gas emissions: **(i)** alternative sources of energy, **(ii)** tree planting and **(iii)** cavity wall and roof insulation in homes.

3 Give supporting information for each of these statements.

   (a) The greenhouse effect is a natural process.

   (b) Global warming already exists.

   (c) Not all countries have equal responsibility for greenhouse gas emissions.

4 **(a)** Why is international action necessary to reduce global warming?

   (b) Will it ever be possible to stop global warming from increasing? Explain your views on this.

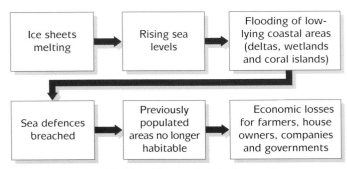

▲ **Figure 4** Worldwide effects of global warming.

Where are the environments that favour tourism?

Although in decline, the coastline remains the most important location for those taking their main summer holiday in the UK. A beach is a necessity – if longshore drift is in the habit of washing it away, the local authority builds groynes. Spectacular coastal scenery close by, such as cliffs, caves, arches and stacks, helps, as also do scenic upland areas inland from the resort for day trips. Another factor of great importance is climate. The greatest concentration of large coastal resorts is along the south coast of England. Here the warmest summer weather and highest number of hours of sunshine are recorded.

Within the past 60 years scenic environments inland have increased in popularity. Pressure of visitors and conflicts between local people and visitors in areas of great scenic beauty led to the setting up of National Parks, made possible by The National Parks and Access to the Countryside Act of 1949. A **National Park** can be defined as 'an area of beautiful and relatively wild countryside'. Creating a National Park has two aims:

*   to preserve and enhance an area's natural beauty
*   to promote people's enjoyment of the countryside.

To achieve both aims is not an easy task – too many visitors destroy the peaceful and beautiful countryside they are all going to see. Rules and regulations are needed and they must be enforced, which is why each park is managed by its own National Park Authority.

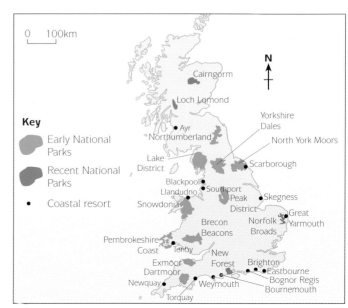

▲ **Figure 1** Coastal resorts and National Parks in the UK.

Management tasks are a mixture of the positive and the negative:

*   managing the land, undertaking conservation work, planting woodland and repairing/re-routing footpaths
*   working with and advising local landowners
*   controlling building and new commercial developments
*   providing access and setting up facilities for visitors, such as information centres, car parks and picnic sites, while controlling where they are located.

▲ **Figure 2** Yorkshire Dales National Park. **A**: Improved footpath using local raw materials near Malham Cove.
  B: Roadside commercial enterprise, which would be discouraged.

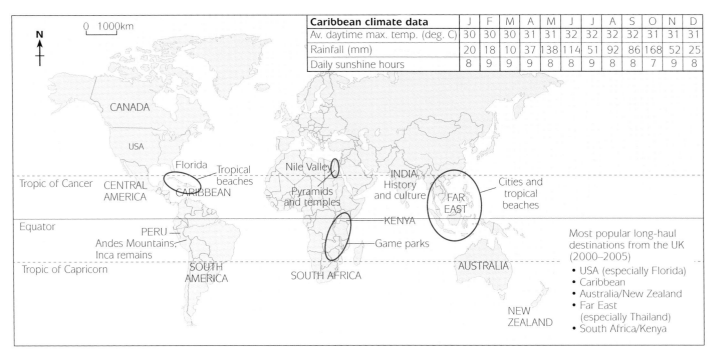

Caribbean climate data	J	F	M	A	M	J	J	A	S	O	N	D
Av. daytime max. temp. (deg. C)	30	30	30	31	31	32	32	32	32	31	31	31
Rainfall (mm)	20	18	10	37	138	114	51	92	86	168	52	25
Daily sunshine hours	8	9	9	9	8	8	9	8	8	7	9	8

▲ **Figure 3** Some popular worldwide destinations.

Worldwide tourism is a great growth industry. Increased leisure time, higher incomes and the growth of cheap and readily available air travel and package holidays have helped turn tourism into one of the world's fastest-growing industries. Tourism accounts for up to 10 per cent of world trade and is one of the world's largest industries. The popularity of long-haul holidays is growing. One purpose may be to visit relatives. Another is the desire to escape from winter weather to a warm tropical beach. This is why the peak season in the Caribbean stretches from December to April, which also coincides with the dry season there (**Figure 3**).

However, people's horizons are broadening and an increasing number wish to see environments and experience cultures different from their own. For many years Kenya dominated the African wildlife and safari market for tourists, but it is now facing stiff competition from South Africa. Central and South America have leapt up the chart of long-haul destinations as people have become aware of the region's natural attractions (mountains, waterfalls, glaciers, jungles and wildlife) as well as the monuments from earlier civilizations (Mayas, Aztecs and Incas).

The desire of people in MEDCs to travel, along with affordability, has opened up new economic opportunities for some LEDCs. Many LEDCs still depend too heavily upon the export of just one or two commodities for their foreign exchange income. Until tourism arrived it was difficult to see how their income could be increased. St Lucia is typical of many small island countries in the Caribbean. Bananas are the main export, but they are grown on smallholdings and the farmers cannot match the low prices of bananas grown by large companies on plantations in Central America. With only 150 000 people on the island, industrialization is not an option, so it is fortunate that visitors from North America and Europe find the climate and scenery there attractive.

Activities

1 Study **Figure 1**.

   (a) Describe the distribution of the ten early National Parks.

   (b) State two ways in which the distribution of the recent National Parks is different.

   (c) Explain why:

   (i) there are more National Parks in the north and west than in the south and east of England

   (ii) the greatest number of coastal resorts is along the south coast of England.

2 Choose one of the National Parks.

   (a) Describe the physical (landscape) features which attract visitors to the Park.

   (b) Choose one of these features and explain its formation.

3 Find information about one of the tourist countries or areas named in **Figure 3** (apart from Kenya) by searching for websites for national tourist organizations and holiday companies. Design an Information Sheet of useful geographical information for someone intending to take a holiday there (such as climate, natural and human attractions, map of locations of places of tourist interest, and transport). Your sheet should fit an A4 page.

Focus on a National Park

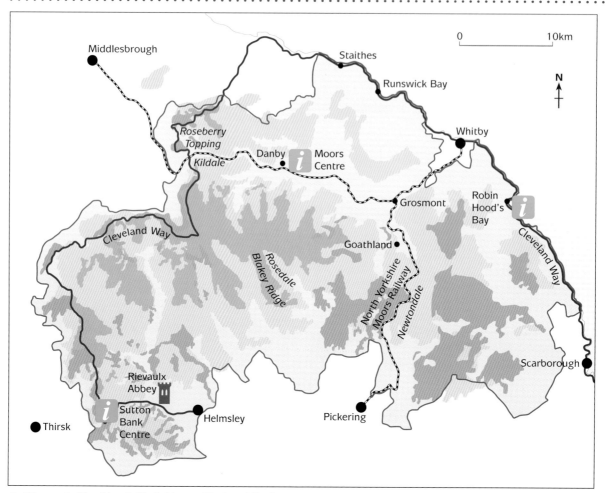

▲ **Figure 1** The North York Moors National Park.

▲ **Figure 2** Inland scene across the North York Moors.

▲ **Figure 3** Rugged coastal scenery, North Yorkshire.

i Information

The North York Moors National Park

Key facts about the Park
- created in 1952
- about 150 000 hectares in area
- population about 25 000
- about 12 million visitors per year.

Physical features and attractions
- plateau-like surface cut by river valleys (dales)
- heather-covered moorlands on the higher parts
- bracken, woods and farmland on the lower slopes
- spectacular coastal scenery of high cliffs with rocky wave-cut platforms and wide bays with sandy beaches
- Jurassic geology with fossils and dinosaur footprints.

Human attractions
- old fishing villages such as Staithes and Robin Hood's Bay
- churches and abbeys such as Rievaulx Abbey
- long-distance footpaths such as the Cleveland Way
- North York Moors Steam Railway from Grosmont to Pickering
- moorland villages in scenic settings, such as Goathland, the location for the TV series *Heartbeat*.

The Park's main problem – pressure from visitor numbers

The Park is close to densely populated urban areas, notably Teesside. On sunny summer weekends as many as 150 000 visitors may arrive in the Park, mostly by car, which leads to congestion on the roads and pressure on car-parking spaces. The visitors are not equally spread throughout the Park, and most make for the main tourist sites such as Robin Hood's Bay, Helmsley and Goathland, which become overcrowded.

Some head for the coast, filling the quaint old fishing settlements which were not built for the era of private transport. Others make for settlements inland, notably Helmsley, close to Rievaulx Abbey, which has the widest range of shops, cafés and restaurants, and Goathland because of its association with the TV series *Heartbeat*. From the top of Roseberry Topping there are fine views over the Tees valley and estuary; it is only a short drive away from highly populated Teesside.

Places such as these which attract many visitors are known as **honeypots**. As with the other National Parks, this is only one of the problems that has resulted from the growth in tourist numbers.

The Threat to *Heartbeat* Country

How the popularity of the TV series is harming Goathland's environment

The popularity of the series, which depicts a policeman's lot on the North York Moors, has caused serious problems for the villagers of Goathland, the moorland community in which it is filmed.

Heartbeat's success brings more than a million fans to the village each year, trampling the grass, churning up the roadsides and causing horrendous traffic congestion with their inconsiderate parking. The tourist hordes are eager to see where their favourite show is set. The low estimate is for 1.25 million visitors a year, which means that on a busy day, with a staggering 700 cars parked in Goathland, it can take three-quarters of an hour to pass through the village.

The response to the problem has been to deter car-parking by significantly expanding the yellow lines in the village and installing kerbs to protect particularly vulnerable roadsides, measures which are starting to prove successful.

David Brewster, the Head of North York Moors National Parks Services, which has a partial responsibility for tackling the problem, said, 'We are not trying to stop *Heartbeat* being filmed here. It does bring income into the area and into Whitby (estimated at £9m in 1996), but there are also significant problems and we have to do something to reduce the impact that flows from them.'

Parish council chairman Keith Thompson said, 'We have a concern for the environment and a duty of care for the village. The village evolved over a long period of time and we should not pass it on bespoiled when problems can be addressed. A lot of people in the middle of the village say they can sit in their gardens and ignore the people milling about outside but there has been some upset to quality of life. However, I don't think people resent the show because the visitors provide a livelihood for a lot of people.'

▲ **Figure 4** Magazine article from 1997.
Source: Adapted from *Yorkshire Life* April 1997

Activities

1 Draw a sketch map of the area of the North York Moors National Park. Locate and label on it some of the physical and human attractions for visitors.

2 Describe with the help of labelled sketches the physical attractions of this National Park.

3 **(a)** Make a chart showing costs and benefits of tourism for Goathland.

   (b) Why will some people in the village object more strongly than others to the great numbers of visitors?

Other problems and conflicts in National Parks

A Footpath erosion

Visitors who walk more than half a mile away from their parked car number fewer than 50 per cent of those who visit the Parks. Dramatic landscape features at the end of a short and easy walk from the car park, such as Malham Cove in the Yorkshire Dales, have footpaths 'almost like motorways' leading up to them (**Figure 1**). There are still many keen walkers who love the hills. The long-distance footpaths such as the Pennine Way, the South Downs Way and the Cleveland Way, which is in the North York Moors, are very popular and well used. The regular pounding of boots soon removes the thin vegetation cover, exposing loose soil which, once eroded by people, is quickly washed away by the rain. As people look for more comfortable walking on the sides of the path, the zone of erosion is widened.

Management in National Parks includes footpath repairs. Steps made of local stone are one solution on steep hillsides (**Figure 2A** on page 214). Slabs of stone are laid in flatter areas. Badly eroded sections may be fenced off and reseeded. Wooden rafts are laid across marshy ground.

◀ **Figure 1** The 'road' to Malham Cove in the Yorkshire Dales National Park. At least most of the erosion from people is limited to the line of the footpath.

B Conflicts between local people and visitors

The basic cause of conflict is that making a living in upland areas, where most of the Parks are located, is never easy. Farming is the activity which covers most land. In the 1980s some farmers in the North York Moors reclaimed moorland into pasture to try to increase their farm income; according to some environmentalists this destroyed the beauty of the wild landscape and reduced wildlife habitats.

Extending the area covered by coniferous forests is even more contentious. Nearly 20 per cent of the open moorland on the North York Moors has been ploughed up and planted with conifers in the past 30 years, and there are some large stretches of forest (**Figure 2**) which dramatically affect the scenery and limit visitor access. The trees have economic value. However, some lessons have been learned over the years. It is now accepted that much can be done to improve the appearance by planting some deciduous trees as well, and to improve the attractions for visitors by developing forest walks, picnic sites and information points. Even so, visitors must be constantly reminded and warned against carelessly starting fires.

The other locally significant conflict, quarrying, was dealt with on pages 32–33. This is an old-established economic activity which can give permanent employment to significant numbers of local people. It does come as a shock to many first-time visitors to a Park to see works in the midst of splendid scenery (**Figure 3**), but there are usually good reasons for the works being there, and their products may be essential to life in the city, from which most Park visitors come.

▲ **Figure 2** Coniferous plantation on the North York Moors.

▲ **Figure 3** Boulby potash mine within the North York Moors Park.

Other problems and conflicts in National Parks

KENYA SAFARI   7 nights

- **Day 1** London to Nairobi with British Airways
- **Day 2** Nairobi to Aberdares
  Transfer to the highlands of the Aberdare range and overnight at the Aberdare Country Club.
- **Day 3** Aberdare National Park
  Transfer to The Ark, situated in the heart of the Aberdare National Park. In the evening enjoy floodlit game viewing.
- **Day 4** Aberdares to Sweetwaters
  Transfer to the Ol Pejeta Rhino Reserve and Sweetwaters tented camp, overlooking one of the busiest waterholes in the reserve.
- **Day 5** Sweetwaters to Mount Kenya
  After an early morning game drive, continue to the famous Mount Kenya Safari Club on the slopes of Africa's second highest mountain.
- **Day 6** Mount Kenya
  A full day at leisure at the Mount Kenya Safari Club. Take an optional excursion to the Animal Orphanage or to Mount Kenya or treat yourself to an 'Aerial Safari', a 20-minute scenic flight over Mount Kenya.

KENYA COAST EXTENSION   7 nights

- **Day 7** Mount Kenya to Masai Mara
  A short flight takes you to Masai Mara and the Mara Safari Club.
- **Day 8** Masai Mara
  Another full day at the Mara Safari Club with morning and evening game drives. You can also take part in the walking safari or take an optional hot-air balloon ride over the Masai Mara Park.
- **Day 9** Masai Mara to Nairobi to Kenya Coast
  After a short-flight transfer to Nairobi, travel by Kenya Airways to Mombasa. Ground transfer to the Turtle Bay Beach Club located 20km south of Malindi on a white sand beach within the Watamu Marine National Park.
- **Days 10–15** Kenya Coast
  Spend a lively and relaxing week next to the sea. Use of the hotel's water sports and sports and fitness facilities is included at no extra cost.
- **Day 16** Kenya Coast to Nairobi to London
  Fly back to Nairobi to connect with the overnight British Airways flight to London.

▲ **Figure 4** A typical holiday in Kenya.

Climatic data

	J	F	M	A	M	J	J	A	S	O	N	D
Nairobi (1800m above sea level)												
Mean daily temperature (°C)	19	20	20	19	18	17	16	16	18	19	18	18
Mean rainfall (mm)	38	64	125	211	158	46	15	23	31	53	109	86
Daily sunshine (hrs)	9	9	9	7	6	6	4	4	6	7	7	8
Mombasa (sea level)												
Mean daily temperature (°C)	28	28	28	27	26	26	25	25	25	26	27	27
Mean rainfall (mm)	25	18	64	196	320	119	89	64	64	86	97	61
Daily sunshine (hrs)	8	9	9	8	6	8	7	8	9	9	9	9

▼ **Figure 5** Kenya.

Activities

1 Using the normal methods for temperature and rainfall (see pages 84–5) and an appropriate method for sunshine hours, plot the climatic data for Nairobi.

2 **(a)** State the main similarities and differences between the climates of Nairobi and Mombasa.

   (b) What is the main reason for the differences in temperature?

3 From the climate point of view, in which month would you advise non-geographical friends to visit Kenya? Explain your answer.

4 From **Figure 5**, calculate:

   (a) The size of Kenya's visible trade gap (difference between total earnings from imports and exports).

   (b) The size of the trade gap after invisible earnings from tourism are taken into account.

   (c) The percentage of total income from exports (visible and invisible) contributed by tourism.

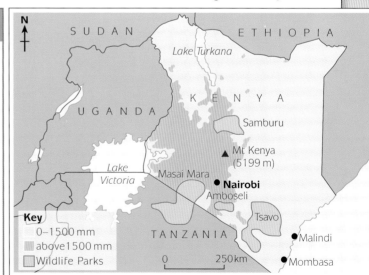

Key
- 0–1500 mm
- above 1500 mm
- Wildlife Parks

0   250km

Key facts 2004

Total population	About 32 million	Total value of exports	US$ 1944 million
Birth rate	34 per 1000		
Death rate	16 per 1000	Main imports	Crude oil and refined products, industrial machinery, motor vehicles and parts
Annual rate of natural increase	1.8 per cent		
GDP per head	US$ 360		
Main exports	Tea, coffee, flowers, fruit and vegetables	Total value of imports	US$ 3192 million
		Earnings from tourism	US$ 544 million

Tourism in Kenya

The information about Kenya on the previous page highlights the fact that Kenya has two different environments attractive to foreign visitors:

* wildlife parks on the plateau
* Indian Ocean coastline.

Although in a typical two-week holiday to Kenya visitors spend one week on safari and one week at the coast (see **Figure 4** on page 219), surveys have shown that for up to 80 per cent of visitors the principal reason for choosing Kenya was the wildlife. Kenya was ahead of most other East African countries in protecting its wildlife within 45 National Parks and Game Reserves, which cover 10 per cent of the country, and in providing luxurious accommodation in safari lodges and clubs. Kenya remained comparatively prosperous when neighbouring countries like Somalia, Ethiopia and Sudan were racked by civil wars.

◄ **Figure 1**
Luxury sarari accommodation.

The benefits

As an LEDC, Kenya relies massively on the inflows of foreign exchange to sustain its economic growth. Earnings from tourism were estimated to have increased to US$ 670 million in 2005, about 12 per cent of total GDP. Over 200 000 people (more than 10 per cent of the official national workforce) are directly employed in tourism, while at least another 300 000 are employed indirectly. This is because tourism as a service industry creates many related job opportunities in a variety of economic sectors, including agriculture, drink, transport, entertainment, textiles and crafts. It generates a large multiplier effect. For example, while staying in a hotel, if tourists spend money in local shops and cafés and on local services such as taxis, this money becomes available to local people for purchasing their daily requirements of food and services. Tourist income spent locally drips down through several levels before it is exhausted.

KENYA

One of the 'Big Five' that tourists can expect to see on safari.

Of all the countries in Africa, Kenya has some of the most prolific and most accessible game parks. Here you can observe some of the greatest examples of wildlife, including the 'Big Five' – elephant, lion, leopard, buffalo and rhino. Scenically stunning with vast expanses of savanna grassland and bush across the plateau of the highlands, it is blessed with beautiful mountains such as Mount Kenya, Africa's second highest peak. A tempting option for the more adventurous visitor is to take the three-day trek complete with porter, cooks and guides to the 5000-metre summit.

As to the beaches, the coastal strip from Malindi to beyond Mombasa has mile upon mile of white coral sand, lapped by the warm waters of the Indian Ocean, sheltered in parts by gently swaying palms. Why not take a trip in a glass-bottomed boat on to the reef where you can see over 240 species of fish and a wide variety of corals? And if you want something different, visit Mombasa – it is hot and dusty but there is a buzz of excitement in its colourful bazaars!

Safaris

Seeing animals in the wild, in their own natural habitat, free of any civilizing influences, is a life-enriching experience. Seeing them in Africa is truly awesome. While a bush safari is one of life's great adventures, we make sure that it is a comfortable one. You will travel in a specially adapted minibus with a guaranteed window seat, be guided by an English-speaking driver and accommodated in comfortable lodges.

▲ **Figure 2**

The problems

These are a mixture of economic, environmental and social.

1 Economic – visitor numbers go up and down

The number of visitors to Kenya crashed in the second half of 1997 with severe consequences for local employment and incomes (**Figure 4**). There was a repeat in 2002 and early 2003 (**Figure 3**) after a missile attack on an aircraft and a car bomb outside a Mombasa hotel that killed 13 people. In early 2003 lost tourist income for Kenya was calculated at US$1 million a day.

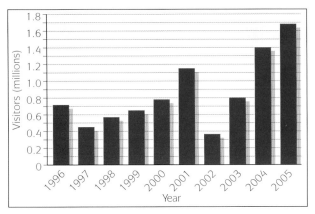

▲ **Figure 3** Number of visitors to Kenya, 1996–2005.

2 Environmental damage

On the reefs, boats drop their anchors into the coral, which is a fragile, living community. People walk on the coral and some tourists take pieces of coral away as souvenirs. In the game parks, drivers are keen to get as close as possible to the animals; the army of minibuses surround and disturb animals (**Figure 5**). They churn up the ground in the wet season and the grassland is changed into a 'dust bowl' in the dry season. Particularly in the Masai Mara, there is too high a concentration of visitors in too small an area.

▲ **Figure 5** Lions and tourists in the Masai Mara.

> When I visited a thatched safari resort in the Masia Mara National Park last week, I was almost the only person staying there. Most of the other 55 rooms were empty. In what should have been the high season, tourism in Kenya has dropped away alarmingly.
>
> All over the interior the game parks and lodges are under half full. On the coast hotels are laying off staff and closing down. Everyone is being hit by the downturn, including the craft sellers along the streets and the hawkers on the beach.
>
> A combination of factors is responsible – some to do with Kenya, some not. Violence has flared up in the approach to Kenya's election. There have been a number of violent incidents in the coastal resorts, including the murder of a member of a band playing in one of the large tourist hotels. Tourists from Europe and North America have been frightened off from travelling to Africa by the massacre of innocent tourists in Luxor – even though this was thousands of miles away in Egypt. For many potential visitors to Kenya, these incidents were too close for comfort.

▲ **Figure 4** Visitor report, 1997.

The Kenya Wildlife Service (KWS) is responsible for the game parks. It is under-resourced for the important work it needs to do, and although many are well motivated, poorly-paid employees with limited experience of the work are more likely to be open to bribery. Some ignore poaching. Others take no action against minibus drivers who go off the roads to get close to the wild animals to earn higher tips from their tourist passengers.

3 Social – conflicts with local people

There are also *conflicts* between local tribespeople, the Masai, and the Kenyan authorities. When the game parks were set up, the Masai were driven off the land to make way for wild animals. Acute shortages of grazing land, coupled with rapid population growth, have forced farmers to move closer to the edges of the parks. Elephants trample their crops. Lions eat cattle. Villagers and tribespeople are injured, or sometimes killed, by wild animals, but they are not allowed to kill them. A survey revealed that less than 2 per cent of the money spent at the world famous Masai Mara Park benefited the local Masai people; even the high US$ 27 daily entry fee went direct to the government in Nairobi.

Activities

Write up and illustrate 'Tourism in Kenya' as a case study using these four headings:

- Tourist attractions and their locations
- Benefits of tourism
- Problems
- Changes (you also need to look ahead to page 223).

Case Study – Tourism in an LEDC (continued)

The impacts of tourism

ADVANTAGES	DISADVANTAGES
A Environmental	**A Environmental**
• Greater awareness of the need for, and interest in, conservation of landscape features, vegetation and wildlife, and preservation of ancient monuments.	• Destruction of the environment, and resulting loss of habitats, in order to build airports, roads, hotels, etc.
• Income from tourism/ entrance fees may pay for management, conservation and repairs.	• Loss of peace and quiet.
	• Pressure on frequently visited landscapes, such as footpath erosion and soil erosion.
	• Pollution problems such as litter, or untreated waste going into rivers or the sea.
B Economic	**B Economic**
• Earns foreign exchange for a country from overseas visitors.	• Country does not gain the full benefits of income from overseas visitors – as little as 15 per cent of money spent on a holiday reaches the host country. A lot of materials and highly-paid staff may be imported. Money for development may be borrowed, increasing debts.
• Increases the size of the domestic economy; people are employed in service occupations as tourism tends to be labour-intensive. There is also a greater market for souvenirs from craft industries and food from farms.	• Numbers of visitors go up and down.
	• Many jobs created by tourism are unskilled, of low status, poorly paid and seasonal.
• The infrastructure (airports, roads, water and electricity supplies), which is improved for tourists, can also benefit local people.	• Some local people, such as farmers or fishermen, may lose their livelihoods in areas where tourist facilities are developed.
C Social/Cultural	**C Social/Cultural**
• Local cultures and traditions may be preserved.	• Local traditions may disappear even faster in favour of copying the visitors.
• Development of tourism may halt migration loss because of new employment opportunities.	• Tourists look down on local people and treat them badly.

▲ **Figure 1** Advantages and disadvantages of tourism.

The case studies on pages 216–17 and 219–21 have shown that tourism, like other economic activities, has both positive and negative impacts upon people and the environment in those areas where it is important. These are summarized in **Figure 1**.

Most of these advantages and disadvantages apply to tourist locations anywhere in the world. However, the disadvantages may be more severe for LEDCs. For example, management and stewardship of the environment are more difficult because they lack the human expertise and money to be as effective as in MEDCs, as the example of Kenya showed.

Also, much of the technology and many of the materials essential for setting up the facilities and services needed for large-scale tourism in LEDCs have to be bought from MEDCs. Think what is needed before a new resort receives one tourist – airport, planes, paved road to the resort, buses, hotels with 'Western' facilities, services such as water, sewerage and electricity. How many of the materials and fittings needed would be manufactured in an LEDC such as Kenya? How much of the money for a holiday to Kenya costing £1500 per person goes to Kenya (**Figure 2**)?

Minimum benefit to an LEDC	Maximum benefit to an LEDC
• Book and pay for holiday in home country.	• Pay for as little as possible in home country.
• Travel on an airline from an MEDC.	• Travel on an airline from an LEDC.
• Stay in luxury hotels owned by American or European companies.	• Stay in small hotels or guest houses and pay locally.
• Eat meals in the hotel.	• Eat in local restaurants.

▲ **Figure 2** Who benefits?

'Green' tourism

The use of 'Green' implies a holiday that is environmentally friendly. It may just be a sales gimmick. It may do no more than just distinguish mass-market package tours to popular destinations from small tours to more distant, environmentally interesting places. One person has described it as 'ordinary tourism dressed up in a politically correct manner' by sticking 'Green' in front of it.

What 'Green tourism', or 'Eco-tourism', really refers to is **sustainable tourism**. To describe it in simple terms, **Green tourism** not only places an emphasis upon protecting the environment, but also involves the local people in making decisions that affect their land and living. The role for the local people is what makes it so different from other types of tourism.

Since banning tourism is an impractical solution, the challenge is to find how it can be used for the benefit of both the environment and local people. For eco-tourism to be effective it needs to be run with the cooperation of the local inhabitants, who need to be able to gain from it.

The need for management and stewardship

The pressures from the growth of tourist numbers have increased the need for management everywhere. In most cases effective management only begins after tourism has begun and some damage has already occurred. However, there is one exception – Antarctica. It is a wilderness and offers tourists magnificent scenery, icebergs and nesting penguins by the million (**Figure 3**).

▲ **Figure 3** Antarctica, an attractive but different tourist location.

Until recently its remoteness had saved it from tourists. By 2005 tourist numbers had passed 30 000 a year. Most arrive by cruise ship, and living on ship reduces the chances of damage to a fragile land environment. Most cruise operators follow the guidelines of IAATO (International Association of Antarctic Tour Operators), but importantly the two with the largest ships do not. They land more than 100 visitors at a time, above the maximum allowed by the guidelines. Supervising so many so that they 'do not go within 5 metres of penguins and other wildlife', 'do not walk on lichens' and 'do not leave litter or waste' becomes more difficult. Will the beauty of the Antarctic wilderness be sustained despite the arrival of tourism?

'Green' tourism in Kenya

Three-quarters of the wildlife in Kenya is found outside the game parks, much of it on land owned by the Masai (**Figure 4**). In colonial times the Masai were driven off the land to make way for wild animals in the parks. The Masai were seen as a nuisance. Now the vegetation is healthier and wildlife more plentiful outside the parks. Three tented camps, owned and run by Kenyans, have been set up in Kimana on an important migration corridor for wildlife between Amboseli and Tsavo National Parks. The Masai are paid a rent of about £1000 a year. The Masai are now seen as vital to the success of these smaller, less environmentally damaging tourist developments.

However, there are problems. Only a small amount of tourist money is trickling down to the Masai. Unable to read or understand leases, they are regularly cheated by tour operators. They need to carry on with their traditional way of life of

planting crops and keeping cattle, activities which do not fit well with encouraging wildlife. Until they can be convinced of the benefits of tourism, the living space for the wildlife will continue to decline.

◄ **Figure 4**
Masai tribesperson.

Activities

1 The government of an LEDC wishes to increase the number of tourist visitors it receives.

   (a) In what ways and why might it do this?

   (b) Explain the disadvantages that might result.

   (c) What should the government do to try to ensure the greatest benefits to the country and its people with the smallest amount of damage to the environment?

2 **(a)** Describe how **Figure 3** shows that Antarctica is a wilderness, 'an area shaped only by the forces of nature'.

   (b) Using the Internet, investigate 'Tourism in Antarctica – attractions and risks' (see Hotlinks, page iv).

Tourism interdependence between MEDCs and LEDCs – does it exist?

Interdependence exists when two or more countries have a shared need to exchange goods and services. It should exist in tourism.

A More and more people in **MEDCs**: are looking for new experiences and new places to visit; want to escape from the cold European and North American winters; have the holiday time and money to travel to distant countries on long-haul holidays.

B **LEDCs** can offer different and unique physical and human attractions. Most people would prefer to see wildlife in its natural habitat than in a zoo or a British 'safari park'. LEDCs contain many of the world's natural and human wonders. Opportunities for activity holidays such as trekking and snorkelling are increased; skiing is possible at any time of year. During winter in the northern hemisphere, it is hot and dry in the Caribbean (see **Figure 3** on page 215) and summer everywhere in the southern hemisphere.

However, as usual the balance is tilted in favour of MEDCs, where the tour operators are based. They strike hard bargains with local tour operators in LEDCs, who have to rely upon them to find customers for their tourist products. European and American tour companies prefer to use their own airlines and international hotel chains such as Hilton and Holiday Inn, allowing only a small percentage of the total holiday cost to trickle down to the host LEDC.

The growth potential for LEDCs from tourism remains enormous. Notice how quickly tourist numbers bounced back in Kenya after the threat to tourists was reduced (see **Figure 3** on page 221) – because Kenya has so much to offer. The tourist tree shows some of the potential spin-offs for LEDCs which are successful in tourism (**Figure 1**). In practice, not all LEDCs can benefit. Wars, political instability, high crime rates and levels of corruption are great deterrents. They exist more widely in African countries than in any other continent.

Exam focus

1 a Study the climatic data for the Caribbean in **Figure 3** on page 215. State two reasons why December to April is the peak tourist season in the Caribbean.   (4 marks)

b (i) Name the tourist-related problem shown in **Figure 2A** on page 214 and **Figure 1** on page 218.   (1 mark)

(ii) Describe how it is being managed.   (2 marks)

(iii) Why was management necessary?   (2 marks)

(iv) How successful is the management? Explain your view.   (2 marks)

c (i) Describe what **Figure 3** on page 221 shows about visitor numbers to Kenya from 1996 to 2005.   (3 marks)

(ii) State and explain the problems for workers in the tourist industry in Kenya in 1997 and 2002.   (5 marks)

d Describe how **Figure 1** on this page shows that tourism in an LEDC can **(i)** increase a country's income and **(ii)** lead to better public services and a higher quality of life for its people.   (6 marks)

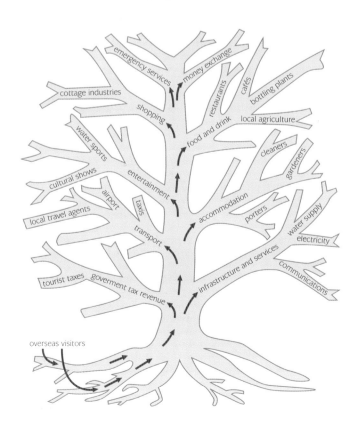

▲ **Figure 1** Tourist tree.

Chapter 13

Development

Rush hour in a country's capital city. What do you think is this country's level of economic development? Is it a low-, medium- or high-income country?

Key Ideas

Contrasts in development are related to economic, environmental, social and political conditions:
- contrasts in level of development between MEDCs and LEDCs are indicated by factors such as population, health, literacy, housing and GNP
- environmental hazards can contribute to lack of development as also can lack of access to clean water, sanitation and adequate food supply
- appropriate technology and sustainable development are needed.

Trade and aid in MEDCs and LEDCs and their consequences:
- the main pattern of world trade is long established and favours MEDCs
- MEDCs in the developed 'North' exchange manufactured goods for raw materials from LEDCs in the developing 'South'
- this should lead to interdependence between MEDCs and LEDCs
- instead there are imbalances in trade leading to debt among LEDCs and the need for different types of aid.

Global contrasts in development between MEDCs and LEDCs

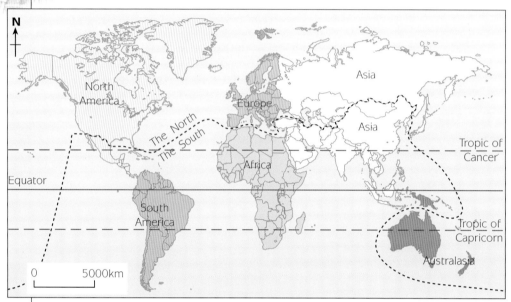

▲ **Figure 1** The world divided into more and less economically developed countries.

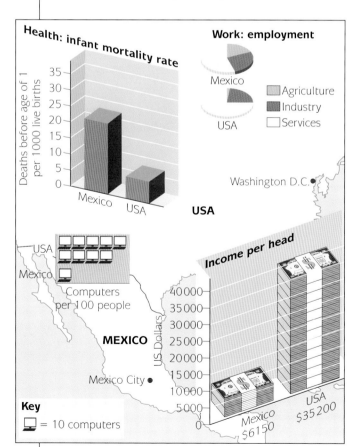

▲ **Figure 2** Contrasts between the USA and Mexico.

Throughout this book there have been repeated references to more economically developed countries (MEDCs) and less economically developed countries (LEDCs). These labels for countries are based upon a world split into two parts – an economically developed 'North' and an economically developing 'South'. The course of the dividing line between the two is shown in **Figure 1**.

Of course, any attempt to place about 180 different countries into one of only two categories of development must have exceptions. There are bound to be countries that do not fit into either category in a satisfactory way. Also, the fortunes of countries change with time and sometimes there are major political changes (as in the former USSR) which affect development, but the dividing line between MEDCs and LEDCs remains in the same position as the one originally fixed in 1980.

One example of a clear-cut divide between 'North' and 'South' is where the line follows the USA–Mexico border. Mexico has achieved higher levels of development than most of the other LEDCs, but it is dwarfed by the size and wealth of the USA. Some of the differences are summarized in **Figure 2**.

The wealth of the USA is almost six times greater than that of Mexico; this means that Americans have more money to spend on consumer goods like cars and computers. Healthcare is better. Many Mexicans are farmers, whereas most Americans are city dwellers working in the service sector. With such a large gap in the level of development on the two sides of a land border, it is not surprising that there are millions of illegal immigrants working in the USA. Mexicans can earn as much in an hour in the USA as in a day in Mexico.

What is meant by 'development'?

'Development' means almost the same as 'wealth'. Although wealth is an economic factor, variations in wealth affect quality of life, health, literacy and housing, which are examples of social conditions. Therefore development can be measured by both economic and social factors.

A *developed country* is a rich country. It may be endowed with natural resources that have been used to create wealth. Most developed countries are industrialized. Incomes are high. After fulfilling the everyday human needs for water, food, shelter and clothing, most people have money left over (disposable income) for

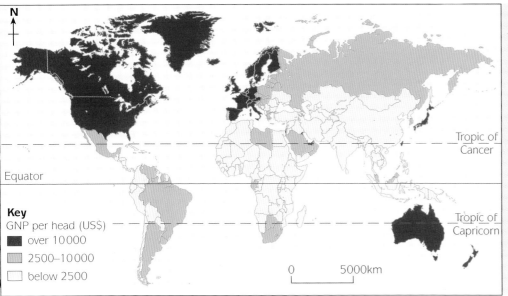

Key
GNP per head (US$)
■ over 10000
▨ 2500–10000
□ below 2500

0   5000km

▲ **Figure 3** GNP per head in the mid-1990s.

- To what extent does plotting the data for GNP support division of the world into two parts as shown in **Figure 1**?

- Can the position of the dividing line in **Figure 1** be justified by the economic data shown?

- How large are the variations in levels of economic development between those countries classified as LEDCs?

- Are variations within Asia greater than those inside the other continents?

buying consumer goods and luxuries for the home and for themselves, or for spending on entertainment, leisure and travel. A large, productive service sector develops.

A *less developed country* is a poor country. In many of these countries there is still a great dependence upon farming, which has not been modernized and from which output is low. Although industry is increasing in some countries, in others it still makes only a small contribution to the economy, while the service sector is dominated by the provision of small services for poor people. For all but the few people who are very rich, there is a constant struggle to achieve even the bare necessities of life. With insufficient food and without clean water, health suffers, particularly that of infants and children. Medical care is sparse. Access to education is limited as well, with resulting low levels of literacy, and the chances of a child improving upon the standard of living and quality of life of its parents are not good. Living in dirty, cramped conditions is the lot of many millions of people in South America, Africa and Asia.

GNP as a measure of wealth

GNP (**Gross National Product**) per head is the most widely used indicator of a country's level of development. It is the US dollar value of a country's final output of goods and services in a year, divided by its population. (US dollars are used for easy comparisons between countries.) When the GNPs for all countries are plotted on a world map (**Figure 3**), a good indication of the world distribution of wealth is given. Countries with medium and low incomes together produce about 20 per cent of the world's goods and services, but house more than 80 per cent of the world's population. Africa, Asia and South America are dominated by them. With the exceptions of Japan, Australia, New Zealand and a few small countries in Asia, such as Singapore and the oil-rich Gulf States, high-income countries are confined to North America and Europe.

GNP is a more reliable indicator of wealth in MEDCs than LEDCs for two reasons.

1 The statistics for working it out are likely to be more readily available and more accurate.

2 Only the values of products sold are included in its calculation. Food produced by farmers for subsistence purposes (to feed themselves and their families) does not have a recorded money value. In many LEDCs, farming remains the main occupation, so it is likely that the GNP represents an undervaluation for poor countries.

Activities

1 The table below shows average income per head by continent.

Continent	Average income per head (US$)
Africa	680
Asia	2100
Australasia	13900
Europe	12100
North America	26900
South America	3080

(a) Show these values on an outline world map using the shading (choropleth) method.

(b) Add the dividing line separating 'North' from 'South'.

(c) How effective is this line for showing world differences in wealth?

2 Try to answer the questions about GNP in the text next to **Figure 3**.

Population and development

Characteristic	Definition	MEDCs	LEDCs
Birth rate (BR)	Live births per 1000 people per year	Low (typical values 8–15)	High (typical values 20–55)
Death rate (DR)	Deaths per 1000 people per year	Low (average 9–10) Trend: Slight rise	Low (average 9–10) Trend: Mainly falling
Natural increase (NI)	Population growth when birth rate is higher than death rate. Replaced by natural decrease when birth rate is lower	Low or non-existent (under 10 per 1000); can be a minus value	High (typical values 15–40)
Stage in DTM (Demographic Transition Model)	Line graphs showing relationship between birth and death rate through time	4 or 5 (5 when DR is higher than BR)	2 or 3 (2 when NI is large)
Population structure	Make up of a population by age and sex, shown by population pyramids	Low % under age 15 (below 25%)	High % under age 15 (30–50%)

Some of the population differences between MEDCs and LEDCs are quite striking. Although these were referred to in Chapter 8, they are repeated in summary form in **Figure 1**, because not all of you will have made a special study of population.

These population characteristics are examples of *social* indicators of development. A fall in a country's birth rate usually accompanies an increase in national and personal wealth. The national government has more funds with which to promote birth control campaigns and to set

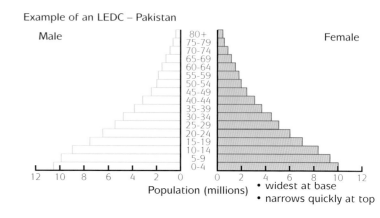

Example of an LEDC – Pakistan

• widest at base
• narrows quickly at top

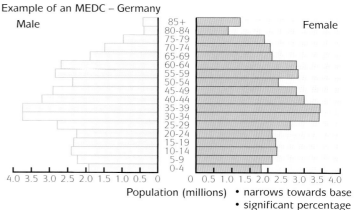

Example of an MEDC – Germany

• narrows towards base
• significant percentage over 65

▲ **Figure 1** Population characteristics.

up family-planning clinics. With greater personal wealth, attitudes towards having children change. No longer are children looked upon as economic assets for old age; instead they are seen as an economic cost that reduces the amount of money left for purchasing consumer goods and services. As birth rates decline the proportion of young people under 15 goes down. Closing the gap between birth and death rates brings down the rate of natural increase; in this way a country progresses from Stage 2 to Stage 3 of the DTM (Demographic Transition Model (see **Figure 1** on page 120), which, if continued, will allow transition into Stage 4. In some (mainly European) countries, birth rates are now lower than death rates. This leads to a natural decrease in

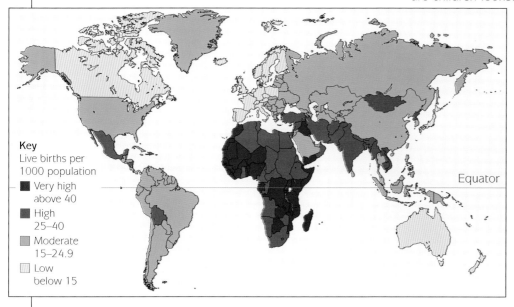

Key
Live births per 1000 population
■ Very high above 40
■ High 25–40
▨ Moderate 15–24.9
▧ Low below 15

Equator

▲ **Figure 2** World distribution of birth rates.

population shown in Stage 5, which is shown in the model by the line for death rate lying above the one for birth rate. When over 40 per cent of a country's population is below the age of 15, as is the case in Pakistan, birth rates will remain high for many years because there are so many young people yet to reach child-bearing and child-producing ages – even if they practise birth control more than their parents did.

Population changes, especially a lowering of the birth rate, go hand in hand with increased levels of economic development. Compare **Figure 2**, showing the world distribution of live births per 1000 people, with **Figure 3** on page 227, showing GNP per head. How closely does the pattern of dark shading on **Figure 2**, showing high birth rates, match up with the pattern of light shading on **Figure 3**, indicating low income per head? One value is low at the same time as the other is high, which suggests a negative relationship between low income and high birth rates.

It is possible to show relationships between two values by drawing a scattergraph. In **Figure 3** on page 227 an almost random selection of twelve countries, with all continents represented, was used. The advantage of a scattergraph is that not only does it highlight whether a general trend exists, it also identifies countries that are anomalous (i.e. ones that do not fit the general trend). This graph confirms the negative relationship with wealth decreasing as birth rates increase. Russia and China can be identified as anomalies; they buck the trend by having low birth rates *and* low incomes per head. To explain those, there must be special factors. In Russia it was economic, political and social upheaval caused by the break-up of the former USSR; in China it was the strict enforcement of the 'one child per family' population policy. Are there other anomalies?

▼ **Figure 3** Income per head and birth rates in 12 countries.

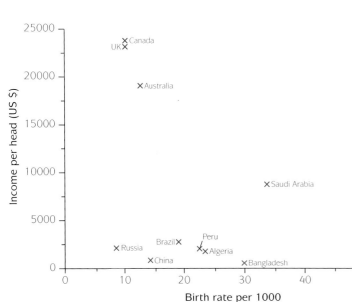

Did you notice in **Figure 1** one population characteristic that cannot be used to differentiate between MEDCs and LEDCs? This is the death rate. Death rates have fallen rapidly all over the world in the last 50 years, due to the spread of improved medical knowledge. Although medical facilities, access to safe water and diet are better in MEDCs than in LEDCs, MEDCs have ageing populations (an increasing percentage of people aged 65 and over). This stops death rates from falling any lower, because there are many people reaching the life expectancy, high though this is. Whereas in LEDCs, the high percentage of young people keeps the death rate low; this is because it is a ratio per 1000 people and many people are at ages well below life expectancy, even though this is usually lower than in MEDCs. Death rates above 20 per 1000 are rare today, almost entirely restricted to a few poor African countries. In 2006 it was reported that average life expectancy had fallen below 40 years in three countries (Sierra Leone, Swaziland and Zimbabwe) due to conflict, poverty and HIV/AIDs, halting (perhaps temporarily) the downward trend in death rates.

Activities

1 (a) From the population pyramids in **Figure 1**, work out the total number of people under 15 and over 64 in **(i)** Germany and **(ii)** Pakistan.

(b) Latest birth and death rates:

	Germany	**Pakistan**
Births (per 1000)	8.2	36.3
Deaths (per 1000)	10.6	9.6

(i) Work out the natural increase or decrease for the two countries.

(ii) Why are they so different?

(iii) The death rate is lower in Pakistan than in Germany. Does this mean that health services in Pakistan are better? Explain your answer.

2 From the summary in **Figure 1**, draw labelled diagrams, graphs or sketches to illustrate the other differences (apart from structure) in population characteristics between MEDCs and LEDCs.

3 (a) Describe the differences between Europe and Africa shown in **Figure 2**.

(b) Using information from Chapter 8, pages 118–19, explain these differences.

Other measures of development

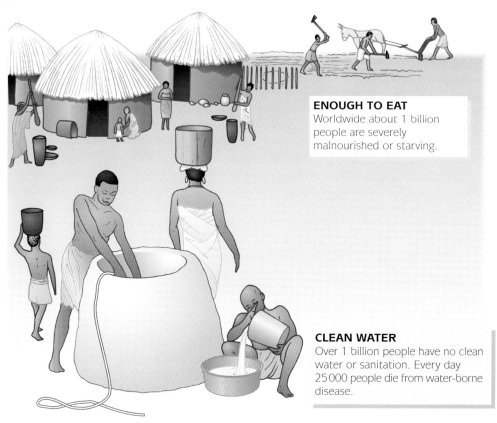

ENOUGH TO EAT
Worldwide about 1 billion people are severely malnourished or starving.

CLEAN WATER
Over 1 billion people have no clean water or sanitation. Every day 25 000 people die from water-borne disease.

Some of the many *social* measures used in the study of development are illustrated Figure 1.

◀ ▼ ▶ **Figure 1** Social measures of development.

	Schooling: percentage not at primary school	
	Sub-Saharan Africa	**South Asia**
1980	42	40
2005	50	22

Housing
- Number of persons per room
- Percentage of houses without access to electricity
- Percentage of houses with running water.

Health
- Infant mortality rate (the number of infants who die within their first year of life per 1000 live births)
- Under-five survival rates
- Life expectancy at birth
- Number of people per doctor
- Number of people per hospital bed
- Percentage of population without access to health services
- Percentage of population without access to safe water
- Average calorie intake per person per day

Infant mortality
Number of babies who die under the age of 1 per 1000 births

■ over 150	▨ 20–49
▨ 100–149	□ 10–19
▨ 50–99	▨ 0–10

UK: 9 deaths per 1000

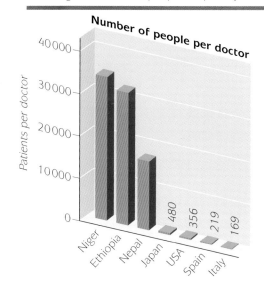

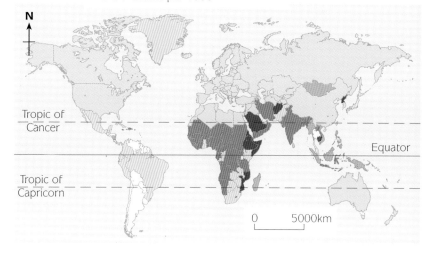

EDUCATION
One billion adults in the world are still unable to read or write, two-thirds of them women.

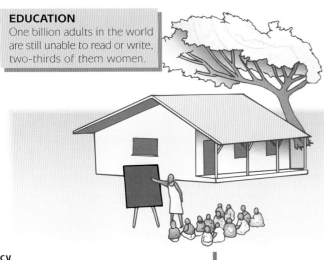

Literacy
- Adult literacy rate (proportion of the adult population who can read and write)
- Average number of years at school per child
- Proportion of children who attend secondary school.

◀ **Figure 2** Shanty housing in Mumbai (Bombay), in which at least half of the city's 16 million inhabitants live, some in even worse conditions than those shown here.

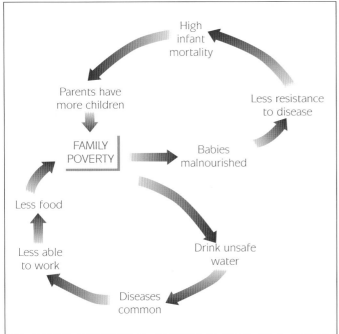

▲ **Figure 3** Poverty cycles.

Clearly, many of the different measures are inter-related in their effects upon people. Children become an infant mortality statistic by not reaching their first birthday, often because nearby clinics and hospitals do not exist or are too expensive. Many diseases, such as dysentery and typhoid, are spread by unsafe water supplies and poor sanitation, to which young children have least immunity. The limited availability of food means that both mother and child are less able to fight off diseases. With better parental education and with clean running water to the houses, fewer children would die. It is estimated that some 35 000 children are still dying every day from preventable diseases in less economically developed countries because of poverty. Many are in the poverty cycle trap (**Figure 3**).

The measures of development in **Figure 1** are labelled *social* measures because they are measuring either personal details or factors that directly affect people's lives. However, it is clear that it is difficult to separate the social from the economic, which is why some measures in **Figure 1** can also be described as *socio-economic*. Other measures of development that are more clearly economic have already been referred to, such as employment structure (**Figure 2** on page 226) and GNP (**Figure 3** on page 227), but they have strong social effects.

Exam focus

1 a (i) Define 'GNP'.   (2 marks)

    (ii) State one advantage and one disadvantage of using GNP to show development.   (2 marks)

b (i) From the world map within **Figure 1**, describe the world pattern of high infant mortality rates (above 100).   (3 marks)

    (ii) Suggest reasons for the world distribution of areas with high infant mortality rates.   (3 marks)

c (i) Describe the ways in which schooling in sub-Saharan Africa is shown in **Figure 1** to be different from that in South Asia.   (2 marks)

    (ii) Suggest how lack of schooling may contribute to family poverty.   (3 marks)

d (i) Draw a labelled sketch to show the main features of the housing in the photograph in **Figure 1**.
                                                (4 marks)

    (ii) Why is it difficult for families living in housing like this to break out of the poverty cycle?   (3 marks)

    (iii) What will be their best hope for breaking out of the poverty cycle? Explain your choice.   (3 marks)

Causes of the lack of development

Measures of development allow rich and poor countries to be identified, but they do not explain why little economic development has occurred in South America, Africa and Asia. Why should LEDCs be concentrated in just three continents? Why are there more than 1 billion people living in absolute poverty (surviving on less than two dollars per day)?

Human factors

Historical – legacy of the colonial past

Many countries in the three continents were colonies of European countries (especially Britain, Spain, France, the Netherlands and Portugal) until well into the second half of the twentieth century. Resources were exploited for the benefit of the colonial power. South American countries, for example, were plundered by the Spanish and Portuguese for their mineral wealth (gold and silver in particular). Native populations were forced off the best farmland so that it could be divided up into large estates and given to Europeans. Land on large estates tends to be used for agricultural exports rather than food crops for local people.

Economic – colonial patterns of trade persist

Exports from South American, African and Asian countries are dominated by primary products: minerals, timber and crops. They are low in added value and are subject to fluctuating world market prices (page 239). Manufacturing in the colonies was not encouraged, but the import of manufactured goods made in Europe was. Industrial goods are high in added value, which makes them worth much more than the raw materials from which they are made. Many LEDCs are caught in the **trade trap** (page 239).

Political – mismanagement of the economy, corruption and wars

Since independence, many African countries have been badly governed, with many leaders more interested in self-enrichment than in economic development of their country. National borders established by European countries in colonial times rarely coincided with tribal boundaries. Tribal conflicts within countries have flared up into armed struggles, resulting in disrupted food production and famine – in Angola, Sudan and Ethiopia, to name just three examples. In Zimbabwe in 2006, after more than 25 years of independence under just one leader, the inflation rate was 900 per cent; instead of exporting agricultural products, over 4 million people were receiving food aid. This situation was blamed on drought and chaotic land redistribution, and life

▲ **Figure 1** Land like this on the steep slopes of the Colca valley in the Andes of Peru is often the only land available to indigenous Indian farmers. The best farmland in valley floors and lowlands is occupied by large estates that were set up in colonial times. Why is farming here associated with a high risk of soil erosion?

expectancy had fallen to the lowest in Africa (36 years compared with 57 years in 1990, when it was one of the highest). In contrast, stable Singapore now has all the characteristics of an MEDC (page 193); the government of China has relaxed total state control over economic activity and welcomed Transnational Corporations, turning China into the world's leading producer of consumer goods such as T-shirts, jeans and trainers (page 195).

There are other factors – **location** may be one of them. LEDCs are located further south than MEDCs and many are within the tropics (see **Figure 1** on page 226), which suggests that environmental factors have a part to play in explaining present low levels of development in South America, Africa and Asia.

Environmental (natural) hazards

Environmental hazards are short-term natural events that threaten life and property. Some result from severe variations from average climatic conditions; **Figure 2** shows that climatic hazards (drought, tropical storms and flood) are responsible for more loss of life than tectonic hazards (earthquakes and volcanoes). Are natural hazards more likely to affect tropical areas, where the majority of LEDCs are located? **Figure 3** gives some clues. Drought and tropical storms in particular, which are responsible for some of the most life-threatening disasters, can be seen to have a greater presence in tropical than in temperate latitudes.

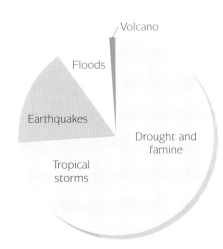

▲ **Figure 2** Percentage loss of life from natural hazards during a 25-year period.

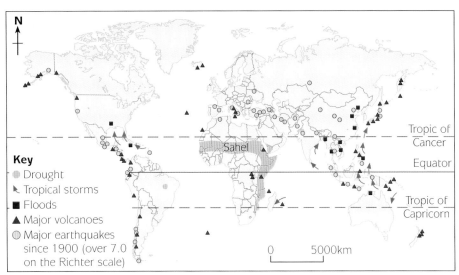

▲ **Figure 3** Areas at risk from natural environmental hazards.

Drought occurs when the rainfall in an area is significantly lower than the average amount expected. The shaded area in Africa in **Figure 3** has been the most drought-affected part of the world since 1970. The drought in Ethiopia in the 1980s eventually received massive media attention (pages 244–5). Rural food supplies were reduced, leading to malnutrition and death, and fewer crops were available for export, affecting both the income of farmers and the national economies. Niger, further west in the Sahel, was the latest country to suffer from severe drought in 2005 (page 96). In comparison, the great British drought of 1995 referred to on pages 96–7 was no more than an inconvenience and a cost to relatively few people; it was not a matter of life and death.

Tropical storms, with their hurricane-force winds and torrential downpours, can destroy settlements and wipe out fields of sugar cane and plantations of bananas (pages 97–8). **Figure 3** shows that only the fringes of three MEDCs are at real risk from tropical storms – which are they? Otherwise all the worst-affected countries are LEDCs, with a high proportion of houses built of flimsy materials and no emergency shelters. Among the worst-hit countries are Caribbean island states, Bangladesh and the Philippines.

Floods occur after heavy or prolonged rainfall causes rivers to burst their banks. There is also coastal flooding due to tropical storms, or tidal surges and big waves that follow earthquakes. Rainfall in the tropics is usually heavy and accompanied by thunder and lightning. The greatest loss of life from flooding is on the densely populated lowlands surrounding the many big rivers in South-East Asia. Bangladesh, with its low-lying coastal location on the Ganges delta, has the highest risk of all (pages 246–7).

Earthquakes and **volcanoes** are restricted to clearly defined zones along plate boundaries. Although many happen in countries lying outside the tropics, when they do strike, poor countries tend to be hit harder than rich ones (**Figure 4**). In wealthy places like California and Tokyo, all new buildings must be constructed to withstand strong earthquake shocks (page 19), but in poor countries like Iran and India many houses are self-built or made of cheap materials; even if regulations exist, they are rarely enforced and widely flouted.

Date	22 December 2003	26 December 2003
Location	California, USA	Bam, Iran
Magnitude (Richter scale)	6.5	6.5
Damage to property	a clock tower was toppled	a large part of the city was flattened
Number of people killed	3	estimated 30,000

◀ **Figure 4** A tale of two earthquakes.

Activities

1 Choose **three** of the five statements below. For each one **(a)** state information (from these two pages) to support it and **(b)** give reasons for it.

   A Colonial history still limits economic development today in many LEDCs.

   B Stable government favours economic development, poor government hinders it.

   C Climatic hazards are responsible for more loss of life than tectonic hazards.

   D Tropical areas are more badly affected by climatic hazards than temperate regions.

   E In MEDCs, environmental hazards are more likely to cause economic losses than loss of life.

2 Which are more important as a cause of lack of economic development – human factors or environmental factors? State and explain your view about the relative importance of human and environmental factors.

Water availability and quality

ℹ️ Information

Facts about water on Earth

- Oceans cover 71% of the Earth's surface

- Oceans and seas contain 97% of all water on Earth

- Only 3% of water on Earth is fresh water

- Most of this 3% is in 'deep freeze' – in ice sheets

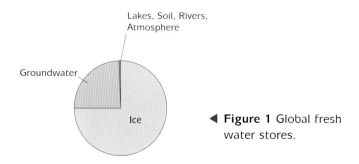

Lakes, Soil, Rivers, Atmosphere

Groundwater

Ice

◀ **Figure 1** Global fresh water stores.

Fresh water is a vital natural resource for all life on Earth (page 202). However, only 3 per cent of the water on Earth is fresh water (**Figure 1**) and most of this is not available for human use.

◀ **Figure 2** Antarctica, where most of the fresh water on Earth is found. Why is it not available for human use?

Water availability

Globally, fresh-water supplies for human use are abundant, despite the low percentage of available fresh water. Precipitation is constantly replenishing natural stores by filling rivers and lakes and topping up underground stores. The main problem is one of uneven distribution among and within countries. In order to assess **fresh-water availability** in a country, two factors are taken into account: (i) the amount of renewable fresh-water resources per year (from precipitation and from rivers flowing in from other countries); (ii) population size.

The majority of water-rich countries are in the hot, wet tropics. The most water-poor countries are located in the desert lands of the Middle East and the northern half of Africa. Many of these countries rely on **aquifers** (underground stores of water trapped in layers of pervious rock) or great dams, such as the Aswan High Dam on the River Nile in Egypt (page 172).

Some water-poor countries are already suffering from severe water shortages, which the UN refers to as **water stress**. By relating water availability in a country to its water consumption, the organization estimates that one in three of the world's people lives in a country already experiencing moderate to high water stress. It believes that the world's thirst for water will become a pressing resource issue during the twenty-first century. Global water consumption is rising at twice the rate of population growth. Much of the new demand is coming from LEDCs, for a variety of reasons:

- higher population growth than in MEDCs

- more water use for irrigation to meet rising demand for food

- traditionally low domestic demand is increasing with economic growth

- industrial and urban expansion is occurring, notably in China and India.

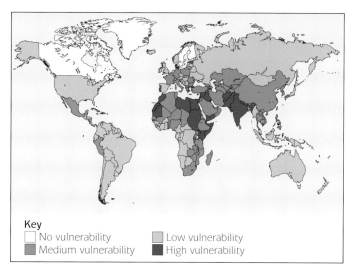

Key
- ☐ No vulnerability
- ☐ Low vulnerability
- ■ Medium vulnerability
- ■ High vulnerability

▲ **Figure 3** Projected water vulnerability in 2025, taking water supplies and income levels into account.
Source: World Resources Institute

By 2025 the UN expects two-thirds of the world population to live in countries with moderate to high water-stress levels. **Figure 3** shows that these countries

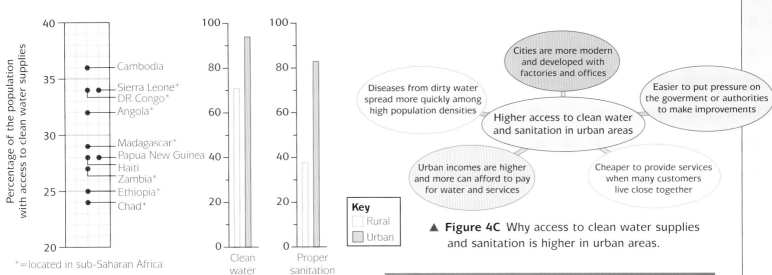

▲ **Figure 4A** The ten countries with lowest access to clean water.

▲ **Figure 4B** Rural and urban areas – world averages for access to clean water and proper sanitation.

Key Rural / Urban

▲ **Figure 4C** Why access to clean water supplies and sanitation is higher in urban areas.

are concentrated in the arid and semi-arid regions of Africa and Asia, where governments are least likely to be able to undertake measures that will reduce or solve problems of water supply such as:

- pollution controls to limit contamination of water supplies in surface and underground stores

- increased efficiency in water use (e.g. by using trickle irrigation that feeds water only to plants' roots)

- effective population policies to control further growth.

Water quality

There are striking differences in water quality and degree of access to clean water supplies between MEDCs and LEDCs. In Europe and North America 100 per cent access to supplies of clean water is taken for granted, whereas in LEDCs access on average is only 70 per cent. As usual, the average masks wide variations between and within countries. The world region with the lowest access is sub-Saharan Africa with just 51 per cent; this is the region where seven of the ten countries shown in **Figure 4A** are located. In rural areas, access to clean water and proper sanitation is almost invariably lower than in the cities (**Figure 4B**, for the reasons given in **Figure 4C**). Indeed, **Figure 4B** shows that it is the exception for proper sanitation to exist in rural areas, where most people use water from surface streams without any treatment. Surface streams have multiple uses, as **Figure 4D** illustrates.

▲ **Figure 4D** A river in India used here as a rubbish dump and for washing clothes and elsewhere as a water supply.

Activities

1 Explain why fresh water is an example of a renewable natural resource.

2 **(a)** Rank the natural stores of fresh water in order of size.

(b) What percentage of the total is unavailable for human use? Explain your answer.

(c) Worldwide, rivers are the main source of fresh water for humans. Make a table and write about the advantages and disadvantages of relying on rivers for water supplies.

3 Study **Figure 3**.

(a) Describe the pattern for **(i)** no water vulnerability and **(ii)** high water vulnerability by 2025.

(b) Describe how the map suggests greater water vulnerability in LEDCs than MEDCs.

4 Draw a spider diagram to show why access to clean water supplies and sanitation is low in rural areas of LEDCs.

Water, health and food supply in LEDCs – problems and solutions

A Water-bornediseases – diseases spread by drinking or by washing food in contaminated water:
Diarrhoea   Typhoid
Dysentery   Hepatitis
Cholera

B Water-relatedinsect-borne diseases – diseases spread by insects that breed in water:
Malaria – carried by mosquitoes
Sleeping sickness – carried by tsetse flies

C Water-baseddiseases – carriers of the disease live in water:
Bilharzia – carried by water snails

◀ **Figure 1** Diseases in tropical regions associated with water.

▶ **Figure 2** Children's health and mortality rates.

- Each day 25000 children die from drinking dirty water.
- Repeated stomach upsets mean that children become malnourished.
- Babies become ill because their mothers only have dirty water to mix with the dried milk.
- Harmful waste water is not always kept away from supplies of fresh drinking water.

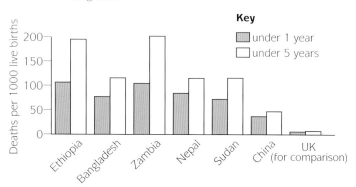

Safe water is an alternative name for clean water. This is because lack of access to clean water supplies has knock-on effects for people's health. Water-related diseases abound in the tropics, where hot, wet climates provide ideal environments for insects and bacteria to thrive and multiply (**Figure 1**). A large proportion of the diseases that make people sick and too weak to work, or that kill the most vulnerable, notably infants and children, are associated in some way with water. Malaria, for example, threatens 40 per cent of the world's population in over 90 countries. It kills four children every minute. The most virulent strains are present in sub-Saharan Africa, the region with 90 per cent of all deaths. Overall, at least 2 million people die from water-related diseases in LEDCs every year (**Figure 2**). Improvements in water availability and quality can bring great social and economic benefits to rural dwellers in LEDCs, as the example from Pakistan in **Figure 3** illustrates.

A

In the past

- Women and children walked up to 6 kilometres to find water, spending 3–6 hours per day
- Most was dirty water taken from stagnant rain water ponds, also used by animals
- Children were often sick with diarrhoea, cholera was a problem in the wet season, no one had money for seeing a doctor
- People were migrating to the cities

B

1995–2002

- Development aid money was used for pumping stations and pipes to bring clean water to 350 villages and almost 1 million rural dwellers
- The approach was community-based with local people involved in planning and construction

C

In 2003

- Drop of over 90% in water-borne diseases
- School enrolment up 80%; children no longer need to fetch water
- Household incomes up 20%; women have more time to generate income
- Attitudes are more commercial; more produce grown that sells well in local markets
- Life styles are changing; women have more dignity
- People who had migrated to the cities are returning

▲ **Figure 3** Rural water supply and sanitation projects in the Punjab in Pakistan.

Food supply

Remarkable progress was made in expanding world food supplies between 1960 and 2000. Although world population doubled from 3 to 6 billion, food production increased even faster see **Figure 1** on page 172. Average calorie consumption per head increased from 2000 in 1960 to over 2500 in 2000. The gains in food production were particularly significant in LEDCs due to a combination of: better seeds (including high-yielding varieties such as IR8 rice seeds); expanded irrigation; higher use of fertilizers and pesticides; Together they became known as the 'Green Revolution' (pages 171 and 173).

Figure 4, however, shows a different picture; many countries, particularly in Africa, still house large populations at risk from malnutrition (lack of nutritious food to keep

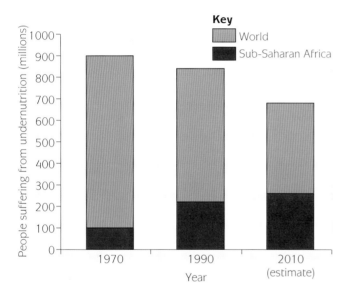

Key
World
Sub-Saharan Africa

▲ **Figure 4** Trends in nutrition.
Source: World Resources Institute

people alive and healthy). It is often said that poverty is the root cause of malnutrition, not insufficient global food production. Poor people lack the resources to produce or buy enough food; the nature of the problems holding back families may be different between rural and urban areas, but they are both poverty-based. Matching global food supplies with actual food needs at the local level has so far proved to be too difficult to overcome. At the same time pressure on the land to produce more is increasing the amount of environmental damage, decreasing the chances of a sustainable future for current numbers of people.

Some of the problems that undermine a family's ability to grow enough food in rural areas of LEDCs are summarized below:

- shortage of farmland; good farmland is owned by large landowners

- soil erosion and land degradation due to forest clearances, overgrazing and overcultivation

- Increasing soil salinity from over-use of irrigation water

- effects of natural hazards, especially drought and floods

- human hazards such as civil wars disrupting production.

In urban areas the problems include low wages, lack of work and high food prices; food is available, but cannot be afforded.

The need for appropriate solutions

Most governments think only of big solutions when faced with major problems; constructing large dams (page 210) is one example. However, some people maintain that alternatives to large dams exist to achieve the same objectives, based on small-scale management works within the local drainage basin. Impressive results from one example in a village in Gujarat (in India) are shown in **Figure 3** on page 176, which is an example of **appropriate technology** – a level of technology suitable for local people to use.

Activities

1 Describe how and why lack of access to clean water supplies:

  (a) is a risk to people's health

  (b) leads to a greater loss of life among infants and young children than adults

  (c) contributes to the cycles of poverty (page 231).

2 Make notes for these two cases studies in a rural area in an LEDC.

  A Case study to illustrate the economic and social benefits after the provision of an assured supply of clean water – using the headings Before and After.

  B Case study to show the benefits of a local management scheme using appropriate technology – under the headings Economic, Social and Environmental.

3 (a) Describe what **Figure 4** shows about hunger in sub-Saharan Africa.

  (b) State other evidence from earlier in this chapter which shows that sub-Saharan Africa is the least developed world region.

  (c) Outline three different reasons for the lack of development of many LEDCs in sub-Saharan Africa.

World trade – why does it favour MEDCs?

◀ **Figure 1** Hotel resort in the tropics. Tourism is a vital source of foreign exchange for some LEDCs.

Trade is the exchange of goods and services between countries. It is called **visible trade** if goods (raw materials, foodstuffs and manufactured goods) are exchanged, because they can be counted, weighed and given a value. It is called **invisible trade** if services are exchanged. Tourism is one example; earnings from workers overseas, and foreign aid, are two more.

The goods exchanged may be **primary products**. These include *raw materials*, such as crude oil, iron ore, cotton or natural rubber, from which other products can be manufactured. They may also be *foodstuffs*. A country may not be able to grow all the food that it needs to keep its population well fed, so it engages in trade. Other countries import foods they cannot grow themselves due to unsuitable physical conditions. The UK diet would be less varied without the import of oranges, bananas, tea, coffee and chocolate.

Secondary products are *manufactured goods*, which may need to be imported into a country because they are not made there. Among MEDCs, however, the same types of goods are traded across borders, such as cars, reflecting freedom of choice and wealth rather than need.

Pattern of trade between LEDCs and MEDCs

Trade between LEDCs and MEDCs forms only a small percentage of total world trade, but it is on the increase, boosted by industrial growth in the Far East in general and in China in particular. Despite the recent growth in exports of consumer goods from Asian countries, the basic pattern of trade established during colonial times largely persists – LEDCs *export* primary products and *import* secondary goods.

Exports from LEDCs to MEDCs include:

- *fuels* such as oil and natural gas: countries in the Middle East have the largest reserves

- *minerals* such as bauxite and iron ore used for making aluminium and steel

- *agricultural raw materials* from tropical plants, such as rubber and cotton

- *foodstuffs* such as coffee, tea, cocoa and bananas, all of them tropical crops.

Many LEDCs depend on the export of a small number of commodities. In Africa, for example, more than 20 countries depend greatly upon earnings from just one or two exports (**Figure 2**). In the other direction, LEDCs import a wide range of manufactured goods produced in factories in MEDCs, from cars to computer software, along with the machinery, technology and knowledge needed to help them develop economically.

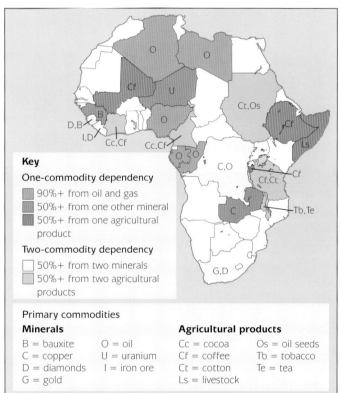

Key

One-commodity dependency

- 90%+ from oil and gas
- 50%+ from one other mineral
- 50%+ from one agricultural product

Two-commodity dependency

- 50%+ from two minerals
- 50%+ from two agricultural products

Primary commodities

Minerals		**Agricultural products**	
B = bauxite	O = oil	Cc = cocoa	Os = oil seeds
C = copper	U = uranium	Cf = coffee	Tb = tobacco
D = diamonds	I = iron ore	Ct = cotton	Te = tea
G = gold		Ls = livestock	

▼ **Figure 2** Many African countries depend upon the export of just one or two primary commodities.

Such over-dependence on just one or two export commodities is dangerous. It falls foul of the old proverb about the dangers of 'keeping all your eggs in one basket'. These dangers are both human and physical. As environmental awareness among people in MEDCs increases, more aluminium cans are recycled, reducing the world demand for bauxite. Which African countries may be affected? There is also product replacement; only half the rubber used in the world is still natural rubber from tropical trees; the rest is synthetic, made from crude oil. Hurricanes in recent years have devastated banana and sugar plantations on several Caribbean islands, but diversification into alternatives is often impossible due to steep relief, lack of capital and inadequate transport infrastructure.

Interdependence between LEDCs and MEDCs

Interdependence describes the relationship that exists between countries that share a need to exchange one another's goods and services. It should exist between LEDCs and MEDCs because:

- LEDCs earn foreign exchange from selling fuels, raw materials and foodstuffs, which MEDCs need to buy to keep their factories working and to supply their people with an assured supply and greater choice of food. The everyday diets of people living in Europe include many foods and drinks that can only be grown in countries with hotter climates and then imported; the long list contains coffee, tea, cocoa and bananas.

- LEDCs use the foreign exchange to buy manufactured goods that they are unable to make for themselves, from which MEDCs benefit by having a larger market for their goods.

In other words, the need for trade between LEDCs and MEDCs clearly exists because each can offer what the other wants. It might be expected to follow from this that the trade will balance, but in practice there is a trade surplus in favour of MEDCs, at the same time as many LEDCs have accumulated large debts.

The pattern of world trade favours MEDCs

Primary products on which LEDCs depend sell at *low* prices, whereas manufactured goods are bought from MEDCs at *high* prices. When you buy bananas or instant coffee, only between 10 and 15 per cent of the money you pay will find its way back to the producing country (**Figure 3**). As raw materials and foodstuffs are processed, value is added. During the past 40 years, the price of manufactured goods has gone up faster than that of raw materials; in other words, to import the same number of tractors as it did ten years ago, an LEDC will need to sell two or three times the amount of raw materials.

Primary products go up and down in price. Prices at which minerals and agricultural commodities are traded between countries throughout the word are fixed by markets in financial centres such as New York and London; the producing countries have little or no control over them. During times of economic recession in industrialized countries, the demand and the price for raw materials go down without any real regard for costs of production. Look at **Figure 4** and see what happened to world coffee prices during a 20-year period. Because coffee beans grow on bushes that take several years to grow, farmers in the

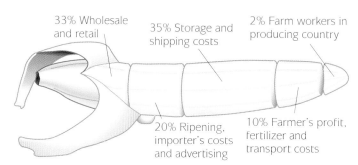

33% Wholesale and retail

35% Storage and shipping costs

2% Farm workers in producing country

20% Ripening, importer's costs and advertising

10% Farmer's profit, fertilizer and transport costs

▲ **Figure 3** Rich pickings – but for whom?

coffee-producing African countries cannot suddenly change crops during times when the world price falls below their costs of production.

The net result is that most LEDCs have a **trade gap** – the value of exports is lower than the value of imports. This leads to debt. Countries are caught in a **trade trap**, especially during times when commodity prices are low. The prospects for many are not good.

▼ **Figure 4** World coffee prices, 1972–2005.

Activities

1 (a) Give a definition of the term 'primary product'.

(b) From **Figure 2** name those primary products exported by more than one African country.

(c) Why do African countries export their primary products mainly to MEDCs in Europe instead of to their African neighbours?

2 State the evidence for each of the following, using labelled illustrations where helpful.

(a) Producers of primary products do not make much money.

(b) The prices of primary products go up and down.

(c) Dependence upon one or two primary products is dangerous.

3 What is interdependence in world trade?

Can poor countries develop economically?

The prospects for most countries are not good. The cartoon below suggests that the **interdependence** between less and more economically developed countries is not a reality.

▲ **Figure 1** Does interdependence between rich and poor countries exist?

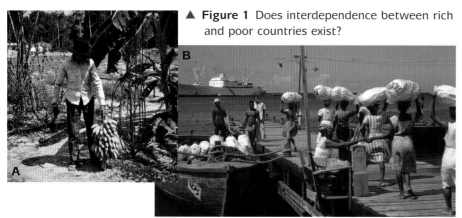

▲ **Figure 2** Banana growing on a small scale. **A:** Farmers harvest bunches of bananas individually. **B:** Caribbean farmers have to carry their bananas to port, where small boats are used to ferry them to the banana boat sitting offshore.

When LEDCs do begin to process and manufacture their own raw materials, trade barriers are the next obstacle they face. Lower wage rates, as well as local raw materials, give LEDCs price advantages over richer, more developed, countries. However, MEDCs have a long history of protecting their own manufacturing industries against cheaper imports, through high tariffs and quotas. The World Trade Organization (WTO) promotes free trade and the removal of import duties. It holds regular international meetings, which have led to removal of some of the obstacles to free trade, but progress tends to be slow because of the vested interests of powerful countries such as the USA and members of the EU.

It is wrong to think that the interests of all LEDCs are the same. The WTO ruled that the favoured treatment given by the EU to banana imports from small-scale producers on Caribbean islands was illegal. The small farmers in the Caribbean (**Figure 2**) know that they cannot hope to compete with bananas grown on plantations in Central and South America by transnational corporations.

Mineral resources have made a few LEDCs wealthy, especially those with oil. The Gulf States of Qatar and United Arab Emirates (which includes Dubai), for example, have GNPs per head higher than the UK. Many of the industrializing countries are in South East Asia (page 193); one of them is Malaysia. **Figure 3** shows how its exports have changed since 1970; can you explain how they indicate that economic development occurred? Judged by its GNP of about US$ 4000 per head, Malaysia is one of the medium-income countries (page 227). The photograph on page 225 shows a scene in Kuala Lumpur, its capital city. Too many people have private forms of transport for it to be a low-income country. In high-income countries more people can afford to replace motorbikes with cars.

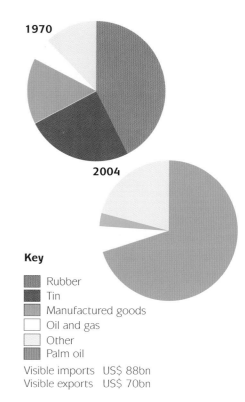

Key
- Rubber
- Tin
- Manufactured goods
- Oil and gas
- Other
- Palm oil

Visible imports   US$ 88bn
Visible exports   US$ 70bn

▲ **Figure 3** Exports from Malaysia in 1970 and 2004.

Fairtrade – can it help poor farmers in LEDCs?

This initiative to give small-scale farmers in LEDCs fair prices for their crops was launched in Europe in the early 1990s. It was born from the realization that many of the farmers in the 'South', who grow our familiar foods and drinks like tea, coffee, chocolate and bananas, are very poor and struggle to survive with a system of world trade with its low and fluctuating prices for primary produce. The FAIRTRADE Mark is an independent label which guarantees that producers in developing countries receive a fair deal. Take a look in your local supermarket: the mark appears on more than 850 varieties of coffee, tea, cocoa, chocolate, honey, snacks, fresh fruit and juices. **Figure 4** records the value of retail sales in the UK – what is the trend? Visit the Fairtrade website (see Hotlinks, page iv) for an update.

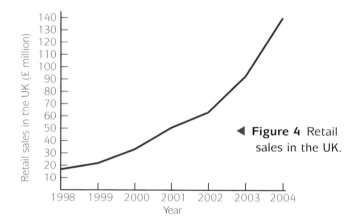

◀ **Figure 4** Retail sales in the UK.

What is a fair deal for farmers in LEDCs? Fairtrade means:

* fair prices to cover the grower's costs of production (no matter how low the market price goes)
* decent working conditions and good health and safety standards
* security from long-term contracts
* respect for the environment to allow sustainable production
* a social premium for community development purposes (so that all the people in an area can benefit).

Fairtrade pays farmers a guaranteed minimum price, even when world prices fall. In addition, a premium is paid, to be spent on community projects. If the world price rises above the Fairtrade price, Fairtrade growers receive the market price plus the premium. Fairtrade crops are bagged and marketed separately so that only fairly-traded produce goes into the manufacture of Fairtrade foods. Nearly 5 million growers and their families in 40 countries benefit from the sale of their goods under the Fairtrade label. For obvious reasons, these are more expensive than non-Fairtrade rival products; they must be of good quality and attractively packaged, because the majority of supermarket shoppers will not buy them just for the cause.

> After Hurricane George destroyed our crops in 1998, we could not find an importer for our bananas. Large companies were not interested in buying from farmers with only two hectares of land. We are grateful to the Fairtrade Foundation which introduced us to a small UK company. The number of farmers in our group has increased from 13 to 70 in four years.

> There was no sanitation in our community until local Fairtrade farmers started donating outside toilets.

> I am earning double what I was earning selling in the local market. I now have regular income and food for my three children.

> I persuaded my fellow farmers to give free uniforms to local schools and baseball teams. Children are better off on the baseball field than on the streets.

▲ **Figure 5** Comments from Fairtrade banana growers in the Dominican Republic (in the Caribbean). They joined Fairtrade in 2000 and live in a poverty-stricken rural area near the border with Haiti.

At the moment Fairtrade is catering for no more than a small, but growing, niche market. Realistically, until pressure from environmentalists and the public is sufficient to hit the profits and reputations of the giant food corporations such as Nestlé and Cadbury-Schweppes, Fairtrade will remain a tiny part of all trade with LEDCs. Big corporations are beginning to show more interest, although some people are suspicious of their motives. Helping small farmers and protecting the environment are not on the traditional route to big profits.

Activities

1 Write down the arguments against free trade in bananas that producers in the Caribbean would use to gain support for their cause.

2 Use **Figure 3**.

   (a) Describe the changes in exports from Malaysia between 1970 and 2004.

   (b) Explain how these changes show that:

   (i) primary industries have declined in importance

   (ii) economic development has occurred.

3 (a) Describe the ways in which Fairtrade is different from normal trade.

   (b) Why do whole communities, not just the growers, benefit from fairly traded produce?

4 Can free trade and fairtrade be solutions to the lack of economic development in LEDCs? Explain your view.

Aid

Aid is money or help given to a country in need. It may include any one or more of the following: food, medical supplies, goods, equipment and people with special skills. Aid may be offered free but sometimes low rates of interest are charged on loans, or conditions are attached to the goods supplied.

Why is aid needed? The previous pages in this chapter provide abundant evidence for its need in many LEDCs.

- great gaps in development exist between poor and rich countries – see pages 226–31
- environmental hazards periodically affect some of the world's poorest countries – see page 233
- imbalances in world trade have created a trade gap and led to debt – see page 239.

The different types of aid can be classified in two different ways, either according to how it is funded and who controls its distribution, or according to whether the need for it is immediate or longer term.

Types of aid according to funding and distribution

1 Political (bilateral aid)

This is aid given by one government to another. Since this is 'two-sided' aid, it is often called bilateral aid. Often this is tied aid with conditions attached to its use; the recipient government is often forced to buy goods and services from the donor country. It is therefore also an example of conditional aid. Indeed, it may be little more than trade promotion – finding an overseas market for the donor country's manufactured products and funding the trade with credits or low-cost loans.

The political nature of this aid raises several problems. The goods may not be best suited to the real development needs of the country and its people. Many bilateral deals focus upon large-scale projects such as dams and power stations, which rank high in prestige but low in value to ordinary people. A second problem is that the choice of donor country is made for political reasons rather than on the basis of need. The UK and France give most bilateral aid to their former colonies. Israel is the single largest beneficiary of aid from the USA, despite not being among the world's poorest countries.

2 Multilateral aid

Governments give money to international agencies which decide how the money is spent. Many of these agencies operate under the United Nations. Each one has a particular area of interest. WHO is the World Health Organization while UNICEF focuses upon helping children, UNESCO upon education and the World Bank upon funding development projects. Although these pay more attention to development needs, as large organizations they are often slow to change and do not always target the real needs of people in poor countries. This is also an example of conditional aid with conditions for its use determined by the organization instead of the government which donated the money.

3 Charitable aid

Charities are examples of what are called non-governmental organizations or NGOs. Most are based and funded in industrialized countries. Money is raised through public appeals, regular voluntary donations and charity shops. They also receive grants from governments, some of which now realize that the NGOs are better able to run projects which are more likely to reach the people in greatest need. Some, such as Oxfam and Christian Aid, have developed a vast international network of operations, but they do work closely with community-based organizations. This allows aid to be more closely targeted on local need, which makes it more effective.

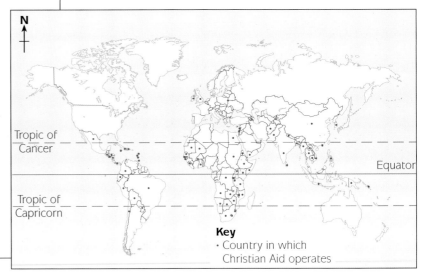

▲ **Figure 1** Christian Aid's global aid programme.

Types of aid according to need

Short-term emergency (relief) aid

This is needed to deal with an emergency situation. It is a lifeline. Often the emergency is caused by a natural disaster such as drought, flood or earthquake, but it may be the consequence of a human disaster such as war. It is hard for people in the Western world to ignore TV pictures beamed by satellite showing the suffering. Food, medical help, clothing, blankets and tents are supplied to help survivors and to relieve the human misery of refugees (hence the alternative name 'relief'). These are only supplied for as long

◀ **Figure 2** Short-term emergency aid in Shindhapalkchock district, Nepal.

as the emergency exists; once something approaching normal life is resumed they are withdrawn (hence 'short-term').

Long-term (development) aid

The purpose of this type of aid is to increase a country's level of development for the future by improving the standards of living and quality of life of its people. This is the type of aid upon which aid agencies prefer to spend their funds. Work for improved health care, much of it done through UN agencies, includes:

- immunization programmes against TB (tuberculosis), polio, measles and diphtheria
- health education programmes in rural areas
- family planning clinics
- clean water supplies and sanitation facilities
- funding research in MEDCs to develop new vaccines against tropical diseases.

◀ **Figure 3** A Bangladeshi woman draws safe, clean water from the village hand-pump at a time of severe flooding. The water pumps are an example of practical long-term aid, useful and appropriate to the local community.

Most of the practical improvements are funded and organized by NGOs, which work through local partners. Their first task is to identify the real needs of the people. The next is to provide practical help. The overall aim is to make changes which are sustainable by the local community so that self-sufficiency will increase and aid will no longer be needed. The types of help given to farmers include:

- agricultural support and technical training
- provision of good-quality seeds suitable for local conditions
- digging ponds and irrigation systems using low-level (appropriate) technology
- community projects for providing loans for tools and business credit
- setting up organizations for marketing surpluses.

Activities

1. Make a full-page copy of the table below. Complete it as fully as you can.

Type of aid	Main features	Advantages	Disadvantages
A Political B Multilateral C Charitable D Short-term E Long-term			

2. Give reasons why:

   (a) governments usually prefer to give bilateral aid

   (b) charities prefer to spend their money on long-term aid rather than emergency aid.

3. Put together information for 'A case study of an aid organization'. Refer either to Christian Aid (see Hotlinks, page iv) or another organization such as Oxfam or Save the Children (see Hotlinks, page iv). Suggested headings are Sources of funding; Countries where they work; Types of work done; Example of a local project.

Ethiopia in the 1980s and 1990s

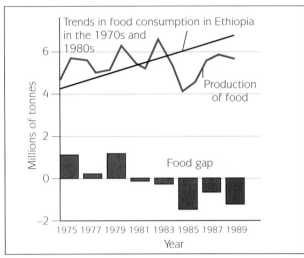

▲ **Figure 1** Food production and consumption trends in Ethiopia.

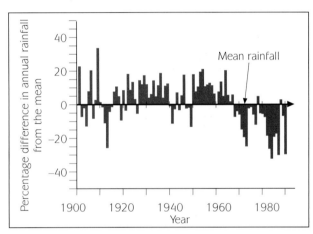

▲ **Figure 2** Variations in annual rainfall in the Sahel region, 1900–1990. Why was drought in the 1980s worse than in the 1970s?

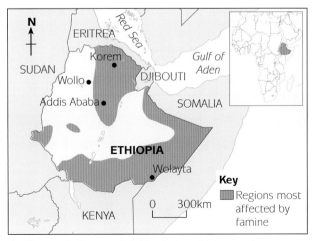

▲ **Figure 3** Areas of Ethiopia in which people were most badly affected by famine in 1984.

Aid in the 1980s

The last year in which Ethiopia was able to feed itself was 1982 (**Figure 1**). The famine was at its greatest in 1984. The natural cause was drought – after years of lower than average annual rainfall, the rains failed again in 1983/84 (**Figure 2**). Human factors worsened the impact of the drought. The military government was engaged in a civil war and spent one-third of Ethiopia's budget on the armed forces. There were food surpluses in some parts of the country, but food could not be moved to areas of need because transport, organization and security were lacking. An estimated 500 000 people died in 1984. People living near the borders in the north and south were most badly affected (**Figure 3**). Some of the first TV pictures seen in the West were of Korem in northern Ethiopia, towards which thousands of people trekked for days in search of food.

Nearly £15 million was raised in 1984 in Britain alone by the UK Disasters Emergency Committee consisting of Christian Aid, Oxfam, Save the Children and Cafod. Nearly all the relief aid was food aid, some transported from overseas and some bought locally with overseas funds. Some food aid was taken by the Ethiopian government for its troops, but most reached the affected areas. Many more people died than was necessary because the military government of Ethiopia concealed news of the famine, which delayed the start of the food aid operation. From 1984 to 1992 the Ethiopian Famine Relief Fund, consisting of Band Aid and Live Aid, raised £110 million. Ethiopia received a massive amount of short-term aid in the 1980s.

Aid in the 1990s

The problem of drought did not go away, but the military government did. It was overthrown in 1991, which gave Ethiopia a chance to start again. However, birth rates remained high and by 1994 there were 2 million more mouths to feed than in 1984. In the first half of 1994 it was estimated that some 5000 people died in Wolayta, southern Ethiopia. A bumper harvest in 1992 had led aid agencies to scale back levels of food aid. A delay in the rains in 1994 meant that people's own food stores were used up. The poor road links limited the speed with which the aid agencies could react to the new emergency. **Figure 4** shows how villagers in Wollo could just about exist without food aid in a year of normal rains, but they remained very vulnerable in dry years. Environmental problems such

Case Study – Aid in Ethiopia from 1984 onwards

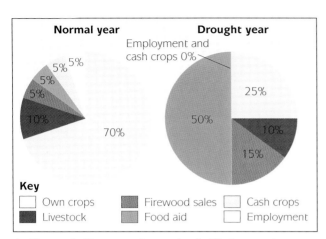

▲ **Figure 4** How people survive in Wollo, northern Ethiopia.

as soil erosion and land degradation continued to be real threats.

During this decade some charities tried to switch away from short-term towards long-term development aid to try to increase the amount of food produced in a manner that was sustainable for the future. For example, the Catholic agency Cafod, a British NGO, worked through local partners to promote integrated agricultural development, helping farmers to identify their own needs. Cafod provided practical help such as:

- finding high-quality seeds which mature even if the rains are poor
- planting drought-resistant trees
- digging ponds and wells
- marketing the surplus produce
- providing loans for buying tools and setting up businesses.

One example of an area where results could be clearly seen by the end of the decade was the plain outside Korem. It was no longer the bare place that it had been in 1984. Crops covered the land; better drought-resistant varieties of seeds and improved irrigation were giving local people more security for the future.

Aid in the new millennium

Some aid agencies complained about needing to commit such a large share of their scarce resources to Ethiopia, leading to the neglect of other countries in great need of development. Three years of failed rains led to famine in 2004, which saw some 14 million Ethiopians relying on food aid. This followed years of population increase (the population doubled between 1974 and 2004) and more or less constant

wars. Much of the aid failed to reach the intended people or regions because of a combination of corrupt officials, inefficient management and transport dislocation during conflicts. Therefore drought and food shortages became famine.

Food aid is expensive – charities would like to concentrate more on long-term aid with its emphasis on sustainable development for the local community. In a drought year in Ethiopia at least 1 million tonnes of food aid are given out. It has been estimated that this could pay for the labour for:

- 167 000 kilometres of access road, or
- 2700 earth dams, or
- over 400 000 kilometres of irrigation channels.

Each of these would bring long-term benefits and make local communities less vulnerable in future emergencies. Additionally there is always the danger of food aid dependency – with regular food handouts, people lose the incentive to produce their own. Too many local officials and 'middle-men' make money by selling what was intended to be free. The response to the famine in 2004 was nothing like as generous as it had been in 1984, suggesting that 'donor fatigue' might be affecting people overseas as far as Ethiopia is concerned.

Activities

Under the heading 'Aid in Ethiopia', write out the answers to the following.

(a) For the 1980s:

- **(i)** Name the type(s) of aid received.
- **(ii)** Give the physical and the human causes of the famine.
- **(iii)** Which was the worst year for famine?
- **(iv)** Where was aid most needed?

(b) For the 1990s:

- **(i)** Name the type(s) of aid received.
- **(ii)** Give the physical and the human reasons why aid was still needed.
- **(iii)** Explain the advantages of development aid over food aid.

(c) Do you think Ethiopia will need aid by 2015? Explain your views.

Country on a delta

Bangladesh, a country of 140 million people, is one of the world's most densely populated countries with about 800 people per square kilometre. It fits all three of the reasons stated on page 242 to explain why some countries need aid.

A *One of the world's poorest countries* – the development indicators listed in **Figure 1** show this. The comparison with the UK emphasizes the enormous size of the gap between the world's rich and poor countries.

	Bangladesh	UK (for comparison)
GNP	330	23 920
Birth rate	29.9	10.6
Infant mortality rate	62.9	5.4
Life expectancy	61	78
Doctors per 1000 people	0.1	1.8

▲ **Figure 1** Development indicators.

B *Natural hazards* – Bangladesh is a delta country crossed by two of Asia's big rivers, the Ganges and the Brahamaputra (**Figure 2**). Most of its flat and low-lying land is less than 10 metres above sea level. The risk of river floods increases in summer during the heavy monsoon rains (**Figure 3**). Cyclones (tropical storms), blown onshore from the Bay of Bengal, bring hurricane-force winds and storm surges that cause coastal flooding in late summer and autumn. 1991, 1998 and 2004 were notable flood years with much cropland destroyed, thousands killed and millions left homeless (pages 48–9).

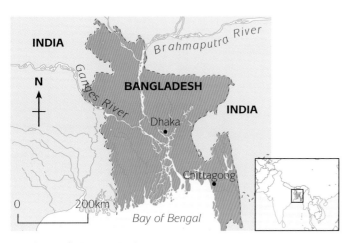

▲ **Figure 2** Bangladesh.

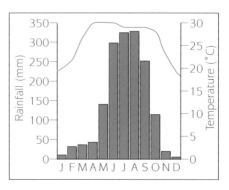

◄ **Figure 3** Climate of Dhaka.

C *Large trade gap* – the visible trade gap in 2003 was wide (**Figure 4**). Why were the contributions from invisibles such as aid so important?

Visible exports (US$ millions)	Visible imports (US$ millions)
Clothing 4900	Machinery and transport equipment 2400
Fish and fish products 400	Fuels 800
Total of all visibles 6100	Total of all visibles 9400

▲ **Figure 4** Bangladesh's trade in the mid-1990s.

Aid from overseas

Most is still political, conditional aid. 'Project aid' would be another label to describe it, because it is mainly used to fund power stations, bridges and roads which the government of Bangladesh cannot afford to provide. The disadvantages associated with this type of aid include:

1 The majority of Bangladeshis, who are poor and rural, benefit little.

2 Much of the funding is in the form of loans; although at low rates of interest, they still have to be repaid.

3 It leads to widespread corruption, because the control of aid projects gives great power to politicians, and fortunes can be made by chosen suppliers and contractors.

Slowly the proportion of aid controlled by charities (NGOs) is increasing, although it is still under 20 per cent of the total. One example of a community-based project is at Manab Mukti, which has Oxfam support. Here there are about 5000 households on a *char* in the Brahmaputra River. *Chars* are low-lying islands which can be farmed by poor families on a seasonal basis.

To help them cope with the regular flooding of the Brahmaputra they were trained in disaster preparedness. They were helped to:

- raise their homesteads above regular flood levels
- make platforms on which to keep their livestock
- build latrines
- install extra piping to raise the community tube wells above flood levels to ensure the availability of clean drinking water during floods.

To enable people to survive extreme floods, two community shelters were built which also had tube wells and latrines. These measures greatly reduce the risk of many water-related diseases.

Measures were also taken to help community development. Agricultural training was given, accompanied by a credit scheme to allow families to take loans for agricultural purposes, especially for growing flood-resistant crops.

Although the people remain poor, Manab Mukti is now a healthier, happier and more prosperous community. This example shows how a small investment of funds in the right place at the right time can have a great effect on people's lives.

Activities

1 Describe and explain the evidence which shows that Bangladesh:

(a) is one of the world's poorest and least developed countries

(b) has a very high flood risk between July and October

(c) has a large trade gap and big debt.

2 (a) State three reasons why aid is important to Bangladesh.

(b) Explain why the government of Bangladesh (like many others in LEDCs) prefers political (bilateral) aid, whereas rural people are more enthusiastic about charitable aid.

Exam focus

1 a World fresh water availability:
1950: 16 000 cubic metres per person
2000: 7000 cubic metres per person

(i) Draw a graph to show these values.
(2 marks)

(ii) State two reasons for the change shown between 1950 and 2000. (2 marks)

b Average water consumption per head per year (cubic metres):

- South America 335
- Europe 625
- Australasia (Oceania) 591
- Asia 542
- North America 1798
- Africa 202

(i) Calculate the difference in average consumption per head per year between continents made up of mainly MEDCs and mainly LEDCs. (2 marks)

(ii) Suggest reasons why consumption is highest in North America (especially the USA) and lowest in Africa (especially countries in sub-Saharan Africa). (5 marks)

c Study **Figures 5A** and **B**, which were both taken in Peru, an LEDC in South America.

▲ **Figure 5A**
Village in the Andes

▲ **Figure 5B**
Residential area in Lima

(i) Describe the differences in housing between these two places in Peru.
(3 marks)

(ii) Describe what the photographs suggest about different levels of wealth between rural and urban areas. (3 marks)

(iii) Name two kinds of aid that are often needed more urgently in rural areas than in cities in LEDCs. (2 marks)

d Choose one case study of an aid project in an LEDC.

(i) Describe what has been done. (4 marks)

(ii) How well does it meet local people's needs? Explain your answer. (2 marks)

Chapter 14

Examination technique

The questions

Each geography GCSE examination question can be broken down into at least two parts:

1 the command words – i.e. what you are being told to do

2 the question theme – i.e. what the question is about.

Some questions specify a location or world region for the question and have a third part:

3 where – the area or areas of the world.

Example of a two-part question:

Name two volcanoes,

command words   question theme

Example of a question with three parts:

State two problems in inner city areas of cities

command words   question theme

in more economically developed countries (MEDCs).

where

Command words

A • Name   • Give   • State   • List

Name one country in which ...

Give two reasons for ...

These are simple and clear command words and need no further comment.

B • Define the term ...   • What is meant by ...?
   • Give the meaning of ...

These command words are asking for definitions. You are likely to be asked to define terms used in the syllabus. Some terms are highlighted in bold in the text, and these terms are explained in the Glossary on pages 253–5. Look at the number of marks – this is vital with this type of question.

Question *What is meant by intensive farming?*   (3 marks)

Answer from Candidate 1 Getting a lot from the land. A lot is grown and sold by the farmer.
same point made

Examiner comment 'The candidate has only attempted to make two points in a three-mark question, and can't gain more than two marks. The answer is only worth one mark because the two statements are making the same point   high inputs.'

Answer from Candidate 2 Intensive means producing as much as possible from the land. To gain a high output, large inputs are needed and the land is carefully farmed.'
same point made

Examiner comment This candidate also begins the answer by making the same point twice, but then extends the answer, making two other valid points. There is just sufficient for the candidate to be given all three marks.

Note that the examiner said *just sufficient* for full marks. If you know and understand the topic and want a high grade, give a little bit more information than you think is needed for full marks, in case part of your answer does not match what is in the mark scheme. Candidate 2's answer could have been extended to say: large inputs, such as using a lot of labour and fertilizers.

C Describe

Describe the features of a delta.

Describe what Figure 1 shows.

(Figure 1 may be a graph, diagram, map or photograph.)

Describe is one of the most commonly used command words in geography examinations. 'Describe' commands you to write about what is there, or its appearance, or what is shown on a graph. You are *not* being asked to explain. The amount of detail expected in the answer is suggested by the number of marks for the question.

Let us take one example. Describing landscapes and landforms is an important part of physical geography. When asked to *describe* a landform (volcano, corrie, etc.), you are really being asked to *say what it looks like.*

* Write about its shape, size and what it is made of.

* Be generous with your use of adjectives, such as wide/narrow, steep/gentle/flat, straight/curved, etc.

Question
Describe the physical features of the landform in the photo.

Approach to the answer	Possible answers
• Name the landform first	volcano (or even composite volcano)
• Describe its shape	cone-shaped
• Describe other features	lava flows on its sides
• Give a more detailed description	steep slopes on the snow-covered top of the cone
	slightly more gentle slopes lower down
	bare rock (lava) on the lower slopes

D • Explain the formation of ...
 • Explain why ...   • Why have ...?
 • Give reasons for ...   • Why does ..?

To answer these questions you need geographical knowledge and understanding. You are being asked to account for the appearance or occurrence of physical and human features of the Earth's surface. These command words do not usually cause problems. What causes problems is giving enough precise information.

Question   *Why is population growth high in LEDCs?*
(4 marks)

Answer   Because birth rates are high and death rates are low.

Examiner comment   This is the basic answer, which has only reached the first level of explanation. It is a one-mark answer. Explanations why birth rates are high and why death rates are low are also needed.

Question   *Explain the formation of spits along some coasts.*
(5 marks)

Answer   A spit forms after longshore drift moves sand along the coast. I have drawn a diagram below to show how it does this.

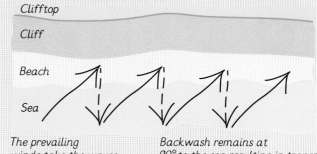

The prevailing winds take the waves on to the beach at an angle   Backwash remains at 90° to the sea resulting in transp. of material in a zig zag fashion

▲ Longshore drift.

Examiner comment   Everything is correct and longshore drift does play a part in the formation of a spit. The question does not ask for a diagram, but relevant points accurately made on the diagram will be credited. But there is only partial explanation here. How and why the spit actually forms after the longshore drift has transported sand is not explained. This answer is only worth two of the five marks.

Questions based on source materials
The source materials used in geography examinations are many and various. However, you will be familiar with:

* the type of source itself, because those most used in examinations will include maps, graphs, diagrams, photographs, tables of data or cartoons.

* the geographical topic covered, because this is part of the syllabus.

There is only a small chance that you will have seen the data before, or the photograph, or the chosen OS map extract. This is why most questions begin with short questions looking at the source materials. **Figure 1** (page 250) takes you through a question based on pie graphs.

Although most questions follow this style, in some you must use knowledge and understanding from the beginning. This makes these questions more difficult and they are likely to be used later in the examination. **Figure 2** (page 250) is a question of this type, based upon photograph interpretation.

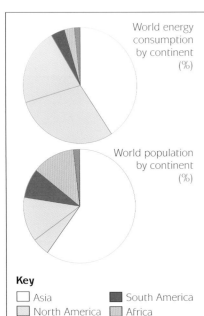

World energy consumption by continent (%)

World population by continent (%)

Key

☐ Asia   ■ South America
☐ North America ■ Africa
☐ Europe   ■ Australasia

Question

a Study the pie graphs.

(i) *Name the continent in which least energy is consumed.* (1 mark)

(ii) *Place the names of the six continents under the appropriate heading in the table below.* (2 marks)

Continents with energy consumption greater than their share of population	Continents with share of world population greater than their energy consumption
1	1
2	2
3	3

(iii) *State one difference between the continents in the two columns of the table.* (1 mark)

b *Suggest reasons why a person living in North America consumes more energy per year on average than a person living in South America.* (6 marks)

> The starter question to give you the chance to look at the data: simple command word and one clear answer

> The data need to be studied more carefully and all the data used. Only the data from the source are needed for the answers.

> Understanding of the difference between continents with MEDCs and those with LEDCs is now needed.

> Knowledge and understanding are now needed. You are being asked to explain the differences. 'Suggest reasons' have been used as the command words instead of 'explain' because the syllabus does not specify study of North and South America. However, you should have learned that rich countries consume more energy resources per head than poor countries and you can be expected to give reasons for this.

▲ **Figure 1**

You need to make a careful study of the photograph before answering part **a** and you should understand what is meant by an urban zone. Do not rush into the answer; take a careful look at the photograph and think the answer through. When answering part **b** use evidence from the photograph to justify your choice. Do not just describe everything that you see.

Questions based upon maps

Many candidates find questions which include the words *distribution* or *pattern* awkward to answer. Remember that in *describing a distribution* you should concentrate upon saying where things are found, but you can also mention where they are not found. Do the same when *describing a pattern*; say where there are many and where there are few. It is important to end with a comment summarizing the main feature of the pattern.

Question

Study the photograph taken in a city in a more economically developed country.

a *Name the urban zone in which the photograph was most likely to have been taken.* (1 mark)

b *Describe the evidence from the photograph which supports your choice of zone.* (4 marks)

▲ **Figure 2**

Question

*Study the world map (**Figure 3** on page 251), which shows rates of illiteracy in the world. From the figure, describe the distribution of countries with high rates of illiteracy (above 50 per cent).* (5 marks)

Approach the answer like this:

- Identify the command word – 'describe'. This is telling you to write about what the map shows.

- Identify the question theme – 'rates of illiteracy above 50 per cent'. Study the key to see how they are shown on the map.

- Study the map.

1 Look first for the locations of countries with above 50 per cent rates of illiteracy. Possible answers would be:

  (i) Most are in Africa.

  (ii) Almost all the countries except for a few countries in the centre and south of Africa are above 50 per cent.

  (iii) A band of very high values above 75 per cent runs from the west coast across to the Red Sea.

  (iv) Some are above 50 per cent in the Middle East and South Asia.

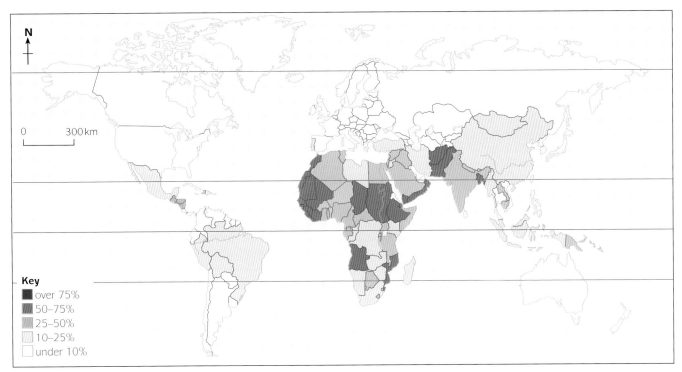

▲ **Figure 3** Percentage of total population unable to read or write.

2 Next look for the location of countries with rates of illiteracy below 50%. Possible brief answers would be:

  (i) In the rest of Asia and in South America illiteracy is lower (mainly 10–50 per cent).

  (ii) Lowest rates, below 10 per cent, are in North America and Europe.

  (iii) These lowest rates are in MEDCs.

Note the following points about these answers:

- They are taken from what can be seen on the map.

- Percentage values are used from time to time.

- The names of continents are used for effective description (think of the world map).

- More than five points are described to try to ensure full marks.

Using case studies

Sometimes the wording of the question makes it clear that you are expected to use your *case study* information. However, the term 'case study' is not normally used in the question. You must look for other wording that suggests that a case study is needed. Examples include:

With reference to an example you have studied ...
Using an example ...   For one farming system in the UK ...
For one industrial area in an LEDC ...

In these questions you have little chance of gaining more than half marks unless you include some case study information in your answer. Look at these examination answers from two candidates.

Question

For one named area with many footloose industries, give the reasons why these industries have been attracted to the area.

(7 marks)

Answer from Candidate 1   There are many footloose industries along the sides of the M4 between London and Bristol. One area is on the industrial estate next to the M4 in Reading where there are food-processing and high-tech industries because it is near a motorway junction. High-tech industries need skilled workers, who are produced in nearby universities such as Oxford and Reading. Rail and road links are very good. I have already said that the footloose industries are along the sides of the M4 motorway, which gives fast road links to London, the biggest market in the country. The M4 also passes Heathrow Airport for markets overseas. High-speed trains run along the rail line between Bristol and London.

Examiner comment   This candidate begins in the best possible way by naming an area. The rest of the answer clearly refers to this chosen area. Some precise information, such as the source of the skilled workers, is used and the significance of some of the points made is stated, such as Heathrow Airport for overseas markets. This answer reaches the highest mark band called Level 3 and is worth six marks. With a little extra detail, or if another reason for location had been referred to in a similarly precise way, the answer would have been worth all seven marks.

Answer from Candidate 2 There are lots of workers who are needed in footloose industries. There are good roads. The railway lines are also good. There are some high-tech industries. These are making lots of money because everyone is now buying computers. High-tech industries are footloose industries. There are lots of these near London near the M4 because of good markets.

Examiner comment Only at the end of the answer is there a named area for footloose industries. This is just 'tagged on' without the other information being related to it. Most of the rest of the answer is very vague with too many 'good's in it. Such general statements could apply to the location of any industry in any place. Although the answer tells us something about the nature of footloose industries, this is not relevant in a question about reasons for their growth. It is a weak and very basic answer to the question, only of Level 1 standard, worth one or at the most two of the seven marks.

These are examples of questions where your case study information must be used to obtain high marks. However, try to use this information in other questions where it fits. For example, the question might be:

What disadvantages can tourism bring to an LEDC?

This question can be answered solely by making general points, but if you have studied Kenya or another LEDC, include relevant information. Specific information about places always enhances the quality of an answer.

Activity – Answers that have gone wrong

What is wrong with these answers? Give your own correct versions.

1 *Name a country which has suffered from drought and been given much food aid.* (1 mark)
   Answer Africa.

2 *With reference to an area you have studied in a less economically developed country (LEDC), explain the disadvantages tourism can bring.* (6 marks)
   Answer Costa del Sol in southern Spain. A line of villas owned by rich Europeans continues for 50 kilometres along the Mediterranean coastline. This has brought wealth to a region which 50 years ago was one of the poorest in Europe. There are many jobs in service industries and young people are attracted from all parts of Spain by the prospects of work.

3 a *Measure the distance along the railway line between stations A and B in Figure 1.* (1 mark)
   Answer 2.5

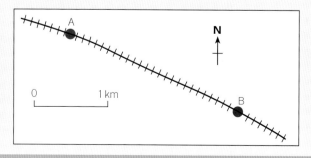

b *In what direction would you be travelling if you took the train from station B to station A?* (1 mark)
   Answer Eastwards.

4 *Study Figure 2 which shows part of the coastline of north-east England. Describe the coastal features shown.* (4 marks)

▲ **Figure 2**

Answer The main features of this coastal area allow us to see where the sea crashes against these rocks. It also allows us to see that the rocks have been severely beaten by the sea in one particular place. Abrasion is one way in which the sea batters rocks. The rocks have obviously been eroded away in one place. This has weakened the rocks themselves making them more vulnerable to more erosion by the sea.

◀ **Figure 1**

Glossary

Contemporary issues new and live issues affecting people around the world, such as recent natural disasters; greenfield versus brownfield sites; or a foot and mouth disease outbreak

Global citizenship to be responsible for ensuring that our individual actions take account of the consequences, not just for ourselves but for the world as a whole, e.g. by reducing air pollution we reduce the risks of acid rain and possible global warming, which would have consequences around the world

Political issues those to do with government, for the local level up to national and international levels

Social and economic issues those that affect people and finances; they are often grouped together as socio-economic issues, e.g. unemployment, quality of life

Sustainable development new developments that ensure there will be at least the same amount of resources in the future as there are today – that is, development does not damage the environment or use up resources in a permanent way; examples include green tourism, renewable energy, tree planting, using brownfield rather than greenfield sites in cities

Values and attitudes the ways in which different groups of people and individuals respond to circumstances or events

Chapter 1

Acid lava thick 'sticky' lava, high silica content, flows short distances

Anticline an upfold, like an arch

Basic lava thin 'runny' lava, low silica content, flows long distances

Composite volcano volcano with alternate layers of lava and ash

Compressional plate margin where two plates are moving together (destructive)

Convection current change in the flow and pressure of the Earth's mantle affecting plate movement

Core the centre of the Earth

Crater circular hole at the top of a volcanic cone

Crust the thin layer at the Earth's surface

Earthquake shaking of the Earth's surface

Epicentre point on the Earth's surface directly above the focus

Focus point within the Earth's crust where an earthquake occurs

Geosyncline a depression between two plates

Landslide rapid movement of rocks and soil under the influence of gravity

Lava name given to molten magma when it erupts at the surface

Magma molten rock from deep within the Earth

Mantle molten rock surrounding the Earth's core

Passive plate margin where two plates are moving alongside each other in the same direction but at different rates (conservative)

Plate section of the Earth's crust

Primary effects (of an earthquake) direct effects of an earthquake, e.g. buildings collapse

Richter scale measure of the strength (intensity) of an earthquake

Secondary effects (of an earthquake) indirect effects, e.g. fire, tsunami, disease

Sedimentary rock formed from sediments laid down under water

Seismograph measures and records the intensity of an earthquake

Shield volcano large volcano formed from basic lava spread over a large area

Subduction zone where a plate is sinking and melting

Syncline a downfold

Tensional plate margin where two plates are moving apart (constructive)

Tsunami huge 'tidal wave' caused by a submarine earthquake

Vent opening through which lava flows from a volcano

Volcano cone-shaped mountain formed by surface eruptions of magma from inside the Earth

Chapter 2

Aquifer underground store of water in permeable or porous rocks

Bourne stream which only flows for part of the year in chalk areas

Cavern underground chamber in limestone, larger than a cave

Chalk escarpment landform with both a steep scarp slope and a gentle dip slope

Chemical weathering break-up of rock by processes such as solution causing changes in the minerals that form the rock

Clay vale large area of low land (larger than a valley) which tends to be wet because of the impermeable clay

Clint flat-surfaced block of rock in a limestone pavement

Dip slope the gentle slope on a chalk escarpment

Dry valley V-shaped valley without any stream flowing in the bottom

Freeze-thaw break-up of rocks by alternate freezing and thawing of water trapped in joints in the rock

Gorge steep-sided rocky valley with a stream flowing in the bottom

Gryke vertical gap between blocks of rock in a limestone pavement

Igneous rock rock that began as magma in the interior of the Earth

Karst scenery area in which several typical Carboniferous limestone landforms are found

Landfill surface sites used for dumping waste materials

Limestone pavement level surface of bare rock broken up into separate blocks

Limestone solution process of chemical weathering that dissolves limestone rock

Metamorphic rock rock whose shape or form has been changed by heat and/or pressure

Non-sustainable activity without a long future, using up or destroying natural resources

Physical weathering break-up of rock by processes such as freeze-thaw without any changes in the minerals that form the rock

Pillar (of limestone) column of lime from floor to roof in a limestone cave or cavern

Reclamation changing land that cannot be used in its present state into useful land (for farming, settlement etc.)

Scarp slope the steep slope on a chalk escarpment

Scree pieces of rock with sharp edges below rock outcrops

Sedimentary rock rock that began as sediments, usually on a sea bed

Sink hole place where there are joints or cracks in limestone rock through which water from a surface stream disappears underground

Spring point where water from underground reaches the surface

Stalactite column of lime hanging down from the roof in limestone caves or caverns

Stalagmite column of lime built up from the floor in caves or caverns

Swallow hole funnel-shaped holes in limestone down which surface streams disappear underground

Tor block of rock outcropping on top of a granite plateau such as Dartmoor

Weathering breakdown of surface rock by weather without any movement of the rock

Chapter 3

Alluvium river-deposited material (sand and silt)

Channel the feature in which a river flows

Confluence where two rivers meet

Delta often triangular-shaped flat land jutting out into the sea at the mouth of a river

Drainage basin area of land drained by a single river

Estuary drowned river mouth in a lowland area

Gorge of recession steep-sided narrow valley created by a waterfall/river retreat

Interlocking spurs spurs of high land which overlap in the upper part of a river valley

Meander bend in the middle and lower courses of a river

Meander scar dried-up ox-bow lake

Mouth where a river enters the sea or a lake

Ox-bow lake semi-circular lake formed by a meander being sealed off from the main course of a river

River cliff steep river bank on the outside of a meander

Saltation small particles 'jumping' along the river bed

Slip-off slope gentle slope on the inside of a meander

Solution a form of chemical weathering

Source the starting point of a river

Suspension small particles of clay and silt carried along in the river

Tributary small river which flows into a larger river

V-shaped valley river valley in its upper course, steep-sided and narrow

Watershed the imaginary line that surrounds a drainage basin

Whinstone a hard, resistant igneous rock

Chapter 4

Abrasion (glacial) process of erosion in which rocks carried in the bottom of a glacier wear away the surface rock

Arête sharp-edged two-sided ridge on the top of a mountain

Boulder clay all materials deposited by ice, usually clay with boulders of different sizes within it

Cirque another name for the circular hollow of a corrie

Corrie circular rock hollow high on a mountainside surrounded by steep, rocky walls

Countryside stewardship managing land and habitats in an environmentally beneficial way

Drumlin egg-shaped small hills made of boulder clay

Erratic large boulder dropped by a glacier, of a different type of rock from the rock now below it

Glacial till another term for boulder clay, especially sandy glacial deposits

Glacial trough flat-floored, steep-sided valley, usually described as a U-shaped valley

Glacier large moving mass of ice

Ground moraine boulder clay deposited by glaciers which gives a hummocky surface

Hanging valley a tributary valley left hanging high above the main valley floor, usually with a waterfall at the bottom end

High-level bench flatter area high on the valley side above the steep sides of a U-shaped valley

Honeypot place that is especially attractive to many visitors/tourists

Ice sheet moving mass of ice that covers all the land surface over a wide area

Lateral moraine line of glacial deposits along the side of a valley

Medial moraine line of glacial deposits down the middle of a valley

Misfit stream small stream in a large glaciated valley

Moraine all material deposited after having been transported by ice

Névé snowfield like those that fill corries with ice and snow

Pleistocene Ice Age time when most of the British Isles were covered by ice, about 20 000 years ago

Plucking process of erosion in which blocks of rock are torn away from the bedrock as the ice moves

Pyramidal peak three-sided slab of rock which forms a mountain peak

Ribbon lake long, narrow lake on the floor of a glaciated valley

Rock bar outcrop of hard rock between rock basins made of softer rocks

Rock basin hollow on the valley floor eroded by glacial abrasion and plucking

Snout furthest point reached by a glacier, where all the ice melts

Snow line above this point the land is covered by snow

Striation deep groove in rock outcrops made by abrasion as the ice flowed over them

Tarn small circular lake that fills the bottom of a corrie after the ice has melted

Terminal moraine ridge of boulder clay dumped by a glacier as it melted

Truncated spur projection in a river valley that has later been eroded away by a valley glacier so that the lower valley sides are now vertical

Valley glacier moving mass of ice confined within a valley

Waterfall (glacial) fall of water down the side of a U-shaped valley from a hanging valley

Chapter 5

Arch a rocky opening through a headland

Backwash backward movement of water down the beach after a wave has broken

Cave area that has been hollowed out by waves at the bottom of a cliff

Coastal management attempts by people to maintain or alter the natural features of the coast to their own advantage

Constructive wave gentle wave with strong swash and weak backwash which deposits material

Destructive wave high wave with strong backwash which breaks frequently, causing erosion

Fetch the length (distance) of water over which the wind has blown, which affects the size and strength of waves

Longshore drift the current that transports material along the coast in a zigzag

Spit long ridge of sand and shingle, attached to the land at one end but in the open sea at the other end

Stack piece of rock surrounded by sea, now standing away from the coast

Swash forward movement of water as a wave breaks

Wave-cut platform area of gently sloping or flat rocks exposed at low tide

Chapter 6

Climate average weather conditions of a place over many years

Climatic hazard short-term weather event that is a threat to life and property

Cold front boundary between a cold air mass and a warm air mass, where the cold air mass moves under the warm air in front of it

Convectional rainfall heavy rain formed by the cooling of moist air as it rises above a heated surface

Drought period of dry weather beyond that normally expected

Front the dividing line (boundary) between two air masses with different temperatures, as a result of which air rises to give rain

Frontal rainfall rain formed by the cooling of moist air as it is forced to rise along a warm or cold front

High pressure air in an anticyclone sinks to create an area of high pressure at the Earth's surface

Hurricane powerful tropical storm affecting the Caribbean and some other tropical ocean areas

Insolation heating of the Earth's surface by the Sun's rays

Latitude distance from the Equator; has an effect on the temperature of a place

Leaching minerals washed out of the soil by high rainfall, making the soil less fertile

Low pressure air in a depression rises, leaving an area of low pressure at the Earth's surface

Mediterranean climate hot, dry summers and warm, wet winters

Occluded front boundary where air is rising after a cold front joins up with a warm front

Precipitation all forms of moisture from the atmosphere: rain, snow, hail, fog, mist

Pressure gradient the difference in pressure between places

Prevailing wind direction of the wind that blows most frequently at a place

Relief rainfall rain formed by the cooling of moist air as it is forced to rise over high land

Temperate maritime climate the warm wet climate of the UK

Tropical storm a deep area of low pressure formed over warm sea surfaces near the Equator, bringing strong winds and heavy rain

Urban heat island when the temperature of a city is higher than the temperature experienced in surrounding rural areas

Warm front boundary between a warm air mass and a cold air mass where the warm air rises up over the cold air in front of it

Weather the day-to-day conditions of temperature, precipitation, cloud, sunshine and wind

Chapter 7

A horizon the top layer in the soil profile

B horizon the second layer in the soil profile, found in soils where the minerals have been washed down from the layer above

Biodiversity great numbers of plants and animals in one ecosystem; greatest in tropical rainforests

Desertification spread of bare desert-like surfaces into surrounding areas, mainly as a result of human actions

Ecosystem the living community of plants and animals and the physical factors upon which they depend, such as climate and soil

Hard pan soil layer in which minerals washed out from higher up the profile, such as iron, have been compressed

Humus layer of organic material at the top of a soil profile

Latosol the red and yellowish-red soil that forms under tropical rainforests

Lower B horizon layer down the soil profile into which minerals have been washed from above

Overcultivation too much crop growing, destroying soil fertility and structure and increasing risks of soil erosion

Overgrazing destruction of grass and other vegetation caused by too many animals on the land

Podsol soil type under coniferous woodland

Savanna tropical grasslands, a natural vegetation zone of mixed grassland and trees

Soil erosion loss of fertile topsoil by wind and water

Sustainable conserved and able to be used for many years into the future

Tropical rainforest dense forest (jungle) growing in hot, wet lowlands near the Equator

Chapter 8

Ageing population increasing percentage of old people (65 and over) in a country

Census collection of data on a country's population characteristics, usually made every 10 years in the form of a questionnaire

Choropleth map map that uses shading to show density, e.g. of population

Densely populated large numbers of people per unit area

Dependants people of non-working age, usually above 65 and below 16

Dependency ratio number of children under 15 and the number of people over 65, relative to the number of adults aged between 15 and 64; the young and old age groups are known as dependants

Dot map map that uses dots to represent numbers, e.g. of people

Economic migrant person who has moved in search of work, e.g. a Mexican moving into California

Emigration when people leave a country

Forced migration people forced to move, e.g. as a result of natural disaster or war

Life expectancy average number of years that a person is expected to live

Migration the movement of people

Natural decrease death rate higher than birth rate: death rate minus birth rate

Natural increase birth rate higher than death rate: birth rate minus death rate

Overgrazing too many animals grazing an area so that the vegetation cover cannot support the number of animals

Population pyramid graph to show age and sex structure of a population

Population structure composition of a country's population by age and sex

Push and pull factors factors that force a person away from one place and to another place

Refugee person forced to move from one country to another, often as a result of war or famine

Rural to urban migration movement from the countryside to the towns

Sparsely populated few people per unit area

Working or independent population income earners and main taxpayers, typically between the ages of 16 and 65

Chapter 9

Bridging point settlement located where a river is forded or bridged

Brownfield area of land, previously built-up, again available to be built on

Burgess model a model of urban structure for towns in MEDCs

Catchment area the area that is served by a particular school, shop or settlement

Central Business District (CBD) the heart of a city where the financial and business interests are (shops, offices, entertainment)

Commuter or dormitory village settlement on the edge of a town or city

Comparison goods goods that are expensive and bought less frequently, e.g. furniture

Convenience goods goods that are cheap and bought frequently, e.g. papers, food

Core (of the CBD) the heart of the CBD where the large department stores are located

Decentralization outward movement of shops and offices from the CBD to the suburbs

Dry-point site site of a settlement which avoids land that is likely to flood, e.g. a gravel mound or side of a valley

Frame (of the CBD) the outer area of the CBD with smaller shops and offices

Gentrification movement of wealthy people into an area of former urban decay (often in inner city zones)

Green belt land around a large town or city which is protected from development, in order to halt the expansion of the town into the countryside

Greenfield open land which has never previously been built on

Hinterland the area surrounding a settlement, port, etc.

Industrial estate zone of light industry, often on the edge of a town

Informal sector in LEDCs many people work in the informal sector as shoe-shine boys, servants, etc.

Inner city the urban zone of MEDCs outside the CBD – a zone of mixed land uses

Morphology arrangement of land uses in an urban area, usually in distinct zones

Periferia housing zone in São Paulo where shanty towns have been upgraded

Route focus where many communication routes (roads, railways) meet

Rural–urban fringe area at edge of an urban area, beyond the suburbs, where there is a mix of rural and urban land uses

Self-help scheme people in shanty towns in LEDCs are given basic services by the government, and some of the means to build their own homes

Site land on which a settlement is built

Situation settlement in relation to its surrounding area

Suburb mainly residential area outside the inner city of an MEDC

Urban land use use of the land in towns and cities

Urban zone area where land uses are similar

Urbanization increase in the percentage of people living in urban areas

Wet-point site place where a settlement is close to a water supply such as a spring on a chalk escarpment

Chapter 10

Agribusiness farms operated by large companies, e.g. Findus, Bird's Eye

Appropriate technology technology that is appropriate to the needs, skills, knowledge and wealth of the people

Arable growing of crops, e.g. wheat, barley

Biodiversity a rich range of plants and animals

Cash crops crops that are grown to be sold for profit

Chagra a clearing in the Amazon rainforest

Climate the average weather conditions of an area

Commercial the sale of farm products for profit

Double or multiple cropping where two or more crops are produced on the same land in one year

Eutrophication the loss of oxygen in streams and lakes, caused by chemical pollution

Extensive farm farm system with low inputs and outputs per land area, usually practised over a large total area, e.g. shifting cultivation in the Amazon Basin or hill sheep farming in Wales

Green Revolution large increase in food production due to the introduction of new, high-yielding varieties of seeds

High-tech (farming) using modern machinery and computers to farm

High-yielding varieties (HYVs) crops that have been specially developed to produce very high yields, e.g. IR8 rice

Inbye the flatter land close to the farm in the valley floor in hill sheep areas

Inputs physical and human raw materials that go into a farming system

Intake enclosed fields (usually with dry-stone walls) on lower slopes of upland farms

Intensive farm farm system with high inputs and outputs per land area, usually practised on a small area of land, e.g. market gardening in Kent, rice growing in Asia

Inundation canal channel used to direct water from a river onto the cropped land nearby

Irrigation the artificial watering of land, using sprinklers, canals, sprays, etc.

Machete simple hand-held axe used by people in the Amazon Basin

Mixed farm single farm where crops are grown and animals also reared

Monsoon season of heavy rainfall in countries such as India and Bangladesh

Outputs the end products of a farm system

Pastoral the rearing of animals

Processes the activities required to turn inputs into outputs

Relief the height and shape of the land

Salinization increase in the salt content in soils due to overuse of irrigation water

Self-sufficient when people can produce all the foodstuffs they need

Soil erosion loss of topsoil through erosion by wind and water

Soil the thin layer of weathered material and organic matter on the land surface

Staple crop main food crop, e.g. rice in Bangladesh

Subsistence farmer produces just enough food to feed the family

Tube well modern deep well

Well means of taking water from the water table below ground-level

Yield amount produced by growing crops or rearing animals

Chapter 11

Brownfield site urban area cleared of old industry and housing and now available for new developments

Business park attractive area of offices, research establishments and light industries, usually on the edge of an urban area

Energy factor such as fuel or electricity, needed for work to be done

Footloose industry light or high-tech industry which has considerable freedom in location

Fuel supply energy source, usually coal, oil or gas

Globalization increasing importance of international operations for people and companies

Government backing political support

Greenfield site rural open land that has not previously been built on

High-tech company firm that researches, develops and makes technological goods, e.g. semiconductors and microchips

Industrial estate area where several manufacturing companies are located, often on the edge of an urban area

Labour the workforce

Light industry industry producing small-sized items that do not require bulky raw materials

Newly industrializing countries (NICs) LEDCs where manufacturing has grown very rapidly leading to economic development

Primary industry one that takes raw materials from the land or sea, e.g. mining, forestry, farming

Raw materials the things needed to make goods in a factory

Secondary industry manufacturing industry that processes raw materials and makes them into other products

Site land upon which a settlement or factory is built

Science park attractive landscaped area for high-tech companies which may be linked with a university

Sunrise industry new growth industry, e.g. high-tech companies

Sunset industry type of manufacturing in decline: many are heavy industries

Tertiary industry activity that supplies services to people and to other industries

Transnational large corporation operating in many countries, but usually with its headquarters in an MEDC

Transport movement of raw materials to the factory, or goods from the factory to market, by road, rail, sea or air

Chapter 12

Acid rain rainwater made more acid by air pollution

Alternative source of energy one that can be used instead of fossil fuels, usually renewable, e.g. wind or solar power

Benefits economic or social advantages for people

Coal fossil fuel made from trees and plants

Consumer-orientated society one in which people are wealthy enough to buy many different goods for themselves and for their homes

Energy efficiency measures to reduce heat loss and use of energy

Energy of the tides generating electricity using changing levels of the sea and waves

Geothermal generating electricity and heat from the ground, usually in volcanic areas

Green tourism form of tourism in which protection of the environment and the way of life of the local people are considered to be important

Hydro electric power electricity generated from the force of moving water

Industrial Revolution period of great industrial growth and change

National Park area of beautiful scenery which is managed to preserve its attractive features, and for visitor recreation

Natural resources naturally occurring materials which can be used by people

Non-renewable natural (or other) resource which can be used only once

Nuclear energy energy released from uranium and plutonium

Primary energy obtained by burning fossil fuels to give heat and energy

Recycling recovery of waste products by converting them into materials that can be used again

Renewable resource that is replaced naturally and so never runs out

Reserves fuel supplies that have been discovered and could be used in the future

Resource substitution using one product in place of another, e.g. aluminium (which is cheaper and can be recycled) instead of tin and steel cans

Secondary energy electricity that is generated from a fuel, e.g. coal, oil, water or wind

Solar energy using the Sun as a source of light and heat

Sustainable activity that has a good future because the environment/resource upon which it depends is not being destroyed

Sustainable tourism tourist activities and locations that have a good future because neither the environment nor the way of life of local people are being destroyed

UK England, Northern Ireland, Scotland and Wales

Wind turbine tall structure with arms turned by the wind to generate power

Worldwide tourism long-distance holidays, including destinations in some LEDCs

Chapter 13

Aid transfer of money, goods and expertise from one country to another either free or at low cost

Appropriate technology level of technology suitable for local people to use

Aquifer underground store of water in permeable rock

Fresh-water availability amount of (non-salty) water that is present for people to use

Gross National Product (GNP) total value of all the goods and services produced by people and companies from one country in one year

Interdependence when two or more countries have a shared need to exchange one another's goods

Invisible trade exchange of services and transfers of money including aid and foreign exchange from tourists

Primary products those obtained from the land or sea without being processed or made into another product, e.g. food, minerals, wood

Secondary products manufactured goods

Trade exchange of goods and services between countries

Trade gap difference in value or amount between exports and imports

Trade trap value of exports consistently lower than the value of imports, resulting in debt

Visible trade exchange of goods that can be counted, measured or weighed

Water stress concern over the amount of fresh water available for human use

Index